AF886632

Springer Proceedings in Mathematics & Statistics

Volume 168

More information about this series at http://www.springer.com/series/10533

Springer Proceedings in Mathematics & Statistics

This book series features volumes composed of select contributions from workshops and conferences in all areas of current research in mathematics and statistics, including OR and optimization. In addition to an overall evaluation of the interest, scientific quality, and timeliness of each proposal at the hands of the publisher, individual contributions are all refereed to the high quality standards of leading journals in the field. Thus, this series provides the research community with well-edited, authoritative reports on developments in the most exciting areas of mathematical and statistical research today.

Regina Y. Liu • Joseph W. McKean
Editors

Robust Rank-Based and Nonparametric Methods

Michigan, USA, April 2015
Selected, Revised, and Extended Contributions

Springer

Editors
Regina Y. Liu
Department of Statistics
Rutgers University
New Brunswick, NJ, USA

Joseph W. McKean
Department of Statistics
Western Michigan University
Kalamazoo, MI, USA

ISSN 2194-1009 ISSN 2194-1017 (electronic)
Springer Proceedings in Mathematics & Statistics
ISBN 978-3-319-39063-5 ISBN 978-3-319-39065-9 (eBook)
DOI 10.1007/978-3-319-39065-9

Library of Congress Control Number: 2016947172

© Springer International Publishing Switzerland 2016
This work is subject to copyright. All rights are reserved by the Publisher, whether the whole or part of the material is concerned, specifically the rights of translation, reprinting, reuse of illustrations, recitation, broadcasting, reproduction on microfilms or in any other physical way, and transmission or information storage and retrieval, electronic adaptation, computer software, or by similar or dissimilar methodology now known or hereafter developed.
The use of general descriptive names, registered names, trademarks, service marks, etc. in this publication does not imply, even in the absence of a specific statement, that such names are exempt from the relevant protective laws and regulations and therefore free for general use.
The publisher, the authors and the editors are safe to assume that the advice and information in this book are believed to be true and accurate at the date of publication. Neither the publisher nor the authors or the editors give a warranty, express or implied, with respect to the material contained herein or for any errors or omissions that may have been made.

Printed on acid-free paper

This Springer imprint is published by Springer Nature
The registered company is Springer International Publishing AG Switzerland

To our mentors

Foreword

This book contains a collection of papers by distinguished researchers that were presented at an international conference to celebrate Joseph McKean's 70th birthday. The conference entitled "Robust Rank-Based and Nonparametric Methods" was held at Western Michigan University on April 9 and 10, 2015. Many papers in this book are contributed by some of Joe's long-standing collaborators, students, and colleagues.

Joe McKean is a truly outstanding scholar. He is internationally recognized as having made fundamental contributions to the theory and practice of nonparametric statistics. Joe has consistently produced very high quality work over the last 40 years resulting in 6 books and more than 100 published papers. Over this time, he has directed 24 Ph.D. dissertations. He personally contributed in the development of rank-based methods for linear models, multivariate models, time series models, experimental designs, mixed models, and nonlinear models. In particular, he is responsible for developing both the theoretical underpinnings and the computational algorithms for these rank-based methods. His contributions are both broad and deep. Joe has developed rank-based methods across a broad range of settings which are competitive in terms of efficiency with parametric methods, when the parametric assumptions hold and at the same time are robust to violations of these assumptions. Put simply, essentially any data set you can analyze with least squares and/or maximum likelihood, you can now do using a robust rank-based method developed and implemented in software by Joe. Joe is a fellow of the American Statistical Association and the 1994 winner of Western Michigan University's Distinguished Faculty Scholar Award.

Joe's contributions to teaching and service are also legendary. In terms of teaching, Joe has taught essentially every graduate statistics class at Western Michigan University. He was the program chair or codirector of five separate Great Lakes Symposiums on Applied Statistics held in Kalamazoo, Michigan. Joe has served as an associate editor for five different journals, including the *Journal of the American Statistical Association*. Joe played a fundamental role in the formation of the Department of Statistics at Western Michigan University, which occurred in

July 2001. Joe was also one of the principal leaders in putting together the Statistical Computing Lab at Western Michigan University.

Describing Joe's long list of outstanding accomplishments speaks to just one aspect of Joe. Joe has a wonderful life outside of academia. He is much loved by his wife Marge, his three daughters, and his four grandchildren. Pour Joe a craft beer and ask him about his international travels and you will be regaled by a tale from one of his visits to Australia or Switzerland.

On a further personal note, we have enjoyed a long-term friendship and collaborative relationship with Joe. We have had the great good fortune to work with him on many research projects. He is the ideal collaborator, always ready to discuss a problem in depth, often coming up with innovative and creative solutions.

In summary, those who have worked with Joe McKean as well as those who have been taught by him have greatly benefited from their interactions with a truly great man.

State College, PA, USA
College Station, TX, USA
January 2016

T.P. Hettmansperger
S.J. Sheather

Preface

This volume of papers grew out of the

International Conference on
Robust Rank-Based and Nonparametric Methods

which was held at Western Michigan University, Kalamazoo, MI, on April 9 and 10, 2015. This conference consisted of 2 days of talks by distinguished researchers in the areas of robust, rank-based, and nonparametric statistical methods. Many of the speakers agreed to submit papers to this volume in areas of their expertise. These papers were refereed by external reviewers, and revised papers were resubmitted for a final review. We thank the referees for their work on these papers.

This collection of papers discuss robust rank-based and nonparametric procedures for many of the current models of interest for univariate and multivariate situations. It begins with a review of rank-based methods for linear and nonlinear models. Many of the succeeding papers extend robust and nonparametric methods to mixed and GEE-type models. Many of the papers develop robust and nonparametric methods for multivariate designs and time series models. Theoretical properties of the analyses, including asymptotic theory and efficiency properties, are developed. Results of simulation studies confirming the validity and empirical efficiency of the methods are presented. Discussion also focuses on applications involving real data sets and computational aspects of these robust procedures. Several R packages for these procedures are discussed, and the URLs for their downloading are cited.

The conference was hosted by the Department of Statistics of Western Michigan University. Our thanks goes to many people who contributed to the success of this conference. In particular, a special thanks goes to Ms. Michelle Hastings, administrative assistant of the Department of Statistics, who was in charge of the local arrangements. Also a special thanks to Dr. Magdalena Niewiadomska-Bugaj, chair of the Department of Statistics; Professor Rajib Paul, Department of Statistics; and Dr. Thomas Vidmar, Biostat Consultants of Portage, for their time spent in organizing this conference. Also, our thanks to Professor Simon Sheather, Texas A&M University, for emceeing the conference banquet.

We deeply appreciate the efforts of the editorial staff of Springer who made the production of this book possible. A special thanks to Ms. Hannah Bracken, associate editor of Statistics at Springer, for her interest and initial guidance on the book, and to Mr. Dhayanidhi Karunanidhi, production coordinator at Springer, for his efforts in the actual production of the book. We acknowledge the help of Dr. Jeff Terpstra on some LaTeXissues. Last but not least, a special thanks to the authors for creating this collection of interesting papers in robust and nonparametric methods.

New Brunswick, NJ, USA — Regina Y. Liu
Kalamazoo, MI, USA — Joseph W. McKean
March 2016

Contents

1 **Rank-Based Analysis of Linear Models and Beyond: A Review** 1
Joseph W. McKean and Thomas P. Hettmansperger

2 **Robust Signed-Rank Variable Selection in Linear Regression** 25
Asheber Abebe and Huybrechts F. Bindele

3 **Generalized Rank-Based Estimates for Linear Models with Cluster Correlated Data** 47
John Kloke

4 **Iterated Reweighted Rank-Based Estimates for GEE Models** 61
Asheber Abebe, Joseph W. McKean, John D. Kloke, and Yusuf K. Bilgic

5 **On the Asymptotic Distribution of a Weighted Least Absolute Deviation Estimate for a Bifurcating Autoregressive Process** 81
Jeff T. Terpstra

6 **Applications of Robust Regression to "Big" Data Problems** 101
Simon J. Sheather

7 **Rank-Based Inference for Multivariate Data in Factorial Designs** ... 121
Arne C. Bathke and Solomon W. Harrar

8 **Two-Sample Rank-Sum Test for Order Restricted Randomized Designs** 141
Yiping Sun and Omer Ozturk

9 **On a Partially Sequential Ranked Set Sampling Paradigm** 163
Douglas A. Wolfe

10 **A New Scale-Invariant Nonparametric Test for Two-Sample Bivariate Location Problem with Application** 175
Sunil Mathur, Deepak M. Sakate, and Sujay Datta

11 Influence Functions and Efficiencies of k-Step Hettmansperger–Randles Estimators for Multivariate Location and Regression 189
Sara Taskinen and Hannu Oja

12 New Nonparametric Tests for Comparing Multivariate Scales Using Data Depth 209
Jun Li and Regina Y. Liu

13 Multivariate Autoregressive Time Series Using Schweppe Weighted Wilcoxon Estimates 227
Jaime Burgos and Jeff T. Terpstra

14 Median Stable Distributions 249
Gib Bassett

15 Confidence Intervals for Mean Difference Between Two Delta-Distributions 261
Karen V. Rosales and Joshua D. Naranjo

Author Index 273

Subject Index 275

List of Contributors

Asheber Abebe Department of Mathematics and Statistics, Auburn University, Auburn, AL, USA

Gib Bassett Department of Finance, University of Illinois at Chicago, Chicago, IL, USA

Arne C. Bathke Fachbereich Mathematik, Universität Salzburg, Salzburg, Austria

Department of Statistics, University of Kentucky, Lexington, KY, USA

Yusuf K. Bilgic Department of Mathematics, SUNY-Geneseo, Geneseo, NY, USA

Huybrechts F. Bindele Department of Mathematics and Statistics, University of South Alabama, Mobile, AL, USA

Jaime Burgos Department of Statistics, Western Michigan University, Kalamazoo, MI, USA

Sujay Datta Department of Statistics, Buchtel College of Arts and Sciences, University of Akron, Akron, OH, USA

Solomon W. Harrar Department of Statistics, University of Kentucky, Lexington, KY, USA

Thomas P. Hettmansperger Department of Statistics, Penn State University, University Park, PA, USA

John Kloke Department of Biostatistics, University of Wisconsin School of Medicine and Public Health, Madison, WI, USA

Jun Li University of California, Riverside, Riverside, CA, USA

Regina Y. Liu Department of Statistics, Rutgers University, New Brunswick, NJ, USA

Sunil Mathur Department of Biostatistics and Epidemiology, Medical College of Georgia, Georgia Regents University, Augusta, GA, USA

Joseph W. McKean Department of Statistics, Western Michigan University, Kalamazoo, MI, USA

Joshua D. Naranjo Department of Statistics, Western Michigan University, Kalamazoo, MI, USA

Hannu Oja Department of Mathematics and Statistics, University of Turku, Turku, Finland

Omer Ozturk The Ohio State University, Columbus, OH, USA

Karen V. Rosales MMS Holdings, Inc., Canton, MI, USA

Deepak M. Sakate Department of Biostatistics and Epidemiology, Medical College of Georgia, Georgia Regents University, Augusta, GA, USA

Simon J. Sheather Department of Statistics, Texas A&M University, College Station, TX, USA

Yiping Sun PTC Therapeutics, Inc., Piscataway, NJ, USA

Sara Taskinen Department of Mathematics and Statistics, University of Jyväskylä, Jyväskylä, Finland

Jeff T. Terpstra Department of Statistics, Western Michigan University, Kalamazoo, MI, USA

Douglas A. Wolfe Department of Statistics, Ohio State University, Columbus, OH, USA

Chapter 1
Rank-Based Analysis of Linear Models and Beyond: A Review

Joseph W. McKean and Thomas P. Hettmansperger

Abstract In the 1940s Wilcoxon, Mann and Whitney, and others began the development of rank based methods for basic one and two sample models. Over the years a multitude of papers have been written extending the use of ranks to more and more complex models. In the late 60s and early 70s Jurečková and Jaeckel along with others provided the necessary asymptotic machinery to develop rank based estimates in the linear model. Geometrically Jaeckel's fit of linear model is the minimization of the distance between the vector of responses and the column space of the design matrix where the norm is not the squared-Euclidean norm but a norm that leads to robust fitting. Beginning with his 1975 thesis, Joe McKean has worked with many students and coauthors to develop a unified approach to data analysis (model fitting, inference, diagnostics, and computing) based on ranks. This approach includes the linear model and various extensions, for example multivariate models and models with dependent error structure such as mixed models, time series models, and longitudinal data models. Moreover, McKean and Kloke have developed R libraries to implement this methodology. This paper reviews the development of this methodology. Along the way we will illustrate the surprising ubiquity of ranks throughout statistics.

Keywords Efficiency • Diagnostics • High breakdown fits • Mixed models • Nonlinear models • Nonparametric methods • Optimal scores • Rank scores • Rfit • Robust

---

J.W. McKean (✉)
Department of Statistics, Western Michigan University, Kalamazoo, MI 49008, USA
e-mail: joseph.mckean@wmich.edu

T.P. Hettmansperger
Department of Statistics, Penn State University, University Park, PA 16002, USA
e-mail: tph@stat.psu.edu

© Springer International Publishing Switzerland 2016

R.Y. Liu, J.W. McKean (eds.), *Robust Rank-Based and Nonparametric Methods*, Springer Proceedings in Mathematics & Statistics 168,
DOI 10.1007/978-3-319-39065-9_1

1.1 Introduction

Our intention in writing the following historical development is to provide our perspective on the evolution of nonparametric methodology (both finite and asymptotic). We will focus on a particular development based on ranks. We will show how beginning with simple rank tests in the 1940s, the area has grown into a coherent group of contemporary statistical procedures that can handle data from increasingly complex experimental designs. Two factors have been essential: theoretical developments especially in asymptotic theory, see Hettmansperger and McKean (2011), and in computational developments, see Kloke and McKean (2014). Statistical inference based on ranks of the data has been shown to be both statistically efficient relative to least squares methods as well as robust. Any history is bound to be selective. We have chosen a line of development that is consistent with the theme of this conference. There is a rich and extensive literature on nonparametric methods. We will confine ourselves to references that directly relate to the history as related to the topics of the conference.

When constructing tests for the median of a continuous population, the simplest nonparametric test is the sign test which counts the number of observations greater than the null hypothesized value of the median. The null and alternative distributions of the sign test statistic are both binomial. In the case of the null hypothesis, the binomial parameter is 0.5, and hence, the null distribution of the sign test statistic does not depend on the population distribution. We call such a test nonparametric or distribution free. The use of the sign test for dichotomous data was first proposed by Arbuthnott (1710).

The modern era for nonparametric or distribution free tests began with the work of Wilcoxon (1945) and Mann and Whitney (1947). Wilcoxon proposed the nonparametric Wilcoxon signed rank test for the median of a symmetric population, and the nonparametric Wilcoxon rank sum test for the difference in population medians. Mann and Whitney (1947) showed that the rank sum test is equivalent to the sign test applied to the pairwise differences across the two samples. Tukey (1949) showed the signed rank test is equivalent to the sign test applied to the pairwise averages from the sample (called the Walsh averages by Tukey). Hence, from the earliest time, we have a connection between rank based methods and the L_1 norm expressed through its derivative, the sign statistic. In what follows we will exploit this connection by considering a rank based norm and its relationship to the L_1 norm. In addition, we will need to include the L_2 norm and least squares for comparison in our discussion.

Noether (1955), based on earlier unpublished work by Pitman (1948), introduced Pitman efficiency for hypothesis tests. Then Hodges and Lehmann (1956, 1960) analyzed the efficiency of various rank tests relative to least squares tests (t- and F-tests) and proved the surprising result that the efficiency of the Wilcoxon tests relative to the t-tests is never less than 0.864, is 0.955 at the normal model, and can be arbitrarily large for heavy tailed model distributions. No longer was a rank test considered quick and dirty with low power. Rank tests now provided a serious alternative to least squares t tests.

Hodges and Lehmann (1963) next developed estimators based on rank test statistics (R-estimates) and showed that they inherit the efficiency of the rank tests that they were derived from. Because of the connection between the Wilcoxon test statistics and the L_1 norm, the Hodges-Lehmann estimate of location is the median of the pairwise averages and the estimate for the difference in locations is the median of the pairwise differences across the two samples. By the mid-sixties rank tests and estimates for location models, including the one-way layout, were available, and they share the excellent efficiency properties. Robustness was introduced during this time by Huber (1964) followed by the work of Hampel (1974). The basic tools of robustness include the influence function and break down point. Ideally we would like to have estimates that have bounded influence and positive breakdown. Indeed, Wilcoxon R-estimates enjoy precisely these good robustness properties in addition to the excellent efficiency properties mentioned above. For example the breakdown value for the Hodges-Lehmann estimate of location, the median of the pairwise averages, is 0.293 while the breakdown of the sample mean is 0.

Hájek and Šidák (1967) published a seminal work on the rigorous development of rank tests. This was followed many years later by a second edition, Hájek et al. (1999) which extends much of the theory and includes material on R-estimates.

Hence, during the 1960s nonparametric and distribution free rank tests and rank-based estimates for location models were well understood and provided excellent alternatives to least squares methods (means, t- and F-tests) from the point of view of both efficiency and robustness. Unfortunately the rank methods did not extend in a straight forward way to the two-way layout with interaction terms. For example, a quick check of text books on nonparametrics written before the mid-seventies did not reference a test for interaction in a two-way layout.

The next step involved the extension of rank methods to linear regression where the two-way layout could be formulated in regression terms and natural rank tests for regression parameters were easy to construct. The rank based statistical methods which require the estimation of nuisance parameters will then be asymptotically distribution free but no longer distribution free for finite samples. The tools for the development of rank regression were provided by Jurečková (1969, 1971) and Jaeckel (1972). Jurečková, in particular, provided the asymptotic theory and Jaeckel provided a rank based dispersion function that when minimized produced R-estimates. McKean (1975) developed corresponding rank tests along with the necessary asymptotic distribution theory for the linear model. In the next several sections we explicitly introduce the linear model and discuss the development of rank based methods and their efficiency and robustness properties. In subsequent sections, we discuss extensions of rank-based analyses to nonlinear models and models with dependent errors.

There is R software available to compute these rank-based analyses. In the examples presented, we discuss some of the R code for the computation of these analyses. The rank-based package for linear models, `Rfit`, (see Kloke and McKean 2012), can be downloaded at CRAN (http://cran.us.r-project.org/). Supplemental packages for the additional models discussed in the examples can be downloaded at the site https://github.com/kloke/.

1.2 Rank-Based Fit and Inference for Linear Models

In this section we will review the univariate linear model and present the rank based norm used to derive the rank based statistical methods along with the basic asymptotic tools. Then we will present the efficiency and robustness results that we mentioned in the introduction, but in more detail. We will also describe some of the rank based methods for residual analysis. For details of this development see Chaps. 3–5 of Hettmansperger and McKean (2011).

Let $\boldsymbol{Y}$ denote the $n \times 1$ vector of observations and assume that it follows the linear model

$$\boldsymbol{Y} = \mathbf{1}\alpha + \boldsymbol{X}\boldsymbol{\beta} + \boldsymbol{e}, \tag{1.1}$$

where $\boldsymbol{X}$ is an $n \times p$ full column rank matrix of explanatory variables, $\mathbf{1}$ is an $n \times 1$ vector of ones, $\boldsymbol{\beta}$ is $p \times 1$ vector of regression coefficients, α is the intercept parameter, and $\boldsymbol{e}$ is the $n \times 1$ vector of random errors. Letting $\boldsymbol{x}_i'$ denote the *ith* row of $\boldsymbol{X}$, we have $y_i = \alpha + \boldsymbol{x}_i'\boldsymbol{\beta} + e_i$. For the theory cited in this section, assume that the random errors are iid with pdf $f(x)$ and cdf $F(x)$, respectively.

A score generating function is a nondecreasing square-integrable function $\varphi(u)$ defined on the interval $(0, 1)$ which, without loss of generality, satisfies the standardizing conditions

$$\int_0^1 \varphi(u)\, du = 0 \text{ and } \int_0^1 [\varphi(u)]^2\, du = 1. \tag{1.2}$$

We denote the scores by $a(i) = \varphi[i/(n + 1)]$.

The basis of traditional analysis of most models in practice is the least squares (LS) fit of the model. This fit minimizes the squared-Euclidean distance between the vector of responses and the estimating region, (subspace if it is a linear model). In the same way, the basis for a rank-based analysis is the fit of the model except that a different norm is used other that the Euclidean norm. This norm leads to a robust fit. For a given score function $\varphi(u)$, the norm is defined by

$$\|\boldsymbol{v}\|_\varphi = \sum_{i=1}^n a_\varphi[R(v_i)]v_i, \quad \boldsymbol{v} \in R^n. \tag{1.3}$$

Note that this is a pseudo-norm; i.e., it satisfies all properties of the norm except it is invariant to constant shifts, i.e., $\|\boldsymbol{v} + a\mathbf{1}\|_\varphi = \|\boldsymbol{v}\|_\varphi$ for all a, where $\mathbf{1}$ is a vector of n ones. The counterpart in LS is the squared-Euclidean pseudo-norm $\sum_{i=1}^n (v_i - \overline{v})^2$.

For convenience, we define the dispersion function $D(\boldsymbol{\beta})$ in terms of the pseudo norm $\| \cdot \|_\varphi$ as

$$D(\boldsymbol{\beta}) = \|\boldsymbol{y} - \boldsymbol{X}\boldsymbol{\beta}\|_\varphi = \sum_{i=1}^n a[R(y_i - \boldsymbol{x}_i'\boldsymbol{\beta})](y_i - \boldsymbol{x}_i'\boldsymbol{\beta}) = \boldsymbol{a}'[R(\boldsymbol{y} - \boldsymbol{X}\boldsymbol{\beta})](\boldsymbol{y} - \boldsymbol{X}\boldsymbol{\beta}), \tag{1.4}$$

where $R(y_i - \boldsymbol{x}_i'\boldsymbol{\beta})$ denotes the rank of $y_i - \boldsymbol{x}_i'\boldsymbol{\beta}$ among $y_1 - \boldsymbol{x}_1'\boldsymbol{\beta}, \ldots, y_n - \boldsymbol{x}_n'\boldsymbol{\beta}$ and $\boldsymbol{a}[R(\boldsymbol{y} - \boldsymbol{X}\boldsymbol{\beta})]$ is the vector with *ith* component $a[R(y_i - \boldsymbol{x}_i'\boldsymbol{\beta})]$. Note that ranks are invariant to constant shifts such as an intercept parameter. The rank-based estimator of $\boldsymbol{\beta}$ is the minimizer

$$\hat{\boldsymbol{\beta}} = \text{Argmin}\, D(\boldsymbol{\beta}). \tag{1.5}$$

Let V_f denote the full model subspace of R^n; i.e., V_f is the range (column space) of $\boldsymbol{X}$. Then $D(\hat{\boldsymbol{\beta}})$ is the minimum distance between the vector of responses $\boldsymbol{Y}$ and the subspace V_f in terms of the norm $\|\cdot\|_\varphi$. For reference, we have

$$D(\hat{\boldsymbol{\beta}}) = \min_{\boldsymbol{\eta} \in V_f} \|\boldsymbol{Y} - \boldsymbol{\eta}\|_\varphi. \tag{1.6}$$

Note that this minimum distance between $\boldsymbol{Y}$ and V_f is unique; i.e., the minimum distance does not depend on the basis matrix of V_f.

Denote the negative of the gradient of $D(\boldsymbol{\beta})$ by

$$\boldsymbol{S}(\boldsymbol{\beta}) = -\bigtriangledown D(\boldsymbol{\beta}) = \boldsymbol{X}'\boldsymbol{a}[R(\boldsymbol{y} - \boldsymbol{X}\boldsymbol{\beta})]. \tag{1.7}$$

Then the estimator also satisfies $\boldsymbol{S}(\hat{\boldsymbol{\beta}}) \doteq \boldsymbol{0}$. Generally, the intercept parameter is estimated by the median of the residuals; i.e.,

$$\hat{\alpha} = \text{med}_i\{y_i - \boldsymbol{x}_i'\hat{\boldsymbol{\beta}}\}. \tag{1.8}$$

Examples of scores functions include: $\varphi(u) = \sqrt{12}[u - (1/2)]$, for Wilcoxon rank-based methods; $\varphi(u) = \text{sgn}[u - (1/2)]$, for L_1 methods; and $\varphi(u) = \Phi^{-1}(u)$, where $\Phi(t)$ is the standard normal cdf, for normal scores methods. In Sect. 1.1, we pointed out that the nonparametric sign and Wilcoxon location estimators are based on minimizers of L_1-norms. This is true also in the regression case for the Wilcoxon and sign scores. First, if sign scores are used then the rank-based estimator of $\boldsymbol{\beta}$ and α, as estimated by the median of the residuals, are the L_1 (least absolute deviations) estimators of α and $\boldsymbol{\beta}$; see page 212 of Hettmansperger and McKean (2011). Secondly, for Wilcoxon scores we have the identity

$$\frac{4(n+1)}{\sqrt{12}} \sum_{i=1}^{n} \sqrt{12}\left(\frac{R(u_i)}{n+1} - \frac{1}{2}\right) u_i = \sum_{i=1}^{n}\sum_{j=1}^{n} |u_i - u_j|, \quad \boldsymbol{u} \in R^n. \tag{1.9}$$

That is, the Wilcoxon estimator of the regression coefficients minimizes the absolute deviations of the differences of the residuals.

In addition to the above defined rank-based estimator of $\boldsymbol{\beta}$, we can also construct an hypothesis test of the general linear hypotheses

$$H_0 : \boldsymbol{M}\boldsymbol{\beta} = \boldsymbol{0} \text{ versus } H_A : \boldsymbol{M}\boldsymbol{\beta} \neq \boldsymbol{0}, \tag{1.10}$$

where $\boldsymbol{M}$ is a $q \times p$ matrix of full row rank. Let V_r denote the reduced model subspace; i.e., the subspace of V_f subject to the null hypothesis. Let W be a $n\times(p-q)$ basis matrix of V_r. Then we write the reduced model as $\boldsymbol{Y} = \alpha\mathbf{1} + \boldsymbol{W\theta} + \boldsymbol{e}$. Let $\hat{\boldsymbol{\theta}}$ denote the rank-based estimator of this reduced model. Then the distance between $\boldsymbol{Y}$ and the subspace V_r is $D(\hat{\boldsymbol{\theta}})$, which is the same for any basis matrix of V_r.

The test statistic of the hypotheses (1.10) is the normalized version of the reduction in distance, $D(\hat{\boldsymbol{\theta}}) - D(\hat{\boldsymbol{\beta}})$, given by:

$$F_{\varphi} = \frac{[D(\hat{\boldsymbol{\theta}}) - D(\hat{\boldsymbol{\beta}})]/q}{\hat{\tau}/2}, \tag{1.11}$$

where $\hat{\tau}$ is an estimator of the scale parameter

$$\tau^{-1} = \int \varphi'(F(t))f^2(t)\,dt = \int f(t)\,d(\varphi(F(t))). \tag{1.12}$$

Koul et al. (1987) developed a consistent estimator of τ. Note that the reduction in distance (dispersion) parallels the least squares reduction in sums of squares.

The approximating distributions of the estimator and the test statistic are determined by a linear approximation of the negative gradient of the dispersion and a quadratic approximation of the dispersion. Let $\boldsymbol{\beta}_0$ denote the true parameter. Then the following approximations can be made asymptotically rigorous under mild regularity conditions:

$$\frac{1}{\sqrt{n}}\boldsymbol{S}(\boldsymbol{\beta}) \asymp \frac{1}{\sqrt{n}}\boldsymbol{X}'\boldsymbol{\varphi}(F(\boldsymbol{y} - \boldsymbol{X\beta}_0)) - \tau^{-1}\frac{1}{n}\boldsymbol{X}'\boldsymbol{X}\sqrt{n}(\boldsymbol{\beta} - \boldsymbol{\beta}_0) \tag{1.13}$$

$$D(\boldsymbol{\beta}) \asymp D(\boldsymbol{\beta}_0) - (\boldsymbol{\beta} - \boldsymbol{\beta}_0)'\boldsymbol{S}(\boldsymbol{\beta}_0) + \frac{1}{2\tau}(\boldsymbol{\beta} - \boldsymbol{\beta}_0)'\frac{1}{n}\boldsymbol{X}'\boldsymbol{X}(\boldsymbol{\beta} - \boldsymbol{\beta}_0).$$

Based on these results the following asymptotic distributions can be obtained:

$$\boldsymbol{S}(\boldsymbol{\beta}_0) \text{ is approximately } MVN(\mathbf{0}, \boldsymbol{X}'\boldsymbol{X})$$

$$\hat{\boldsymbol{\beta}} \text{ is approximately } MVN(\boldsymbol{\beta}_0, \tau^2(\boldsymbol{X}'\boldsymbol{X})^{-1}) \tag{1.14}$$

$$F_{\varphi} \text{ is approximately } F(q, n-p-1), \text{ under } H_0; \tag{1.15}$$

($qF_{\varphi} \to \chi^2$-distribution with q degrees of freedom, under H_0).

Based on (1.14), an approximate $(1-\alpha)100\,\%$ confidence interval for the linear function $\boldsymbol{h}'\boldsymbol{\beta}$ is

$$\boldsymbol{h}'\hat{\boldsymbol{\beta}} \pm t_{\alpha,n-p}\hat{\tau}\sqrt{\boldsymbol{h}'(\boldsymbol{X}'\boldsymbol{X})^{-1}\boldsymbol{h}}, \tag{1.16}$$

where $t_{\alpha,n-p}$ is the upper $\alpha/2$ quantile of a t-distribution with $n-p$ degrees of freedom. The use of t-critical values and F-critical values for tests and confidence procedures is supported by numerous small sample simulation studies; see McKean and Sheather (1991) for a review of such studies.

1.2.1 Diagnostics

After fitting a model, a residual analysis is performed to check for quality of fit and anomalies. A standard diagnostic tool for a LS fit is the scatterplot of residual versus fitted values. A random scatter indicates a good fit, while patterns in the plot indicate a poor fit and, often, lead to the subsequent fitting of more adequate models. For example, suppose the true model is of the form

$$\boldsymbol{Y} = \mathbf{1}\alpha + \boldsymbol{X}\boldsymbol{\beta} + \boldsymbol{Z}\boldsymbol{\gamma} + \boldsymbol{e}, \tag{1.17}$$

where $\boldsymbol{Z}$ is an $n \times q$ matrix of constants and $\boldsymbol{\gamma} = \boldsymbol{\theta}/\sqrt{n}$, $\boldsymbol{\theta} \neq \mathbf{0}$. We fit, though, Model (1.1) using LS; i.e., the model has been misspecified. A straight forward calculation yields

$$\hat{\boldsymbol{Y}}_{LS} = \alpha\mathbf{1} + \boldsymbol{X}\boldsymbol{\beta} + \boldsymbol{H}\boldsymbol{e} + \boldsymbol{H}\boldsymbol{Z}\boldsymbol{\gamma} \text{ and } \hat{\boldsymbol{e}}_{LS} = \boldsymbol{e} - \boldsymbol{H}\boldsymbol{e} + (\boldsymbol{I} - \boldsymbol{H})\boldsymbol{Z}\boldsymbol{\gamma},$$

where $\boldsymbol{H}$ is the projection matrix onto the range of $\boldsymbol{X}$. If range($\boldsymbol{Z}$) $\not\subset$ range$(\boldsymbol{X})^{\perp}$ then both the fitted values and residuals are functions of $\boldsymbol{Z}\boldsymbol{\gamma}$ and, hence, there will be information in the plot concerning the misspecified model. Note that the function $\boldsymbol{H}\boldsymbol{e}$ is unbounded; so, based on this representation, outliers in the random errors are diffused throughout the residuals and fitted values. This leads to distortions in the residual plot which, for example, may even mask outliers.

For the rank-based fit of Model (1.1) when Model (1.17) is the true model, it follows from the above linearity results that

$$\hat{\boldsymbol{Y}}_{rb} = \alpha\mathbf{1} + \boldsymbol{X}\boldsymbol{\beta} + \tau_{\varphi}\boldsymbol{H}\boldsymbol{\varphi}[F(\boldsymbol{e})] + \boldsymbol{H}\boldsymbol{Z}\boldsymbol{\gamma} \text{ and } \hat{\boldsymbol{e}}_{rb} = \boldsymbol{e} - \tau_{\varphi}\boldsymbol{H}\boldsymbol{\varphi}[F(\boldsymbol{e})] + (\boldsymbol{I} - \boldsymbol{H})\boldsymbol{Z}\boldsymbol{\gamma}.$$

Thus, as with LS, there is information in the residual plot concerning misspecified models. Note from the rank-based representation, the function $\boldsymbol{H}\boldsymbol{\varphi}[F(\boldsymbol{e})]$ is bounded. Hence, the rank-based residual plot is less sensitive to outliers. This is why outliers tend to standout more in residual plots based on robust fits than in residual plots based on LS fits. Other diagnostic tools for rank-based fits are discussed in Chaps. 3 and 5 of Hettmansperger and McKean (2011). Among them are the Studentized residuals. Recall that the *ith* LS Studentized residual is $\hat{e}^*_{LS,i} = \hat{e}_{LS,i}/[\hat{\sigma}\sqrt{1 - (1/n) - h_i}]$, where h_i is the *ith* diagonal entry of the projection matrix $\boldsymbol{H}$ and $\hat{\sigma}$ is the square root of *MSE*. Note that $\hat{e}^*_{LS,i}$ is corrected for both scale and location. Rank-based Studentized residuals are discussed in Sect. 3.9.2 of Hettmansperger and McKean (2011). As with LS Studentized residuals, they are corrected for both location and scale. The usual outlier benchmark for Studentized residuals is ± 2, which we use in the examples below.

1.2.2 Computation

Kloke and McKean (2012, 2014) developed the R package `Rfit` for the rank-based fitting and analysis of linear models. This package along with its auxiliary package `npsm` can be downloaded from the site CRAN; see Sect. 1.1 for the url. The Wilcoxon score function is the default score function of `Rfit` but many other score functions are available in `Rfit` including the normal scores and the simple bent scores (Winsorized Wilcoxons). Furthermore, users can easily implement scores of their choice; see Chap. 3 of Kloke and McKean (2014) for discussion. In subsequent examples we demonstrate how easy `Rfit` is to use.

Analogous to least squares, the rank-based analysis can be used to conduct inference for general linear models, i.e., a robust ANOVA and ANCOVA; see Chaps. 3–5 of Hettmansperger and McKean (2011). As an illustration, we end this section with an example that demonstrates how easy the rank-based analysis can be used to test for interaction in a two-way design.

1.2.3 Example

Hollander and Wolfe (1999) provide an example on light involving a 2×5 factorial design; see, also, Kloke and McKean (2014) for discussion. The two factors in the design are the light regimes at two levels (constant light and intermittent light) and five different dosage levels of luteinizing release factor (LRF). Sixty rats were put on test under these ten treatments combinations (six repetitions per combination). The measured response is the level of luteinizing hormone (LH), nanograms per ml of serum in the resulting blood samples.

We chose Wilcoxon scores for our analysis. The full model is the usual two-way model with main and interaction effects. The right panel in Fig. 1.1 shows the mean profile plots based on the full model rank-based estimates. The profiles are not parallel indicating that interaction between the factors is present. These data comprise the `serumLH` data set in `Rfit` and hence is loaded with `Rfit`. The `Rfit` function `raov`, (robust ANOVA), obtains the rank-based analysis with one line of code as indicated below. The reduction in dispersion test, (1.11), of each effect is adjusted for all other effects analogous to Type III sums of squares in SAS; see Sect. 5.5 of Kloke and McKean (2014). Further, the design need not be balanced. Here is the code and resulting (with some abbreviation) rank-based ANOVA table:

```
> raov(serum~light.regime+LRF.dose+
       light.regime*LRF.dose, data=serumLH)

Robust ANOVA Table
                      DF        RD        F  p-value
light.regime           1 1642.3332 58.03844 0.00000
LRF.dose               4 3027.6734 26.74875 0.00000
light.regime:LRF.dose  4  451.4559  3.98850 0.00694
```

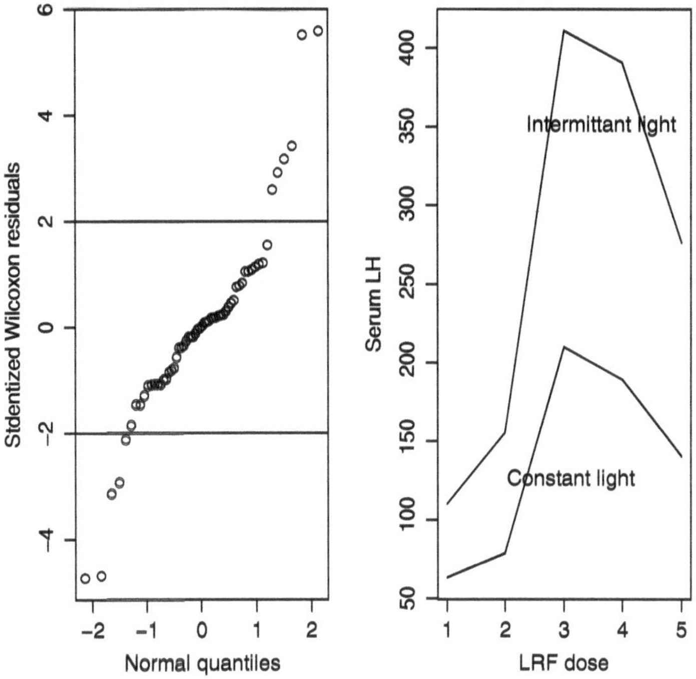

Fig. 1.1 Plots for the serum LH data

As the figure suggested, the factors light regime and LRF dose interact, $p = 0.00694$. In contrast, the LS analysis fails to reject interaction at the 5 % level, $p = 0.0729$. For this example, at the 5 % level of significance, the rank-based and LS analyses would lead to different interpretations. The left panel of Fig. 1.1 displays the q–q plot of the Wilcoxon Studentized residuals. This plot indicates that the errors are drawn from a heavy tailed distribution with numerous outliers, which impaired the LS analysis. The estimate of the ARE between the rank-based and least squares analyses is the ratio

$$\widehat{ARE} = \frac{\hat{\sigma}^2}{\hat{\tau}_{\varphi}^2} \tag{1.18}$$

where $\hat{\sigma}^2$ is the MSE of the full model LS fit. This is often thought of as a measure of precision. For this example, this ratio is 1.88. So, the rank-based analysis cuts the LS precision by a factor of $1/1.88 = 0.53$.

In a two-way analysis when interaction is present often subsequent inference involves contrasts of interest. To demonstrate how easy this is accomplished using `Rfit`, suppose we consider the contrast between the expected response at the peak (factor LRF dose at 3, factor light at the intermittent level) minus the expected response at the peak (factor LRF dose at 3, factor light at the constant level). Of course this is after looking at the data, so we are using this in a confirmatory mode. Our confidence interval is of the form (1.16). Using the following code, it computes to 201.16 ± 65.63. The difference is significant.

```
#  full model fit
fitmod <- rfit(serum~factor(light.regime)+
   factor(LRF.dose)+ factor(light.regime)*
   factor(LRF.dose), data=serumLH)
#  hvec picks the contrast
hvec <- rep(0,60); hvec[27] <- -1; hvec[57] <- 1
# estimate of contrast
contr <- hvec%*%fitmod$fitted.values
# error in CI
se2 <- t(hvec)%*%mat%*%vc%*%t(mat)%*%hvec
# error term in the confidence interval.
err <- qt(.975,50)*sqrt(se2)
```

1.3 Efficiency and Optimality

In general, the relative efficiency of one statistical method to another, in estimating or testing, is the squared ratio of the slopes in their respective linear approximations. Restricting ourselves to the linear model, (1.1), and rank-based procedures, in light of the asymptotic linearity results, (1.14), such ratios involve the scale parameter τ. In particular, suppose we consider two rank-based methods using the respective score functions $\varphi_1(u)$ and $\varphi_2(u)$, and, hence, the norms $\|\cdot\|_{\varphi_1}$ and $\|\cdot\|_{\varphi_2}$. Then the aysmptotic efficiency of method 1 to method 2 is

$$e(\|\cdot\|_{\varphi_1}, \|\cdot\|_{\varphi_2}) = \frac{\tau_2^2}{\tau_1^2}. \tag{1.19}$$

Note by (1.14) that this is the same as the ratio of the asymptotic variances of the associated estimators of the regression coefficients. Values greater than 1 indicate that methods based on $\|\cdot\|_{\varphi_1}$ are superior. The larger slope indicates a more sensitive method, where the slope is τ^{-1}.

For LS, $e(\|\cdot\|_{\varphi_1}, \text{LS}) = \sigma^2/\tau_1^2$, where σ^2 is the variance of the random errors. For comparison with Wilcoxon methods, we already mentioned for location models that $e(\text{Wilcoxon}, \text{LS})$ is 0.955 when the error distribution is normal. Hence, $e(\text{Wilcoxon}, \text{LS})$ is the same for both linear and location models. Another striking result, due to Hodges and Lehmann (1956), shows that this efficiency is never less than 0.864 and may be arbitrarily large for heavy tailed distributions. In the case of the normal scores methods, the efficiency relative to least squares is 1 at the normal model and never less than 1 at any other model!

An optimality goal is to select a score function to minimize τ_φ; i.e., maximize τ_φ^{-1}. We can write expression (1.12) as

$$\tau^{-1} = \int_0^1 \varphi(u)\varphi_f(u)\,du, \tag{1.20}$$

where

$$\varphi_f(u) = -\frac{f'[F^{-1}(u)]}{f[F^{-1}(u)]}. \tag{1.21}$$

Recall that the scores have been standardized so that $\int \varphi^2(u)\,du = 1$. Hence τ^{-1} can be expressed as

$$\begin{aligned} \tau^{-1} &= \frac{\int_0^1 \varphi(u)\varphi_f(u)\,du}{\sqrt{\int_0^1 \varphi^2(u)\,du}\sqrt{\int_0^1 \varphi_f^2(u)\,du}} \left\{ \sqrt{\int_0^1 \varphi_f^2(u)\,du} \right\} \\ &= \rho \left\{ \sqrt{\int_0^1 \varphi_f^2(u)\,du} \right\}. \end{aligned} \tag{1.22}$$

The first factor on the right in the first line is a correlation coefficient which we have indicated by ρ. Thus τ^{-1} is maximized if we select the score function to be $\varphi_f(u)$, (standardized form). This makes the correlation coefficient 1 and τ_φ^{-1} equal to the term in the braces. This term, though, is the square-root of Fisher information. Therefore, by the Rao-Cramér lower bound, the choice of $\varphi_f(u)$ as the score function leads to an asymptotically efficient (optimal) estimator. For the score functions discussed in earlier sections, it follows that the optimal score function for normally distributed errors is the normal score function, for logistically distributed errors is the Wilcoxon score function, and for Laplace distributed errors is the sign score function.

Of course this optimality only can be accomplished provided that the form of f is known. Evidently, the closer the chosen score is to φ_f, the more optimal the rank based analysis is. A Hogg-type adaptive scheme where the score function is selected based on initial (Wilcoxon) residuals has proven to be effective; see Sect. 7.6 of Kloke and McKean (2014). McKean and Kloke (2014) successfully modified this scheme for fitting a family of skewed normal distributions.

1.3.1 Monte Carlo Study

To illustrate the optimality discussed above, we conducted a small simulation study of a proportional hazards model. Consider a response variable T with a $p \times 1$ vector of covariates $\boldsymbol{x}$. Assume T has a $\Gamma(1, \zeta)$ distribution where $\zeta = \exp\{\boldsymbol{x}'\boldsymbol{\beta}\}$ and $\boldsymbol{\beta}$ is a $p \times 1$ vector of parameters. It follows that

$$\log T = \boldsymbol{x}'\beta + \epsilon, \tag{1.23}$$

where ϵ has an extreme-valued distribution; see Chaps. 2 and 3 of Hettmansperger and McKean (2011). The optimal scores for this model are the log-rank scores generated by $\varphi(u) = -1 - \log(1 - u)$.

We simulated this model for the following situation: sample size is $n = 20$; the covariates are $(1, x_i)$, where $x_i = i/21$; and $\alpha = -2$ and $\beta = 5$. The methods involved are LS and the three rank-based methods: optimal scores (log-rank), Wilcoxon scores, and normal scores. These scores are intrinsic to `Rfit`. The simulation size is 10,000. To show how simple the coding is, here is the gist of the program's loop portion for a simulation:

```
mu <- exp(alpha + beta*x)
y <- rgamma(20,1,1/mu); ly <- log(y);
fitw <- rfit(ly~x)                              # Wil.
fitly <- rfit(ly~x,scores=logrankscores)        # opt.
fitns <- rfit(ly~x,scores=nscores)              # ns
fitls <- lm(ly~x)                               # LS
```

Table 1.1 presents the empirical relative efficiencies (ratios of mean square errors) for the parameter β. The efficiencies are relative to the optimal rank-based score procedure. The log-rank score procedure is most efficient followed by the normal scores procedure and then the Wilcoxon. Least squares (LS) performed the worst.

Table 1.1 Empirical AREs of estimators of the slope parameter β relative to the rank-based estimator based on the optimal log-rank scores

Method			
Optimal	Wilcoxon	Normal scores	LS
1	1.215	1.175	1.380

The simulation size is 10,000

1.4 Influence and High Breakdown

1.4.1 Robustness Properties

In the 1960s and 1970s new tools to assess robustness properties of estimators were developed beginning with Huber (1964) and Hampel (1974). The breakdown value of a location estimator is the (limiting) proportion of the data that must be contaminated in order to carry the value of the estimator beyond any finite bound. In the one sample location model with score function $\varphi^+(u)$, the breakdown for the rank based estimator is ϵ where

$$\int_0^{1-\epsilon} \varphi^+(u)\, du = \frac{1}{2}\int_0^1 \varphi^+(u)\, du.$$

A simple computation shows that the least squares estimate, the mean, has 0 breakdown point, worst possible. The median has breakdown 0.5, the best possible. The median of the pairwise averages (Wilcoxon score) has breakdown 0.293, while the normal scores estimate has breakdown 0.239.

Another robustness tool, the influence function, is a measure of how fast the estimator changes when an outlier is moved out beyond the edges of the sample. It is provided by the linear approximation of the negative gradient, $\boldsymbol{S}(\boldsymbol{\beta})$, (1.7). Consider the linear model (1.1). For the rank-based estimator, using the score function $\varphi(u)$, the influence function is given by:

$$\Omega(\boldsymbol{x}, y) = \tau \left(\frac{1}{n}\boldsymbol{X}'\boldsymbol{X}\right)^{-1} \varphi(F(y))\boldsymbol{x},$$

where $(\boldsymbol{x}, y)$ is the value at which we evaluate the influence. When the score function is bounded the influence is bounded in the $\boldsymbol{y}$-space. However, note that influence is unbounded in factor space. In the case of the location model, the influence functions for the median and the median of the pairwise averages are both bounded. Note also that least squares estimators have unbounded influence functions in both the $\boldsymbol{y}$- and the $\boldsymbol{X}$-spaces.

For most designed experiments and for designs with predictors which are well behaved, the rank-based estimators offer a robust and highly efficient alternative to LS for fitting and analyzing linear models. In the case of messy predictors, though, a robust alternative with bounded influence in factor space and positive breakdown is most useful. In fact a primary use of such fits is to highlight the difference between its fit and that of a highly efficient robust fit and, thus, alerting the user to possible anomalies in factor space. We next discuss a high breakdown rank-based (HBR) fit and its accompanying diagnostics which serve this purpose.

1.4.2 High-Breakdown and Bounded Influence Rank-Based Estimates

For the linear model, Chang et al. (1999) developed a rank-based estimator that has bounded influence and can achieve a 50 % breakdown point. It is a weighted version rank-based Wilcoxon fit. By the identity (1.9), the Wilcoxon estimator minimizes the least absolute deviations of the differences of the residuals. Let $\{b_{ij}\}$ be a set of nonnegative weights. Consider estimators which minimize

$$\hat{\boldsymbol{\beta}} = \text{Argmin} \sum_{i<j} b_{ij} |(y_i - \boldsymbol{x}_i'\boldsymbol{\beta}) - (y_j - \boldsymbol{x}_j'\boldsymbol{\beta})|. \tag{1.24}$$

If the weights are all 1, then this is the Wilcoxon estimator.

Chang et al. (1999) proposed weights which are both functions of factor space and residual space. For factor space, it uses robust distances based on the high breakdown minimum covariance determinant (MCD) which is an ellipsoid in p-space that covers about half the data. For residual space, it uses the high breakdown least trim squares (LTS) fit for an initial fit. See Rousseeuw and Van Driessen (1999).

A brief description of the weights are given next. These are the weights defined for the R function `hbrfit` which are in the R package `npsmReg2` and are discussed in Sects. 7.2 and 7.3 of Kloke and McKean (2014). This package can be downloaded at the github site indicated at the end of Sect. 1.1.

Let $\hat{\boldsymbol{e}}_0$ denote the residuals from the initial LTS fit. Let $\boldsymbol{V}$ denote the MCD with center $\boldsymbol{v}_c$. Define the function $\psi(t)$ by $\psi(t) = 1,\ t,\ \text{ or } \ -1$ according as $t \geq 1$, $-1 < t < 1$, or $t \leq -1$. Let σ be estimated by the initial scaling estimate MAD $=$ $1.483\ \text{med}_i|\hat{e}_i^{(0)} - \text{med}_j\{\hat{e}_j^{(0)}\}|$. Letting $Q_i = (\boldsymbol{x}_i - \boldsymbol{v}_c)'\boldsymbol{V}^{-1}(\boldsymbol{x}_i - \boldsymbol{v}_c)$, define

$$m_i = \psi\left(\frac{b}{Q_i}\right) = \min\left\{1, \frac{b}{Q_i}\right\}.$$

Consider the weights

$$\hat{b}_{ij} = \min\left\{1, \frac{c\hat{\sigma}}{|\hat{e}_i^{(0)}|}\frac{\hat{\sigma}}{|\hat{e}_j^{(0)}|} \min\left\{1, \frac{b}{\hat{Q}_i}\right\} \min\left\{1, \frac{b}{\hat{Q}_j}\right\}\right\}, \tag{1.25}$$

where b and c are tuning constants. We set b at the upper $\chi^2_{.05}(p)$ quantile and c is set as

$$c = [\text{med}\{a_i\} + 3MAD\{a_i\}]^2,$$

where $a_i = \hat{e}_i^{(0)}/(MAD \cdot Q_i)$. From this point-of-view, it is clear that these weights downweight both outlying points in factor space and outlying responses. Note that the initial residual information is a multiplicative factor in the weight function. Hence, a good leverage point will generally have a small (in absolute value) initial residual which will offset its distance in factor space.

In general, the HBR estimator has a 50 % breakdown point, provided the initial estimates used in forming the weights have 50 % breakdown. Further, its influence function is a bounded function in both the Y and the $\boldsymbol{x}$-spaces, is continuous everywhere, and converges to zero as the point $(\boldsymbol{x}^*, Y^*)$ gets large in any direction. The asymptotic distribution of $\hat{\boldsymbol{\beta}}_{HBR}$ is asymptotically normal. As with all high breakdown estimates, $\hat{\boldsymbol{\beta}}_{HBR}$ is less efficient than the Wilcoxon estimates but it regains some of the efficiency if the weights depend only on factor space.

McKean et al. (1996) developed diagnostics to detect differences in highly efficient and high breakdown robust fits. Their diagnostic TDBETA measures the total difference in fits of the regression coefficients, standardized by the variance-covariance of the Wilcoxon fit. The benchmark is similar to the classic diagnostic DFFITS. A second diagnostic CFITS measures the difference of predicted values at each case. This diagnostic is useful for data sets where TDBETA exceeds its benchmark. Section 7.3 of Kloke and McKean (2014) give a full discussion of these diagnostics with examples. McKean et al. (1999) extended these diagnostics to differences between robust and LS fits. Next, we present an example which illustrates HBR fits and these diagnostics.

1.4.2.1 Stars Data

This data set is drawn from an astronomy study on the star cluster CYG OB1 which contains 47 stars; see Chap. 3 of Hettmansperger and McKean (2011) for discussion. The response is the logarithm of the light intensity of the star while the predictor is the logarithm of the temperature of the star. The scatterplot of the data is in the left panel of Fig. 1.2. Four of the stars, called giants, form a cluster of outliers in factor space while the rest of the stars fall in a point cloud. The panel includes the overlay plot of the Wilcoxon and HBR linear fits. The four giants form a cluster of high leverage points, exerting a strong influence on the Wilcoxon fit while having a minor influence on the HBR fit. The diagnostic TDBETAS between the Wilcoxon and HBR fits has the value 67.92 which exceeds the benchmark of 0.340, indicating a large difference in the fits. The right panel of Fig. 1.2 shows the values of the diagnostic CFITS versus case. The benchmark for this diagnostic is 0.34. The four largest values are the four giant stars. Hence, for this data set, the diagnostics work. The diagnostic TDBETAS alerts the user to the large difference between the fits and CFITS indicates the major points contributing to this difference. The next two largest CFITS values are of interest to astronomers, also. These are stars between the giants and the main sequence stars. Although not shown, the least squares fit is similar to the Wilcoxon fit. The fits and diagnostics can be computed with the following code:

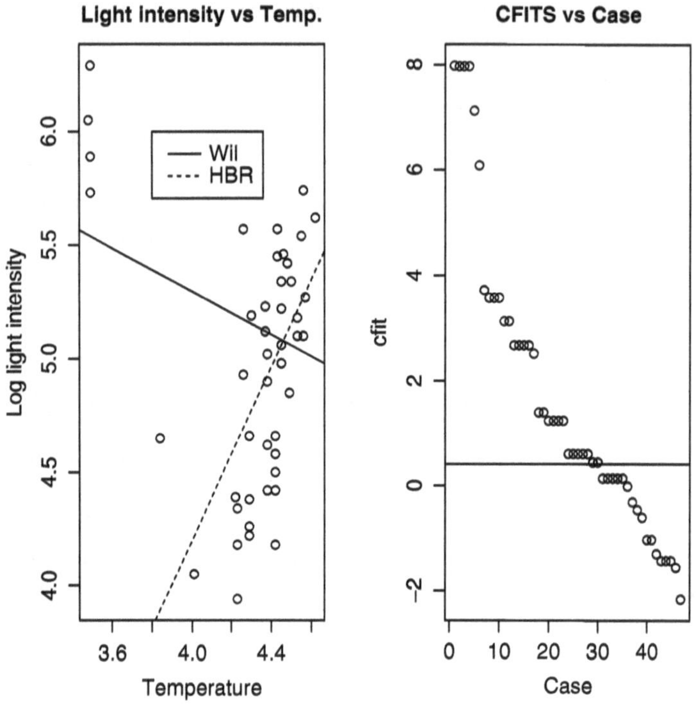

Fig. 1.2 The *left panel* displays the scatterplot of the log of light intensity versus the temperature of the star, overlaid with the Wilcoxon and HBR fits. The *right panel* displays the values of CFITS versus the case numbers. The *horizontal line* is the benchmark value

```
fitw <- rfit(lintensity ~ temp)
fith <- hbrfit(lintensity ~ temp)
fitls <- lm(y ~ x)
fitsdiag <- fitdiag(temp,lintensity,est=c("WIL",
                                          "HBR"))
```

1.5 Extensions to Mixed and Nonlinear Models

In the past 20 years, there have been extensions of the rank-based analysis to many other models. This includes nonlinear models and models with dependency among the responses. In this section, we briefly discuss a few of these models, ending with an example involving a mixed model.

For traditional least squares-based methods for these models, the geometry essentially remains the same in that least squares estimation is based on minimizing the squared-Euclidean distance between the vector of responses and the region of estimation. This is true of the rank-based approach, also, except that the rank-based norm, (1.3), replaces the squared-Euclidean norm.

1.5.1 Multivariate Linear Models

Davis and McKean (1993) extended the linear model rank-based procedures for general score functions to the multivariate linear model. They developed asymptotic theory for the estimators and tests of linear hypotheses of the form $\boldsymbol{A\beta K}$ for the matrix of regression coefficients $\boldsymbol{\beta}$. See Sect. 6.6 in Hettmansperger and McKean (2011) for discussion and examples. These are component wise estimators, so computations can be based on the package `Rfit`; see the web site indicated at the end of Sect. 1.1 to download a preversion of the package `Rfitmult`. These methods are regression equivariant but they are not affine invariant. Oja (2010) and his collaborators developed affine invariant rank-based procedures for Wilcoxon scores using a transformation retransformation procedure. Nordhausen and Oja (2011) developed the R package `MNM`, downloadable at CRAN, to compute these affine procedures.

1.5.2 Nonlinear Linear Models

For responses y_i, consider a nonlinear model of the form $y_i = g(\boldsymbol{\theta}, \boldsymbol{x}_i) + e_i$, $i = 1, \ldots, n$, where g is a specified nonlinear function, $\boldsymbol{\theta}$ is a $k \times 1$ vector of unknown parameters, and $\boldsymbol{x}_i$ is a $p \times 1$ vector of predictors. Let $\boldsymbol{y}$ and $g(\boldsymbol{\theta}, \boldsymbol{x})$ denote the corresponding $n \times 1$ vectors. Given a rank score function $\varphi(u)$, the associated rank-based estimator of $\boldsymbol{\theta}$ is

$$\hat{\boldsymbol{\theta}}_{\varphi} = \text{Argmin}\|\boldsymbol{y} - g(\boldsymbol{\theta}, \boldsymbol{x})\|_{\varphi},$$

where $\|\cdot\|_{\varphi}$ is the norm defined in expression (1.3). Abebe and McKean (2007) obtained asymptotic theory for $\hat{\boldsymbol{\theta}}_{\varphi}$ for the case of Wilcoxon scores. The efficiency properties of the Wilcoxon estimator are the same as in the linear model case; so, the estimator is highly efficient for the nonlinear model. Abebe and McKean (2013) extended this development to high breakdown rank-based nonlinear estimators of $\boldsymbol{\theta}$ which have bounded influence in both the response and factor spaces. The R package `npsmReg2` contains the R function `wilnl` which computes these estimators; see Chap. 7 of Kloke and McKean (2014) for further discussion.

1.5.3 Time Series Models

Consider the autoregressive model of order p, $Ar(p)$:

$$\begin{aligned} X_i &= \phi_0 + \phi_1 X_{i-1} + \phi_2 X_{i-2} + \cdots + \phi_p X_{i-p} + e_i \\ &= \phi_0 + \boldsymbol{Y}'_{i-1}\boldsymbol{\phi} + e_i, \quad i = 1, 2, \ldots, n \end{aligned} \qquad (1.26)$$

where $p \geq 1$, $\boldsymbol{Y}_{i-1} = (X_{i-1}, X_{i-2}, \ldots, X_{i-p})'$, $\boldsymbol{\phi} = (\phi_1, \phi_2, \ldots, \phi_p)'$, and $\boldsymbol{Y}_0$ is an observable random vector independent of $\boldsymbol{e}$. Let $\boldsymbol{X}$ and $\boldsymbol{Y}$ denote the corresponding $n \times 1$ vector and the $n \times p$ matrix with components X_i and $\boldsymbol{Y}'_{i-1}$, respectively. For the score function $\varphi(u)$, the rank-based estimator of $\boldsymbol{\phi}$ is given by

$$\hat{\boldsymbol{\phi}}_\varphi = \text{Argmin}\|\boldsymbol{X} - \boldsymbol{Y}'\boldsymbol{\phi}\|_\varphi,$$

where $\boldsymbol{Y}$ is the matrix with rows $\boldsymbol{Y}'_{i-1}$. Koul and Saleh (1993) developed the asymptotic theory for these rank-based estimates. Because of the structure of the $AR(p)$ model, outliers in the random errors become ensuing points of high leverage. As a solution to this problem, Terpstra et al. (2000, 2001) proposed estimating $\boldsymbol{\phi}$ using the HBR estimators of Sect. 1.4. They obtained the corresponding asymptotic theory for these HBR estimators and showed their validity and empirical efficiency in several large simulation studies. Section 7.8 of Kloke and McKean (2014) discusses the computation of these estimates using the R package `Rfit`.

1.5.4 Cluster Correlated Data

Frequently in practice data are collected in clusters. Common examples include: repeated measures on subjects, experimental designs involving blocks, clinical studies over multiple centers, and hierarchical (nested) designs. Generally, the observations within a cluster are dependent.

For discussion, suppose we have m such clusters. Let y_{ki} denote the *ith* response within the *kth* cluster, for $i = 1, \ldots, n_k$ and $k = 1, \ldots, m$, and let $\boldsymbol{x}_{ki}$ denote the corresponding $p \times 1$ vector of covariates. For cluster k stack the n_k responses in the vector $\boldsymbol{y}_k$ and let $\boldsymbol{X}_k$ denote the $n_k \times p$ matrix with rows $\boldsymbol{x}'_{ki}$. Assume a linear model of the form

$$\boldsymbol{y}_k = \beta_0 \mathbf{1}_{n_k} + \boldsymbol{X}_k\boldsymbol{\beta} + \boldsymbol{e}_k, \quad k = 1, \ldots, m, \tag{1.27}$$

where $\boldsymbol{e}_k$ follows a n_k-multivariate distribution and the vectors $\boldsymbol{e}_1, \ldots, \boldsymbol{e}_m$ are independent. We then stack the vectors $\boldsymbol{y}_k$ and matrices $\boldsymbol{X}_k$ into the vector $\boldsymbol{Y}$ and matrix $\boldsymbol{X}$, respectively.

There are several rank-based analyses available for these models. Abebe et al. (2016) develop a rank-based analysis for a general estimating equations (GEE) model which includes models of the form (1.27). This allows for very general dependency structure. For a general score function $\varphi(u)$, Kloke et al. (2009) developed the asymptotic theory for the rank-based estimator defined in expression (1.5), i.e., the minimizer of the norm $\|\boldsymbol{Y} - \boldsymbol{X}\boldsymbol{\beta}\|_\varphi$. Their development includes consistent estimators of standard errors and consistent test statistics of general linear hypotheses. The theory requires the additional assumption that the univariate marginal distributions of $\boldsymbol{e}_k$ are the same. This is true for many of the usual models in practice such as simple mixed models (compound symmetry covariance structure) and stationary times series models for the clusters.

The R package `jrfit` computes the analysis developed by Kloke et al. (2009); see Chap. 8 of Kloke and McKean (2014) for discussion and examples. This includes the fit and several options for the estimation of the covariance structure, including compound symmetry and two general estimators, (a sandwich-type estimator and a general nonparametric estimator). We conclude this discussion with an example which illustrates the use of `jrfit` on cluster data.

1.5.4.1 Example

For an example, we consider the first base study presented in Hollander and Wolfe (1999). This study investigated three methods of rounding first base for baseball players who are running from home plate to second base. The response is the player's total running time. The three methods are narrow angle (NA), round out (RP), and wide angle (WA); see Hollander and Wolfe for details. Twenty-two ball players participated in the study and each ran six times, two repetitions for each method. The average time of the two repetitions are the response times available. The data can be found in the `firstbase` data set in the R package `npsmReg2`; see Chap. 8 of Kloke and McKean (2014).

Let y_{ij} denote the running time for the *jth* player on method i and consider the randomized block design given by

$$y_{ij} = \mu + \alpha_i + b_j + \epsilon_{ij}, \tag{1.28}$$

where α_i denotes the *ith* treatment fixed effect; b_j denotes the random effect for the *jth* player; and e_{ij} denotes the random error. We assume that the random errors are iid and the random effects are iid with different distributions. We further assume that the random errors and the random effects are independent.

Although, finite variance is not required for the asymptotic theory, for the discussion, we assume finite variances. The variance-covariance structure of Model (1.28) is compound symmetric. Besides fixed effects analyses, we are interested in the estimation of the variance components given by σ_b^2 the variance of b_j, σ_ϵ^2 the variance of ϵ_{ij}, and the intraclass correlation coefficient $\rho = \sigma_b^2/(\sigma_b^2 + \sigma_\epsilon^2)$. Kloke et al. (2009) provided robust estimates of these components based on the rank-based fit of Model (1.28). These estimates have been incorporated into the package `jrfit`.

The null hypothesis for the fixed effects is $H_0 : \alpha_1 = \alpha_2 = \alpha_3 = 0$. The traditional nonparametric test of this hypothesis is based on Friedman's test statistic, which for this example results in the value of 11.14 with p-value 0.003. The comparative rank-based analysis is the Wald-type test on the rank-based fit. As shown below, the value of the test statistic is 19.31 with a p-value of 0.0001. As with the Friedman test, the Wald-type test is highly significant. An experimenter, though, wants a much more in depth analysis than just this test of the fixed effects. The left

Table 1.2 Rank-based estimates of fixed effects and variance components, firstbase data

	Fixed effects		Variance
	Effect	SE	components
$\mu_{RO} - \mu_{NA}$	0.000	0.016	$\hat{\sigma}_B^2 = 0.0124$
$\mu_{WA} - \mu_{NA}$	−0.053	0.012	$\hat{\sigma}_\epsilon^2 = 0.0049$
$\mu_{WA} - \mu_{RO}$	−0.057	0.023	$\hat{\rho} = 0.715$

panel of Fig. 1.3 displays the comparative boxplots of the methods. Note that, based on this plot, it appears that the wide-angle method results in the quickest times. Such a judgement is easily confirmed by considering the estimates and confidence intervals for the three pairwise comparisons. These are shown in Table 1.2 based on the rank-based fit. They do confirm that the wide-angle method results in significantly faster times than the other two methods. Furthermore, the estimated effects provide the experimenter with an estimate of how much faster the wide-angle method is than the other two methods.

Table 1.2 also displays the robust estimates of the variance components. Note that the estimate of the intraclass correlation coefficient is 0.715 indicating a strong correlation among the times of a runner. The right panel of Fig. 1.3 shows the normal $q-q$ plot of the Studentized residuals of the rank-based fit. The horizontal lines at ± 2 are the usual benchmark for potential outliers. This plot confirms the outliers in the boxplots and indicates a heavy tailed error structure. The three largest positive outliers correspond to Runner 22 who had the slowest times in all three methods.

The results in Table 1.2 and Fig. 1.3 are based on computations using the package `jrfit`. Some of the code for the computations is given by:

```
#  The data are in the data set firstbase in the
#  package npsmReg2. More discussion of the
#  computations can be found in Chapter 8
#  of Kloke McKean (2014).
#
#  times is the vector of running times; player is
#  the indicator of the player; method is the
#  indicator of the method.
   xmat <- model.matrix(~as.factor(method))[,2:3]
   fit <- jrfit(xmat,times,player)
   stud <- rstudent(fit)       #Studentized residuals
   vee(fit$resid,fit$block,method='mm')      #Var comp.
   h1 <-c(0,1,0); h2<-c(0,0,1);  hmat<-rbind(h1,h2)
   mid <- solve(hmat%*%fit$varhat%*%t(hmat))
   tst <- t(hmat%*%fit$coef)%*%mid%*%hmat%*%fit$coef
   19.31442
   1-pchisq(tst,2)
   6.396273e-05
```

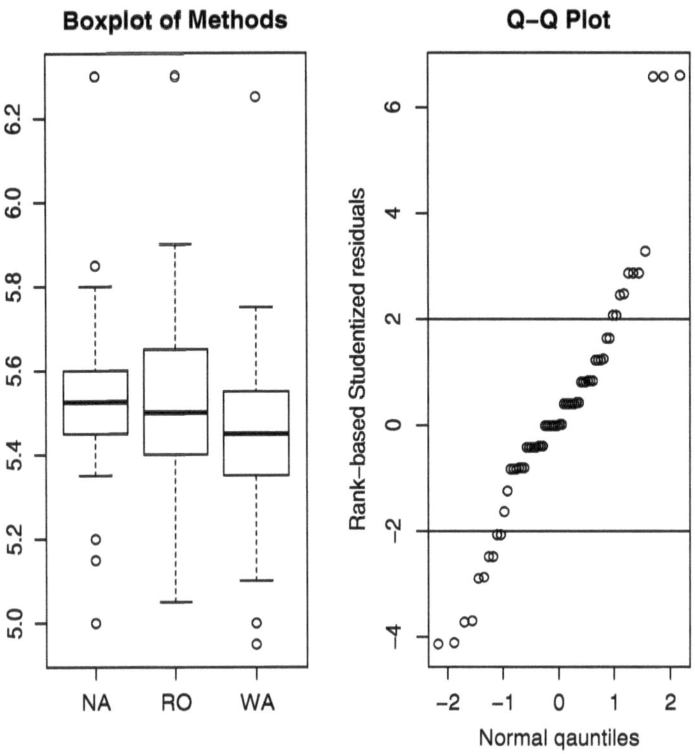

Fig. 1.3 The *left panel* displays the comparative boxplots for the three methods of rounding first base, while the right panel shows the $q-q$ plot of the rank-based Studentized residuals

1.6 Conclusion

As we indicated in Sect. 1.1, the rank tests for simple location problems were initially used because of their quick calculation in the pre computer age. These methods were further found to be highly efficient and robust by Hodges and Lehmann in the mid 1950s. The traditional t-tests for these problems, though, are based on least squares (LS) fitting which easily generalizes to much more complicated models, including linear, nonlinear, and models with dependent error structure. For all such models, the LS fitting is based on minimizing the squared-Euclidean distance between the vector (or matrix) of responses and the region (space) of estimation. Further, LS testing of general linear hypotheses is based on a comparison of distances between the vector of responses and full and reduced model subspaces. Also, there are ample diagnostic procedures to check the quality of the LS fit of a model. As in the location problems, though, LS procedures are not robust. Hence, a generalization was needed for the robust nonparametric methods.

As briefly outlined in Sect. 1.2, the extension of nonparametric methods to linear models came about in the late 1960s and early 1970s, with the robust estimation procedures developed by Jurečková and Jaeckel. In particular, as shown by McKean and Schrader (1980), Jaekel's estimation involves minimizing a distance between the responses and the full model subspace. This distance is based on the norm defined in expression (1.3). For general linear hypotheses, McKean and Hettmansperger (1976) developed an accompanying analysis based on a comparison of distances between the vector of responses and full and reduced model spaces where distance is based on the norm (1.3). Diagnostics for this rank-based analysis were developed by McKean et al. (1990). This rank-based analysis is as general as the traditional LS analysis. As with LS, for any linear model, it offers a complete procedure including fitting, diagnostic checking of the fit, confidence regions, and tests of general linear hypotheses. Details of this analysis are discussed in Chaps. 3–5 of Hettmansperger and McKean (2011).

As discussed in Sect. 1.3, the rank-based analysis is highly efficient. For example, the rank-based procedure based on Wilcoxon scores has efficiency 0.955 relative to LS procedures when the random errors are normally distributed and is much more efficient when the distribution of the random errors has heavy tails. Further, if the form of the error distribution is known, then optimal scores can be used which results in fully asymptotically efficient procedures. Rank-based procedures based on minimizing the norm (1.3) are robust in response space but, similar to LS procedures, are not robust in factor space. A simple weighting scheme, based on robust distances in factor space and residuals from an initial robust fit, leads to a robust rank-based procedure which is robust in both response and factor space as well as having a 50 % breakdown point.

As reviewed in Sect. 1.5, these rank-based procedures have been extended to nonlinear models and models in which the errors have dependencies. For these models, LS fitting is still based on minimizing squared-Euclidean distance between the responses and the space of estimation. In the same way, the rank-based fitting of these models is obtained by minimizing the distance based on the norm (1.3). In recent years, asymptotic theory has been developed for these rank-based procedures. Hence, besides linear models, robust rank-based procedures exist for diverse models, including nonlinear models, autoregressive times series models, multivariate regression models, mixed models, and hierarchical models.

The easy computation of rank-based analyses is performed with R software. For linear models, the package `Rfit` offers a complete computation for the rank-based analysis. A wide variety of scores functions are intrinsic to the package with an option for user-supplied scores. For models other than linear there are accompanying R packages for computations. We have discussed the computation based on these packages throughout this paper. See Kloke and McKean (2014) for discussion of these packages.

References

Abebe, A., & McKean, J. W. (2007). Highly efficient nonlinear regression based on the Wilcoxon norm. In D. Umbach (Ed.), *Festschrift in Honor of Mir Masoom Ali on the Occasion of his Retirement* (pp. 340–357).

Abebe, A., & McKean, J. W. (2013). Weighted Wilcoxon estimators in nonlinear regression. *Australian & New Zealand Journal of Statistics, 55*, 401–420.

Abebe, A., McKean, J. W., Kloke, J. D., & Bilgic, Y. (2016). Iterated reweighted rank-based estimates for GEE models. In R. Y. Liu & J. W. McKean (Eds.), *Robust rank-based and nonparametric methods*. New York: Springer

Arbuthnott, J. (1710). An argument for divine providence taken from the constant regularity observed in the birth of both sexes. *Philosophical Transactions, 27*, 186–190.

Chang, W., McKean, J. W., Naranjo, J. D., & Sheather, S. J. (1999). High breakdown rank-based regression. *Journal of the American Statistical Association, 94*, 205–219.

Davis, J., & McKean, J. W. (1993). Rank based methods for multivariate linear Models. *The Journal of the American Statistical Association, 88*, 241–251

Hájek, J., & Šidák, Z. (1967). *Theory of rank tests*. New York: Academic.

Hájek, J., Šidák, Z., & Sen, P. K. (1999). *Theory of rank tests* (2nd ed.). New York: Academic.

Hampel, F. R. (1974). The influence curve and its role in robust estimation. *Journal of the American Statistical Association, 69*, 383–393.

Hettmansperger, T. P., & McKean, J. W. (2011). *Robust nonparametric statistical methods* (2nd ed.). Boca Raton, FL: Chapman-Hall.

Hodges, J. L., Jr., & Lehmann, E. L. (1956). The efficiency of some nonparametric competitors of the t-test. *Annals of Mathematical Statistics, 27*, 324–335.

Hodges, J. L., Jr., & Lehmann, E. L. (1960). Comparison of the normal scores and Wilcoxon tests. In *Proceedings 4th Berkeley Symposium* (Vol. 1, pp. 307–317).

Hodges, J. L., Jr., & Lehmann, E. L. (1963). Estimates of location based on rank tests. *Annals of Mathematical Statistics, 34*, 598–611.

Hollander, M., & Wolfe, D. (1999). *Nonparametric statistical methods* (2nd ed.). New York: Wiley.

Huber, P. J. (1964). Robust estimation of a location parameter. *Annals of Mathematical Statistics, 35*, 73–101.

Jaeckel, L. A. (1972). Estimating regression coefficients by minimizing the dispersion of the residuals. *Annals of Mathematical Statistics, 43*, 1449–1458.

Jurečková, J. (1969). Asymptotic linearity of rank statistics in regression parameters. *Annals of Mathematical Statistics, 40*, 1449–1458.

Jurečková, J. (1971). Nonparametric estimate of regression coefficients. *Annals of Mathematical Statistics, 42*, 1328–1338.

Kloke, J. D., & McKean, J. W. (2012). Rfit: Rank-based estimation for linear models. *The R Journal, 4*, 57–64.

Kloke, J. D., & McKean, J. W. (2014). *Nonparametric statistical methods using R*. Boca Raton, FL: Chapman-Hall.

Kloke, J., McKean, J. W., & Rashid, M. (2009). Rank-based estimation and associated inferences for linear models with cluster correlated errors. *Journal of the American Statistical Association, 104*, 384–390.

Koul, H. L., & Saleh, A. K. M. E. (1993). R-estimation of the parameters of autoregressive [AR(p)] models. *The Annals of Statistics, 21*, 534–551.

Koul, H. L., Sievers, G. L., & McKean, J. W. (1987). An estimator of the scale parameter for the rank analysis of linear models under general score functions. *Scandinavian Journal of Statistics, 14*, 131–141.

Mann, H. B., & Whitney, D. R. (1947). On a test of whether one or two random variables is stochastically larger than the other. *Annals of Mathematical Statistics, 18*, 50–60.

McKean, J. W., Jr. (1975). *Tests of Hypotheses Based on Ranks in the General Linear Model*. Ph.D. dissertation, University Park, Penn State University.

McKean, J. W., & Hettmansperger, T. P. (1976). Tests of hypotheses of the general linear model based on ranks. *Communications in Statistics, Part A-Theory and Methods, 5*, 693–709.

McKean, J. W., & Kloke, J. D. (2014). Efficient and adaptive rank-based fits for linear models with skewed-normal errors. *Journal of Statistical Distributions and Applications, 1*, 18. http://www.jsdajournal.com/content/1/1/18.

McKean, J. W., Naranjo, J. D., & Sheather, S. J. (1996). Diagnostics to detect differences in robust fits of linear models. *Computational Statistics, 11*, 223–243.

McKean, J. W., Naranjo, J. D., & Sheather, S. J. (1999). Diagnostics for comparing robust and least squares fits. *Journal of Nonparametric Statistics, 11*, 161–188.

McKean, J. W., & Schrader, R. (1980). The geometry of robust procedures in linear models. *Journal of the Royal Statistical Society, Series B, Methodological, 42*, 366–371.

McKean, J. W., & Sheather, S. J. (1991). Small sample properties of robust analyses of linear models based on r-estimates. In W. Stahel & S. Weisberg (Eds.), *Directions in robust statistics and diagnostics, part II* (Vols. 1–20). New York: Springer.

McKean, J. W., Sheather, S. J., & Hettmansperger, T. P. (1990). Regression diagnostics for rank-based methods, *Journal of the American Statistical Association, 85*, 1018–1028.

Noether, G. E. (1955). On a theorem of Pitman. *Annals of Mathematical Statistics, 26*, 64–68.

Nordhausen, K., & Oja, H. (2011). Multivariate L1 methods: The package MNM. *Journal of Statistical Software, 43*(5), 1–28. http://www.jstatsoft.org/v43/i05/.

Oja, H. (2010). *Multivariate nonparametric methods with R*. New York: Springer.

Pitman, E. J. G. (1948). *Notes on nonparametric statistical inference* (Unpublished notes).

Rousseeuw, P., & Van Driessen, K. (1999). A fast algorithm for the minimum covariance determinant estimator. *Technometrics, 41*, 212–223.

Terpstra, J., McKean, J. W., & Naranjo, J. D. (2000). Highly efficient weighed Wilcoxon estimates for autoregression. *Statistics, 35*, 45–80.

Terpstra, J., McKean, J. W., & Naranjo, J. D. (2001). GR-estimates for an autoregressive time series. *Statistics and Probability Letters, 51*, 172–180.

Tukey, J. W. (1949). *The simplest signed rank tests*. Princeton University Stat. Res. Group, Memo Report no. 17.

Wilcoxon, F. (1945). Individual comparisons by ranking methods. *Biometrics, 1*, 80–83.

Chapter 2
Robust Signed-Rank Variable Selection in Linear Regression

Asheber Abebe and Huybrechts F. Bindele

Abstract The growing need for dealing with big data has made it necessary to find computationally efficient methods for identifying important factors to be considered in statistical modeling. In the linear model, the Lasso is an effective way of selecting variables using penalized regression. It has spawned substantial research in the area of variable selection for models that depend on a linear combination of predictors. However, work addressing the lack of optimality of variable selection when the model errors are not Gaussian and/or when the data contain gross outliers is scarce. We propose the weighted signed-rank Lasso as a robust and efficient alternative to least absolute deviations and least squares Lasso. The approach is appealing for use with big data since one can use data augmentation to perform the estimation as a single weighted L_1 optimization problem. Selection and estimation consistency are theoretically established and evaluated via simulation studies. The results confirm the optimality of the rank-based approach for data with heavy-tailed and contaminated errors or data containing high-leverage points.

Keywords Adaptive Lasso • Wilcoxon estimation • Oracle property • Penalized least squares • LAD regression

2.1 Introduction

The growing need for dealing with 'big data' has made it necessary to find ways of determining the few important factors to consider in the statistical modeling. In the linear and generalized linear models, this translates to identifying the covariates

A. Abebe (✉)
Department of Mathematics and Statistics, Auburn University, 221 Parker Hall, Auburn, AL 36849, USA
e-mail: ash@auburn.edu

H.F. Bindele
Department of Mathematics and Statistics, University of South Alabama, 411 University Blvd. N., ILB 316, Mobile, AL 36688-0002, USA
e-mail: hbindele@southalabama.edu

© Springer International Publishing Switzerland 2016

R.Y. Liu, J.W. McKean (eds.), *Robust Rank-Based and Nonparametric Methods*, Springer Proceedings in Mathematics & Statistics 168,
DOI 10.1007/978-3-319-39065-9_2

that are most needed in the prediction of the outcome. In this regard, the Lasso method introduced in Tibshirani (1996) has garnered significant attention in the past two decades. The Lasso method takes advantage of the singularity of the L_1 penalty to effectively select variables via the penalized least squares procedure. This work has been refined and extended in various directions. See, for example, Fan and Li (2001), Zou and Hastie (2005), Zou (2006), Wang and Leng (2008), and references therein. Much of the focus has been in establishing the so-called "Oracle" property Fan and Li (2001) that consists of selection consistency and estimation efficiency. These are both asymptotic properties where selection consistency refers to ones ability to correctly identify the zero regression coefficients while estimation efficiency refers to ones ability to provide a $\sqrt{n}$-consistent estimator of the non-zero coefficients.

However, there are not too many results that address the lack of optimality of these variable selection procedures when the model errors are not Gaussian and/or when the data contain gross outliers. An approach based on penalized Jaeckel-type rank-regression was discussed in Johnson and Peng (2008), Johnson et al. (2008), Johnson (2009), Leng (2010) and Xu et al. (2010). The computation is complicated and, as in unpenalized rank-regression, the approach used in these papers will only result in robustness in the response space. For variable selection, however, getting a handle on leverage is crucial. One paper that discussed this issue and tried to address the influence of high leverage points is Wang and Li (2009), where they considered penalized weighted Wilcoxon estimation. Our proposed approach based on minimization of a penalized weighted signed-rank norm is much simpler to compute and provides protection against outliers and high-leverage points. It also allows one flexibility through choice of score generating functions. One limitation of our proposed approach is that it requires symmetry of the error density. In this case, the estimates are equivalent to Jaeckel-type rank-regression estimates.

Consider the linear regression model given by

$$y_i = \boldsymbol{x}_i'\boldsymbol{\beta}_0 + e_i, \qquad 1 \leq i \leq n, \tag{2.1}$$

where $\boldsymbol{\beta}_0 \in \mathscr{B} \subset \mathbb{R}^d$ is a vector of parameters, $\boldsymbol{x}_i$ is a vector of independent variables in a vector space $\mathbb{X}$, and the errors e_i are assumed to be i.i.d. with a distribution function F. Let $\mathbf{V}_n = \{(y_1, \boldsymbol{x}_1), \ldots, (y_n, \boldsymbol{x}_n)\}$ be the set of sample data points. Note that $\mathbf{V}_n \subset \mathbb{V} \equiv \mathbb{R} \times \mathbb{X}$. We shall assume that $\mathscr{B}$ is a compact subspace of $\mathbb{R}^d$, $\boldsymbol{\beta}_0$ is an interior point of $\mathscr{B}$.

Rank-based approaches have been shown to possess a high breakdown property resulting on robust and efficient estimators. The rank-based approach considered in this paper is based on the so-called the weighted signed-rank (WSR) norm proposed in Bindele and Abebe (2012) for estimation of coefficients of general nonlinear models. Here we consider WSR with added penalty for simultaneous estimation and variable selection in linear models. That is, we obtained an estimator $\hat{\boldsymbol{\beta}}_n$ of $\boldsymbol{\beta}_0$ satisfying

$$\hat{\boldsymbol{\beta}}_n = \operatorname*{Argmin}_{\boldsymbol{\beta} \in \mathscr{B}} Q(\boldsymbol{\beta}), \tag{2.2}$$

where $Q(\boldsymbol{\beta})$ is a penalized WSR objective function

$$Q(\boldsymbol{\beta}) = D_n(\mathbf{V}_n, w, \boldsymbol{\beta}) + n\sum_{j=1}^{d} P_{\lambda_j}(|\beta_j|). \tag{2.3}$$

and $D_n(\mathbf{V}_n, w, \boldsymbol{\beta})$ is the WSR dispersion function defined by

$$D_n(\mathbf{V}_n, w, \boldsymbol{\beta}) = \sum_{i=1}^{n} w(\boldsymbol{x}_i)a_n(i)|z(\boldsymbol{\beta})|_{(i)}. \tag{2.4}$$

Here $z_i(\boldsymbol{\beta}) = y_i - \boldsymbol{x}_i'\boldsymbol{\beta}$, $|z(\boldsymbol{\beta})|_{(i)}$ is the ith ordered value among $|z_1(\boldsymbol{\beta})|, \ldots, |z_n(\boldsymbol{\beta})|$, and the numbers $a_n(i)$ are scores generated as $a_n(i) = \varphi^+(i/(n+1))$, for some bounded and non-decreasing score function $\varphi^+ : (0,1) \to \mathbb{R}^+$ that has at most a finite number of discontinuities. The function $w : \mathbb{X} \to \mathbb{R}^+$ is a continuous weight function. The penalty function $P_{\lambda_j}(\cdot)$ is defined on $\mathbb{R}^+$. When the penalty function is the Lasso penalty Tibshirani (1996) $P_{\lambda_j}(|t|) = \lambda|t|$ for all j, we will refer to the resulting estimator as the WSR-Lasso (WSR-L), and when the penalty function is the adaptive Lasso Zou (2006) $P_{\lambda_j}(|t|) = \lambda_j|t|$, we will refer to the estimator as WSR-Adaptive Lasso (WRS-AL) estimator. We should point out that for $\varphi^+ \equiv 1$, the objective function in (2.3) reduces to the WLAD-Lasso discussed in Arslan (2012). If additionally $w \equiv 1$, then we obtain the LAD-lasso discussed in Wang et al. (2007). While these LAD based estimators are easy to compute and provide robust estimators, they lack efficiency especially when the error density at zero is small (Hettmansperger and McKean 2011; Leng 2010). Note that, while not stressed in our notation, $\check{\boldsymbol{\beta}}_n$ depends on the tuning parameter $\boldsymbol{\lambda} = (\lambda_1, \ldots, \lambda_d)'$.

Using the same idea in Wang et al. (2007), either under WSR-L or WSR-AL, one can write $Q(\boldsymbol{\beta})$ as

$$Q(\boldsymbol{\beta}) = \sum_{i=1}^{n+d} \nu_i |z_i^*(\boldsymbol{\beta})|, \tag{2.5}$$

where $z_i^*(\boldsymbol{\beta}) = y_i^* - \boldsymbol{x}_i^{*\prime}\boldsymbol{\beta}$ with

$$(y_i^*, \boldsymbol{x}_i^*)' = \begin{cases} (y_i, \boldsymbol{x}_i)', & \text{for } 1 \le i \le n, \\ (0, n\lambda_i \mathbf{e}_i)', & \text{for } n+1 \le i \le n+d. \end{cases} \tag{2.6}$$

and

$$\nu_i = \begin{cases} w(\boldsymbol{x}_i)\varphi^+\left(\frac{R(z_i(\boldsymbol{\beta}))}{n+1}\right), & \text{for } i \le n, \\ 1, & \text{for } i > n. \end{cases}$$

Here $\mathbf{e}_i$ is the d-dimensional vector with ith component equal to 1 and all the others equal to 0. To this end, Eq. (2.5) can be seen as the weighted L_1 objective function. In Eq. (2.6) the WSR-L objective function is obtained by putting $\lambda_i = \lambda$ for all i. To avoid any possible confusion, we will use $Q_\ell^w(\cdot)$ and $Q_{a\ell}^w(\cdot)$ for WSR-L and WSR-AL objective functions, respectively.

Remark 2.1. Considering the unpenalized objective function $D_n(\mathbf{V}_n, w, \boldsymbol{\beta})$ defined in Eq. (2.4), asymptotic properties (consistency and $\sqrt{n}$-asymptotic normality) of the WSR estimator with $w \equiv 1$ were established under mild regularity conditions in Hössjer (1994). Considering the weighted case, analogous asymptotic results were obtained by Bindele and Abebe (2012) for general nonlinear regression model.

2.2 Asymptotics

In this section, we provide the asymptotic properties of the WSR-AL estimator defined in (2.2) under regularity conditions. Consider the following assumptions

(I_1) $P(\boldsymbol{x}'\boldsymbol{\beta} = \boldsymbol{x}'\boldsymbol{\beta}_0) < \alpha$ for all $\boldsymbol{\beta} \neq \boldsymbol{\beta}_0$, $0 < \alpha \leq 1$, and $E_G[|\boldsymbol{x}|^r] < \infty$ for some $r > 1$, G being the distribution of $\boldsymbol{x}$.

(I_2) The density f of ε is symmetric about zero, strictly decreasing on $\mathbb{R}^+$, and absolutely continuous with finite Fisher information. Its derivative f' is bounded and $E_F(|\varepsilon|^r) < \infty$ for some $r > 1$.

These two assumptions ensure the strong consistency of $\tilde{\boldsymbol{\beta}}_n = \text{Argmin}_{\boldsymbol{\beta}}\, D_n(\mathbf{V}_n, w, \boldsymbol{\beta})$.

2.2.1 *Consistency and Asymptotic Normality*

We shall assume that $p_0 \leq d$ of the true regression parameters are nonzero. Thus, without loss of generality, we assume $\beta_{0j} \neq 0$ for $j \leq p_0$ and $\beta_{0j} = 0$ for $j > p_0$. Thus $\boldsymbol{\beta}_0$ can be partitioned as $\boldsymbol{\beta}_0 = (\boldsymbol{\beta}_{0a}', \boldsymbol{\beta}_{0b}')'$ with $\boldsymbol{\beta}_{0b} = \mathbf{0}$. Also, $\hat{\boldsymbol{\beta}}_n$ can be similarly partitioned as $\hat{\boldsymbol{\beta}}_n = (\hat{\boldsymbol{\beta}}_{na}', \hat{\boldsymbol{\beta}}_{nb}')'$ with $\hat{\boldsymbol{\beta}}_{na} = (\hat{\beta}_{n,1}, \ldots, \hat{\beta}_{n,p_0})'$, and $\hat{\boldsymbol{\beta}}_{nb} = (\hat{\beta}_{n,p_0+1}, \ldots, \hat{\beta}_{n,d})'$.

Following Johnson and Peng (2008), we define

$$H_{\lambda_j}(|t|)sgn(t) = \frac{d}{dt}P_{\lambda_j}(|t|) \quad and \quad \dot{H}_{\lambda_j}(|t|)sgn(t) = \frac{d}{dt}H_{\lambda_j}(|t|).$$

Also, under Eq. (2.5), taking the negative gradient with respect to $\boldsymbol{\beta}$, we obtain

$$S(\boldsymbol{\beta}) = \nabla_{\boldsymbol{\beta}} Q(\boldsymbol{\beta}) = \sum_{i=1}^{n+d} v_i \boldsymbol{x}_i sgn(z_i^*(\boldsymbol{\beta})) = S_n(\boldsymbol{\beta}) + n\sum_{j=1}^{d} H_{\lambda_j}(|\beta_j|)sgn(\beta_j),$$

where $S_n(\boldsymbol{\beta}) = -\nabla_{\boldsymbol{\beta}} D_n(\mathbf{V}_n, w, \boldsymbol{\beta})$. In addition to $(I_1) - (I_2)$, we will need the following assumption:

(I_3) Define $a_n = \max_{1\le j\le p_0} H_{\lambda_j}(|t|)$ and $b_n = \min_{j>p_0} H_{\lambda_j}(|t|)$, $\forall\, t$ fixed, and assume that

(*i*) $\sqrt{n}a_n \to 0$ and $\sqrt{n}b_n \to \infty$ as $n \to \infty$
(*ii*) $\lim_{n\to\infty} \inf_{|t|\le c/\sqrt{n}} \{\lambda_n^{-1} H_{\lambda_j}(|t|)\} > 0$ for any $c > 0$.

Remark 2.2. Note that for the adaptive Lasso case where $P_{\lambda_j}(|t|) = \lambda_j|t|$, and in assumption (I_3), a_n and b_n are reduced to $a_n = \max_{1\le j\le p_0} \lambda_j$ and $b_n = \min_{p_0+1\le j\le d} \lambda_j$, as $H_{\lambda_j}(|t|) = \lambda_j$. It is worth pointing out the Lasso penalty does not satisfy assumption (I_3) which is not surprising as it is well-known that the Lasso estimator does not have the oracle property, and (I_3) is key to ensuring the oracle property of the resulting estimator.

Theorem 2.1. *Under assumptions* $(I_1) - (I_3)$, $\hat{\boldsymbol{\beta}}_n$ *exists and is a* $\sqrt{n}$*-consistent estimator of* $\boldsymbol{\beta}_0$.

The proof this theorem is provided in Appendix.

Next consider the following assumption commonly imposed in the framework of signed-rank estimation, see Hössjer (1994) and Abebe et al. (2012):

(I_4) $\varphi^+ \in C^2((0,1)\setminus E)$ with bounded derivatives, where E is a finite set of discontinuities.

Following Hössjer (1994), set

$$\gamma_{\varphi^+} = \int_0^1 \left(\varphi^+(t)\right)^2 dt \quad \text{and} \quad \zeta_{\varphi^+} = \int_0^1 \varphi^+(t) h_F(t) dt = -\int_{-\infty}^{\infty} \varphi^+(F^{-1}(u)) f'(u) du,$$

where $h_F(u) = -f'(F^{-1}(u))/f(F^{-1}(u))$. As it is pointed out in Hössjer (1994), (I_1) and (I_2) imply that $\zeta_{\varphi^+} > 0$. Also, letting J denote the joint distribution of $(y, \boldsymbol{x})$ and by symmetry of f, one can define a corresponding symmetric distribution as follows:

$$\begin{aligned} H_{\boldsymbol{\beta}}(t) &= \frac{1}{2}\left[P_J(z_i(\boldsymbol{\beta}) \le t) + P_J(-z_i(\boldsymbol{\beta}) \le t)\right] \\ &= \frac{1}{2}\left[E_G\{F(t) + \boldsymbol{x}^\tau \boldsymbol{\beta}\} + E_G\{F(t - \boldsymbol{x}^\tau \boldsymbol{\beta})\}\right]. \end{aligned} \tag{2.7}$$

Now setting $F_{\boldsymbol{\beta},i}(t) = \frac{1}{2} E_G\{\boldsymbol{x}_i F(t + \boldsymbol{x}^\tau \boldsymbol{\beta})\}$ and $\boldsymbol{\xi}(\boldsymbol{\beta}) = (\xi_1(\boldsymbol{\beta}), \ldots, \xi_n(\boldsymbol{\beta}))^\tau$, where

$$\xi_i(\boldsymbol{\beta}) = 2\int_{-\infty}^{\infty} \varphi^+(H_{\boldsymbol{\beta}}(t)) dF_{\boldsymbol{\beta},i}(t),$$

it is shown under $(I_1) - (I_3)$ in Hössjer (1994) that $S_n(\boldsymbol{\beta}) - \boldsymbol{\xi}(\boldsymbol{\beta}) \to 0$ *a.s.* as $n \to \infty$. Let $W(\boldsymbol{x}) = diag\{w_1(\boldsymbol{x}), \ldots, w_n(\boldsymbol{x})\}$ and define the expected weighted Gram matrix $\Sigma = E_G[\boldsymbol{x}'W(\boldsymbol{x})\boldsymbol{x}]$. Now partition $\boldsymbol{x}$ as $\boldsymbol{x} = (\boldsymbol{x}_a, \boldsymbol{x}_b)$, according to nonzero and zero coefficients, and let Σ_a denote the top left $p_0 \times p_0$ sub-matrix of Σ. We will assume that Σ_a is positive definite. The following main result gives the asymptotic properties (oracle property) of the penalized WSR estimator given in (2.2). Its proof is provided in Appendix.

Theorem 2.2. *Under assumptions* (I_1) *to* (I_4), *we have* $\lim_{n\to\infty} P(\hat{\boldsymbol{\beta}}_{nb} = \mathbf{0}) = 1$, *and*

$$\sqrt{n}(\hat{\boldsymbol{\beta}}_{na} - \boldsymbol{\beta}_{0a}) \xrightarrow{\mathscr{D}} N(0,\ \zeta_{\varphi+}^{-2}\gamma_{\varphi+}\Sigma_a),$$

where Σ_a *is a* $p_0 \times p_0$ *positive definite matrix.*

Remark 2.3. From the two theorems above, (i) and (ii) in assumption (I_3) together with (I_1), I_2 and (I_4) are imposed to ensure the $\sqrt{n}$-consistency, the oracle property and the $\sqrt{n}$-asymptotic normality of the proposed estimator. Note that although Theorem 2.2 is similar to that of Johnson and Peng (2008), the definitions of a_n and b_n given here are more general and the assumptions needed for the asymptotic normality of the gradient function $S_n(\boldsymbol{\beta})$ are very different.

2.3 Some Practical Considerations

2.3.1 Estimation of the Tuning Parameter $\boldsymbol{\lambda}$

Another important issue in the estimation of $\boldsymbol{\beta}_0$ in model (2.1), is the choice of the λ_j's in Eq. (2.3). As proposed by Johnson et al. (2008) $\boldsymbol{\lambda}$ can be estimated as follows

$$\hat{\boldsymbol{\lambda}} = \underset{\boldsymbol{\lambda}}{\text{Argmin}} \frac{D_n(\mathbf{V}_n, w, \hat{\boldsymbol{\beta}}_n(\boldsymbol{\lambda}))/n}{\{1 - e(\boldsymbol{\lambda})\}^2}, \tag{2.8}$$

where $e(\boldsymbol{\lambda}) = tr\big[\mathbf{X}\{\mathbf{X}'\mathbf{X} + \Sigma_{\boldsymbol{\lambda},\hat{\boldsymbol{\beta}}_n(\boldsymbol{\lambda})}\}^{-1}\mathbf{X}'\big]$ and $\mathbf{X}$ is the $n \times d$ matrix with column vectors $\boldsymbol{x}_i$ and $\Sigma_{\boldsymbol{\lambda},\hat{\boldsymbol{\beta}}_n(\boldsymbol{\lambda})}$ a diagonal matrix with entries

$$H_{\lambda_j}(|\hat{\beta}_{nj}(\lambda)|)\text{sgn}(\hat{\beta}_{nj}(\lambda)).$$

This cross validation procedure was considered by Johnson et al. (2008) and was shown to have advantage over the least squares cross valuation criterion that is obtained by replacing the numerator of the right hand side of Eq. (2.8) by the least squares objective function. Note that although the idea similar, the objective function $D_n(\mathbf{V}_n, w, \boldsymbol{\beta})$ considered in this paper is very different to the one considered in Johnson et al. (2008). If we restrict ourselves to WSR-AL, another alternative to estimating $\boldsymbol{\lambda}$ is to consider the AIC and BIC approaches discussed in Wang et al. (2007) based on the considered objective function. That is, obtain $\hat{\boldsymbol{\lambda}}$ as,

$$\hat{\boldsymbol{\lambda}} = \underset{\lambda}{\text{Argmin}} \left\{ Q_{a\ell}^{w}(\tilde{\boldsymbol{\beta}}_n) - \sum_{j=1}^{d} \log(n\lambda_j) \right\} \text{ the for AIC approach,} \tag{2.9}$$

which leads to $\hat{\lambda}_j = 1/(n|\tilde{\beta}_{nj}|)$, and

$$\hat{\boldsymbol{\lambda}} = \underset{\lambda}{\text{Argmin}} \left\{ Q_{a\ell}^{w}(\tilde{\boldsymbol{\beta}}_n) - \sum_{j=1}^{d} \log(n\lambda_j) \log n \right\} \text{ the for BIC approach,} \tag{2.10}$$

which leads to $\hat{\lambda}_j = \log n/(n|\tilde{\beta}_{nj}|)$, where $\tilde{\boldsymbol{\beta}}_n = \underset{\boldsymbol{\beta} \in \mathscr{B}}{\text{Argmin}}\, D_n(\mathbf{V}_n, w, \boldsymbol{\beta})$.

2.3.2 Choice of Weights

In our analysis, we choose the weight function $w(\boldsymbol{x})$ to be

$$w(\boldsymbol{x}) = \min\left[1, \frac{\eta}{d(\boldsymbol{x})}\right],$$

where $d(\boldsymbol{x}) = (\boldsymbol{x} -_x)' \mathbf{C}_x^{-1} (\boldsymbol{x} -_x)$ is a robust Mahalanobis distance, with $_x$ and $\mathbf{C}_x$ being robust estimates of location and covariance of $\boldsymbol{x}$, respectively and η being some positive constant usually set at $\chi^2_{0.95}$ in practice. Under this choice, it is shown in Bindele and Abebe (2012) that the resulting estimator has a bounded influence function.

2.3.3 Computational Algorithm

For computation purposes, the following steps can be followed:

1. Obtain the unpenalized (W)SR estimator $\hat{\boldsymbol{\beta}}_{\varphi^+}$.
2. Use $\hat{\boldsymbol{\beta}}_{\varphi^+}$.
   - Estimate $\hat{v}_i$ as $v_i(\hat{\boldsymbol{\beta}}_{\varphi^+})$.
   - Use AIC/BIC in Eq. (2.9) or (2.10) with $\tilde{\boldsymbol{\beta}}_n = \hat{\boldsymbol{\beta}}_{\varphi^+}$ to estimate $\boldsymbol{\lambda}$, say $\hat{\boldsymbol{\lambda}}$.
3. Form $z^*(\boldsymbol{\beta}, \hat{\boldsymbol{\lambda}}) = y^* - \boldsymbol{x}_{\hat{\lambda}}^{*\prime}\boldsymbol{\beta}$, where $\boldsymbol{x}_{\hat{\lambda}}^*$ is as defined in Eq. (2.6) with $\boldsymbol{\lambda} = \hat{\boldsymbol{\lambda}}$.
4. Find

$$\underset{\boldsymbol{\beta}}{\text{Argmin}} \sum_{i=1}^{n+d} \hat{v}_i |z_i^*(\boldsymbol{\beta}, \hat{\boldsymbol{\lambda}})|$$

   using any weighted LAD software (e.g. `quanteg, rfit` in `R`).

2.4 Simulation and Real Data Studies

To demonstrate the performance of our proposed method, several simulation scenarios and a real data set are considered.

2.4.1 *Low Dimensional Simulation*

The setting for the low-dimensional simulation is taken from Tibshirani (1996). We take a sample of size $n = 50$ where the number of predictor variables is $d = 8$ and $\boldsymbol{\beta}_0$ is set at $\boldsymbol{\beta}_0 = (3, 1.5, 0, 0, 2, 0, 0, 0)'$. Thus $p_0 = 3$. To study the effect of tail thickness, contamination, and leverage, we considered three different scenarios:

Scenario 1: The vector of predictor variables $\boldsymbol{x}$ is generated as $\boldsymbol{x} \sim N_8(\mathbf{0}, V)$, where $V = (v_{ij})$ and $v_{ij} = 0.5^{|i-j|}$. The error distributions are t and contaminated normal. That is, the errors are generated as $e \sim t_{df}$ for several degrees of freedom (df) and $e \sim (1-\epsilon)N(0,1) + \epsilon N(0, 3^2)$ for several levels of contamination ϵ. These distributions allow us to investigate the effect of tail thickness and the rate of contamination on the proposed method.

Scenario 2: The vector of predictors $\boldsymbol{x}$ is generated as $\boldsymbol{x} \sim (1-\epsilon)N_8(\mathbf{0}, V) + \epsilon N_8(\mathbf{1}\mu, V)$, with $\mu = 5$ and the errors are generated as $e \sim N(0, 1)$. This enables us to study the effect of contamination (such us gross outliers and leverage points) in the design space.

Scenario 3: This scenario considers a partial model misspecification similar to the one in Arslan (2012). In this case, we take $\boldsymbol{\beta}_0 = (3, 1.5, 0, 0, 2, 0, 0, 0)'$ and $\boldsymbol{\beta}_0^* = (3, \ldots, 3)'$. Then $\boldsymbol{x}$ and y are generated as follows: for $i = 1, \ldots, n - [n\epsilon]$, $\boldsymbol{x}_i \sim N_8(\mathbf{0}, V)$ and $y_i = \boldsymbol{x}_i'\boldsymbol{\beta}_0 + N(0, 1)$, for $i = n - [n\epsilon] + 1, \ldots, n$, $\boldsymbol{x}_i \sim N_8(\mathbf{1}\mu, V)$, $\mu = 5$, and $y_i = \boldsymbol{x}_i'\boldsymbol{\beta}_0^c + N(0, 1)$. Varying ϵ in $[0, 1)$ allows us to study the effect of various levels of model contamination.

In all cases, we considered the adaptive lasso penalty where the tuning parameter is computed using the BIC criterion. The estimators studied were least squares (LS-AL), least absolute deviations (LAD-AL), signed-rank (SR-AL), weighted LAD (WLAD-AL), and weighted SR (WSR-AL). The weights were computed as discussed above using minimum covariance determinant (MCD) of Rousseeuw (1984). We performed 1000 replications and calculated the average number of correct zeros (true negatives), the average number of incorrect zeros (false negatives), the percentage of correct models identified, and relative efficiencies versus LS-AL of the proposed estimators for estimating β_1 based on estimated MSEs. The results of Scenario 1 are given in Figs. 2.1 and 2.2 while the results of Scenarios 2 and 3 are given in Figs. 2.3 and 2.4, respectively.

Figure 2.1 shows that LAD-AL and SR-AL (unweighted) estimators are not very good at identifying zeroes (left panels) compared to WLAD-AL and WSR-AL.

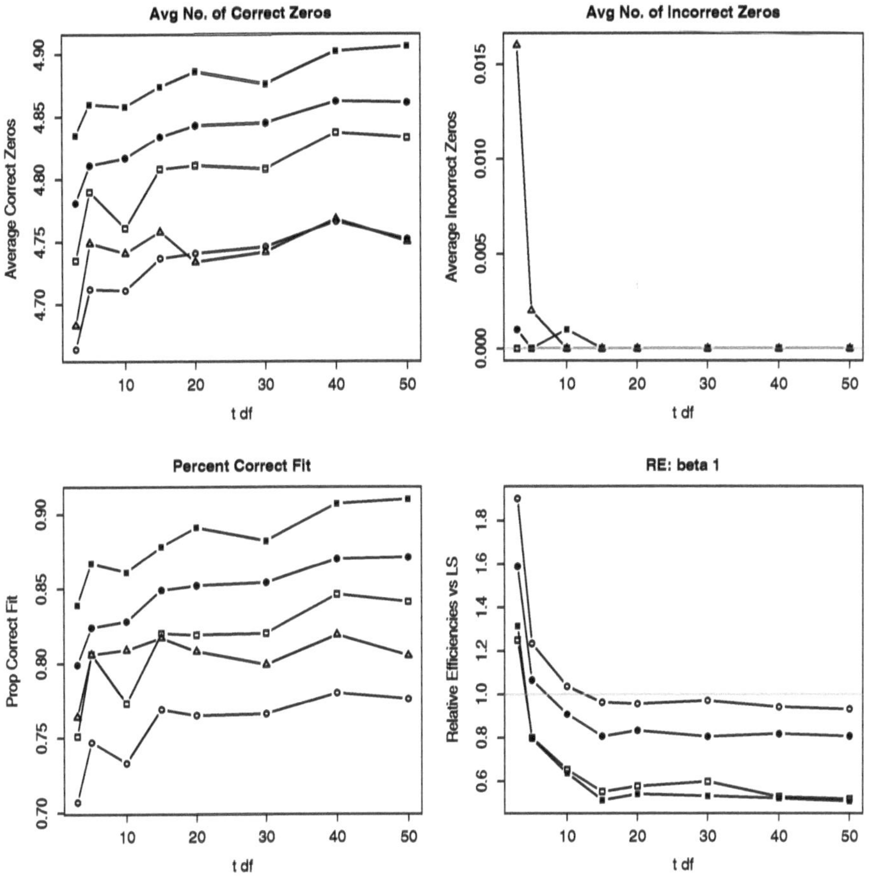

Fig. 2.1 Average number of correct and incorrect zeroes, Relative model error, Percentage of correct fit, relative efficiencies (RE) against t distribution *df* (Scenario 1). The symbols in the plots are LS-AL (*open triangle*), LAD-AL (*open square*), SR-AL (*open circle*), WLAD-AL (*filled square*) and WSR-AL (*filled circle*)

They are, slightly more efficient than their weighted counterpart in estimating nonzero coefficients. Their relative efficiencies versus LS-AL stabilize towards the theoretical relative efficiencies of 0.955 and 0.63 as the tails of the t distribution approach the tails of the standard normal distribution.

Figure 2.2 shows that with the exception of LS-AL, the performance in detecting true zeroes of all other estimators deteriorates as the proportion of contamination increases (left panels). On the other hand, the false negatives of LS-AL increase with increasing contamination (top right panel). Taken together, these indicate that LS-AL increasingly over-penalizes when the proportion of outliers in the data increases. and SR-AL (unweighted) estimators are not very good at identifying zeros (left panels) compared to WLAD-AL and WSR-AL. Once again the unweighted

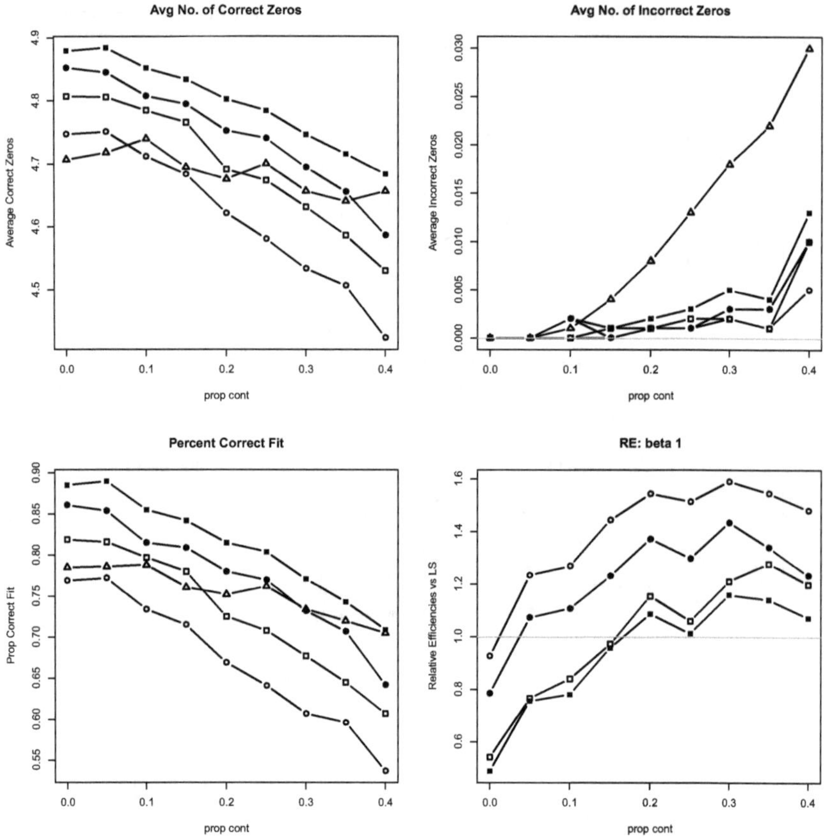

Fig. 2.2 Average number of correct and incorrect zeroes, Relative model error, Percentage of correct fit, relative efficiencies (RE) against contamination proportion (ϵ) of the contaminated normal distribution (Scenario 1). The symbols in the plots are LS-AL (*open triangle*), LAD-AL (*open square*), SR-AL (*open circle*), WLAD-AL (*filled square*) and WSR-AL (*filled circle*)

LAD and SR are slightly more efficient in estimating nonzero coefficients than their weighted counterparts while the relative efficiencies of both weighted and unweighted estimators increases with increasing proportion of error contamination.

Figure 2.3 shows that, even when the model is correctly specified, high leverage points have a detrimental effect on model selection. While the number of true positives decrease, the weighted cases appear to provide some resistance for low percentage of high-leverage points. With respect to the estimation of nonzero coefficients, the false negative rates of LS-AL increase sharply compare to all other estimators (top right panel). Once again, LS-AL is increasingly over-penalizing the model with increasing proportion of high-leverage points. It is not surprising that LS-AL is also inefficient in the estimation of nonzero coefficients, especially

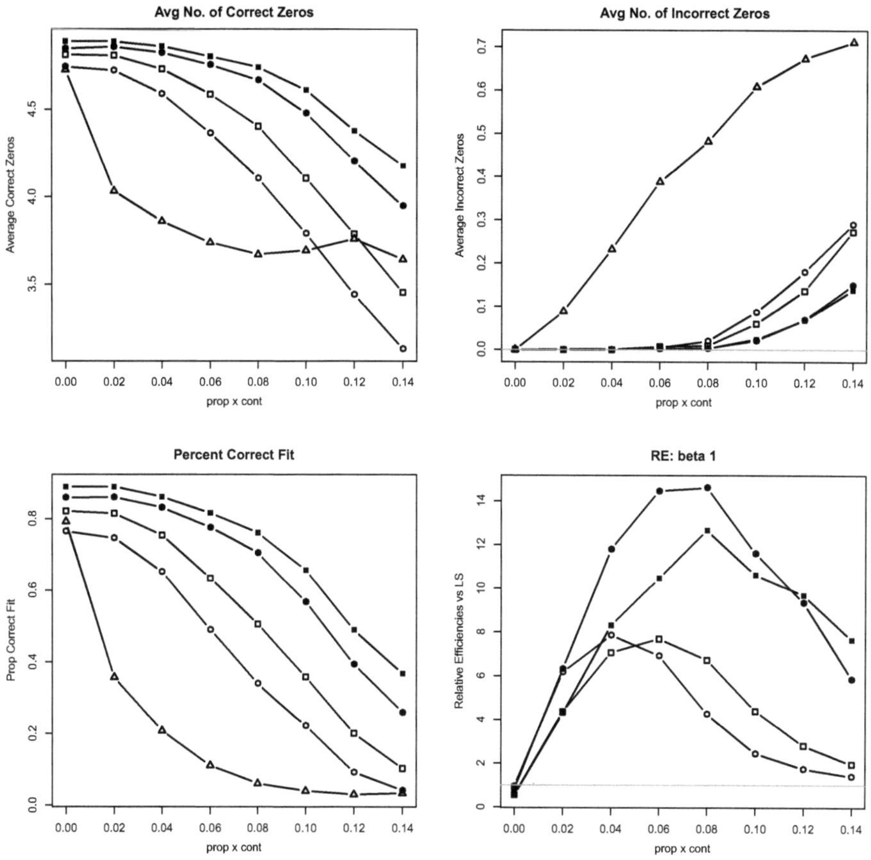

Fig. 2.3 Average number of correct and incorrect zeroes, Relative model error, Percentage of correct fit, relative efficiencies (RE) against contamination proportion (ϵ) of the distribution of the predictor $\boldsymbol{x}$ (Scenario 2). The symbols in the plots are LS-AL (*open triangle*), LAD-AL (*open square*), SR-AL (*open circle*), WLAD-AL (*filled square*) and WSR-AL (*filled circle*)

compared to WLAD-AL and WSR-AL, especially for moderate proportion (4–8 %) of high-leverage points.

Our observations remain similar to the above for model misspecification (Scenario 3). In this case, the performance of all the estimators deteriorates quite rapidly with increasing contamination. LS-AL is once again the worst offender and WLAD-AL and WSR-AL provide the highest relative efficiency. The unweighted forms are much less efficient in comparison.

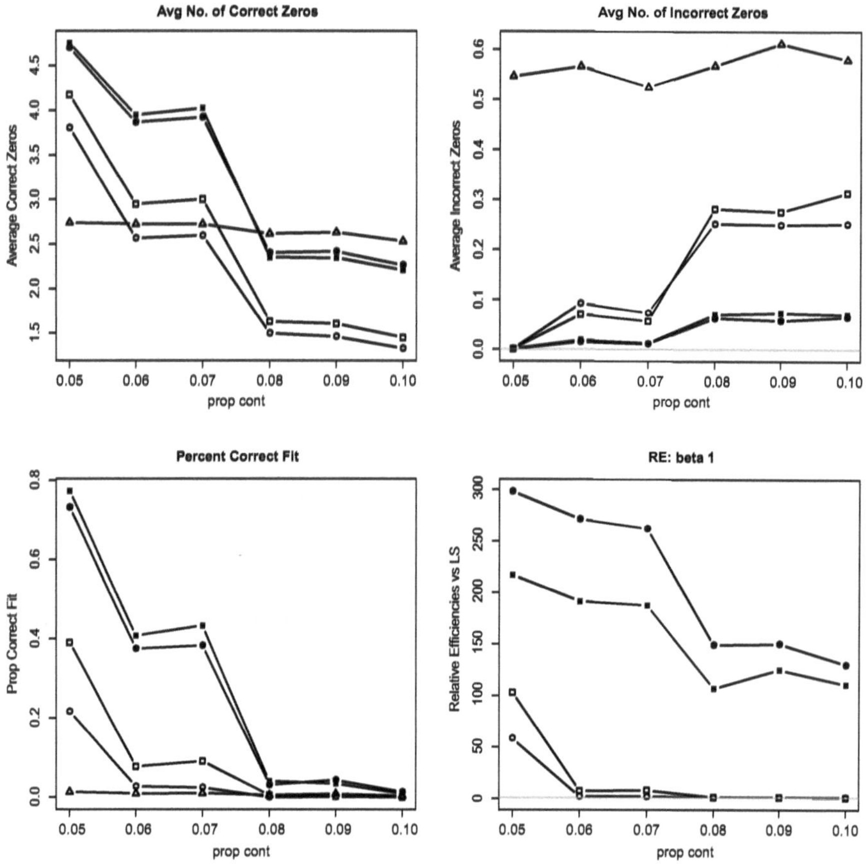

Fig. 2.4 Average number of correct and incorrect zeroes, Relative model error, Percentage of correct fit, relative efficiencies (RE) against contamination proportion (ϵ) of the model contamination (Scenario 3). The symbols in the plots are LS-AL (*open triangle*), LAD-AL (*open square*), SR-AL (*open circle*), WLAD-AL (*filled square*) and WSR-AL (*filled circle*)

2.4.2 High-Dimensional Simulation

Again as in Tibshirani (1996), consider the linear model (2.1), where $\boldsymbol{x}$ is a 100×40 matrix with entries $x_{ij} = z_{ij} + z_i$ such that z_{ij} and z_i are independent and generated from standard normal distributions. This setting makes the x_{ij}'s to be pairwise correlated with correlation coefficient of about 0.5. The random error in Eq. (2.1) is generated from two different distributions: the contaminated normal distribution with different rates of contamination and the t distribution with different degrees of freedom. The regression coefficient vector is set at $\boldsymbol{\beta} = (0, \ldots, 0, 2, \ldots, 2, 0, \ldots, 0, 2, \ldots, 2)$, where there are ten repeats in each block. From 1000 replications, average numbers of correct zeroes, average number of incorrect zeroes and percentage of correct fit are reported. The simulation results are displayed in Fig. 2.5, where for clarity of presentation we only report results of LS-AL, SR-AL, and WSR-AL fits.

Our observations are quite similar to the low-dimensional case. LS-AL over-penalizes with increasing proportion of high leverage points, even when the model is correctly specified. SR-AL and WSR-AL provide superior performance in high leverage situations (rows three and four of Fig. 2.5). WSR-AL is clearly the best among the three for heavier tailed errors (top row). The percentage of correctly estimated models deteriorates with increasing error contamination (second row) for all the methods.

2.4.3 Boston Housing Data

The data considered here is the Boston Housing dataset which contains median values of housing in 506 census tracts and 13 predictors comprised of characteristics of the census tract. The full description of the data can be found in Leng (2010) and the dataset is available in the R library MASS. So, for sake of brevity, the description will not be included here. We first fit unpenalized regression models using the LS and SR procedures. The results are given in Table 2.1. We then fit penalized regression models using LS-AL, SR-AL, and WSR-AL. These results are displayed in Table 2.2.

The results in Table 2.1 indicate that both LS and SR find the variables INDUS and AGE insignificant while ZN is marginally significant. However, the LS and SR estimated coefficients are quite different in some cases outside of two standard errors of each other. Also, the residual plot given in Fig. 2.6 indicates the presence of heavy tails casting doubt on the LS results. In fact, observing the plot of studentized residuals of LS and SR in Fig. 2.6 plotted on the same scale, it is clear that the SR fit identifies many more outlying observations than the LS estimator. The results of penalized regressions given in Table 2.2 show that LS-AL eliminates the two insignificant variables (INDUS, AGE) from the model while SR-AL and WSR-AL eliminate a third variable (ZN) from the model. Thus, our observations are in line with those of Leng (2010).

The obvious question is if this reduction in model is associated with loss in prediction accuracy. To evaluate this, we performed cross validation where we randomly split the data into a training set containing approximately 90% of the data and a testing set containing the remaining 10 %. We fit the models using the training sets and calculated the absolute error for the test sets $|y - \hat{\alpha} - \boldsymbol{x}'\hat{\boldsymbol{\beta}}|$, where $\hat{\alpha}$ is estimated using the mean (for LS) and median (for LAD and SR) of the training set residuals $y - \boldsymbol{x}'\hat{\boldsymbol{\beta}}$. Table 2.3 gives the mean absolute error and the median model size over 100 iterations. The estimators considered all use the adaptive lasso penalty. Weights were computed using three different versions of the Mahalanobis distance: classic (Mah), minimum volume ellipsoid (MVE) of Rousseeuw (1984), and minimum covariance determinant (MCD) of Rousseeuw (1984).

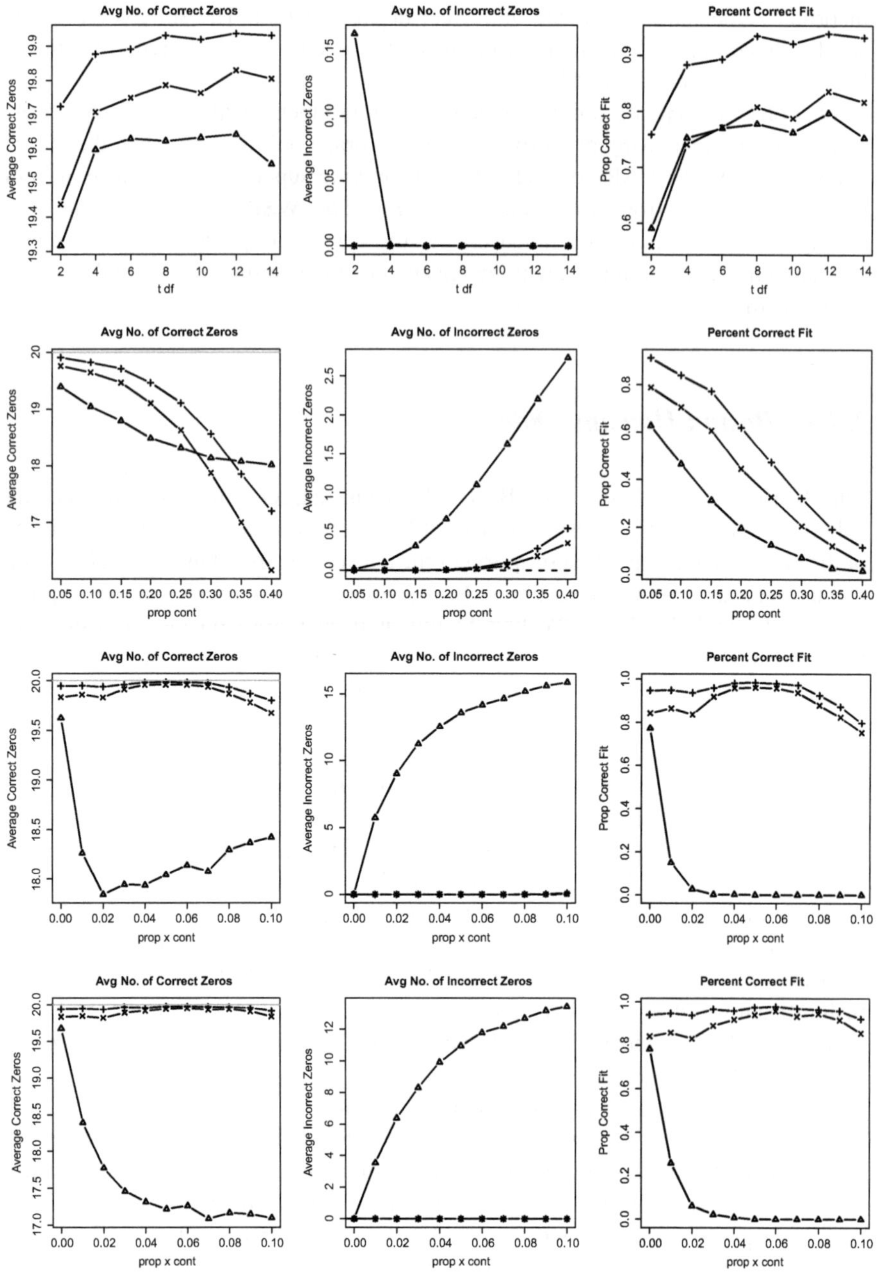

Fig. 2.5 Average number of correct and incorrect zeroes and percentage of correct fit for the high-dimensional simulation. The symbols in the plots are LS-AL (*open triangle*), SR-AL (*times*) and WSR-AL (*plus*). *First row* represents t distributed errors, *second row* represents contaminated normal, *third row* represents high-leverage points, and the *last row* represents model misspecification

Table 2.1 Estimated coefficients using LS and SR

	LS			SR		
	Coef	se	t	Coef	se	t
CRIM	−0.0103	0.0013	−7.8083	−0.0089	0.0010	−8.8709
ZN	0.0012	0.0005	2.1338	0.0008	0.0004	2.0147
INDUS	0.0025	0.0025	1.0022	0.0023	0.0019	1.2355
CHAS	0.1009	0.0345	2.9255	0.0781	0.0263	2.9691
NOX	−0.7784	0.1529	−5.0912	−0.3925	0.1166	−3.3662
RM	0.0908	0.0167	5.4300	0.1766	0.0128	13.8376
AGE	0.0002	0.0005	0.3983	−0.0006	0.0004	−1.5251
DIS	−0.0491	0.0080	−6.1486	−0.0359	0.0061	−5.9025
RAD	0.0143	0.0027	5.3725	0.0094	0.0020	4.6518
TAX	−0.0006	0.0002	−4.1574	−0.0005	0.0001	−4.6360
PTRATIO	−0.0383	0.0052	−7.3086	−0.0300	0.0040	−7.5140
B	0.0004	0.0001	3.8468	0.0006	0.0001	7.5509
LSTAT	−0.0290	0.0020	−14.3036	−0.0229	0.0015	−14.8047

Table 2.2 Estimated regression coefficients using LS, SR, LS-AL, SR-AL, and WSR-AL

	LS	LS-AL	SR	SR-AL	WSR-AL
CRIM	−0.0103	−0.0101	−0.0089	−0.0077	−0.0088
ZN	0.0012	0.0009	0.0008	0.0000	0.0000
INDUS	0.0025	0.0000	0.0023	0.0000	0.0000
CHAS	0.1009	0.0975	0.0781	0.0506	0.0546
NOX	−0.7784	−0.6990	−0.3925	−0.3238	−0.3171
RM	0.0908	0.0911	0.1766	0.1696	0.1708
AGE	0.0002	0.0000	−0.0006	0.0000	0.0000
DIS	−0.0491	−0.0489	−0.0359	−0.0251	−0.0263
RAD	0.0143	0.0126	0.0094	0.0056	0.0053
TAX	−0.0006	−0.0005	−0.0005	−0.0003	−0.0003
PTRATIO	−0.0383	−0.0376	−0.0300	−0.0317	−0.0329
B	0.0004	0.0004	0.0006	0.0005	0.0006
LSTAT	−0.0290	−0.0288	−0.0229	−0.0249	−0.0245

It is evident from Table 2.3 that while the model performances remain relatively similar, the median model sizes of the MCD and MVE weighted adaptive lasso estimation required far fewer variables. For comparable model sizes, SR-AL estimator provides lower absolute error than LS-AL, LAD-AL, WLAD-AL (Mah), and WSR-AL (Mah). Also a comparison of WLAD-AL (Arslan 2012) and WSR-AL shows that on average WSR-AL achieves a lower mean absolute error using a slightly smaller model.

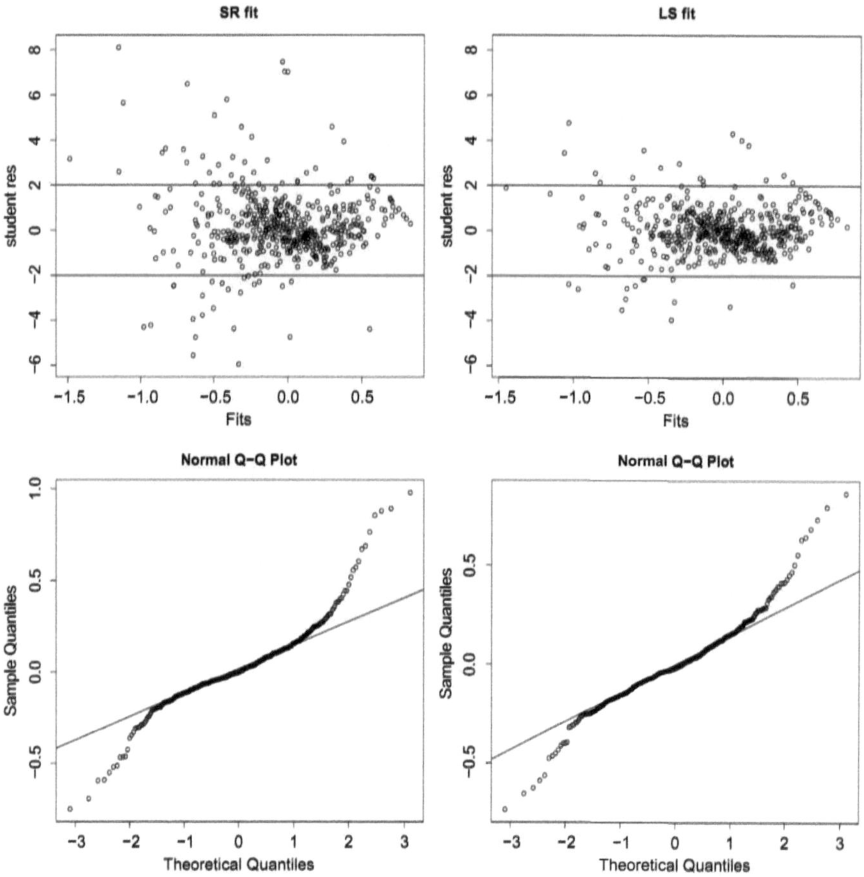

Fig. 2.6 Plots of studentized residuals versus fitted values as well as residual Q-Q plots of LS and SR fits

Table 2.3 Results of cross validation

Method	Mean absolute error (St dev)	Median model size
LS-AL	0.1408 (0.0184)	11.0
LAD-AL	0.1368 (0.0226)	11.0
SR-AL	0.1356 (0.0214)	11.0
WLAD-AL (Mah)	0.1365 (0.0213)	11.0
WSR-AL (Mah)	0.1360 (0.0210)	11.0
WLAD-AL (MVE)	0.1474 (0.0232)	10.0
WSR-AL (MVE)	0.1452 (0.0219)	10.0
WLAD-AL (MCD)	0.1523 (0.0244)	8.5
WSR-AL (MCD)	0.1490 (0.0224)	8.0

2.5 Discussion

This paper considered variable selection for linear models using penalized weighted signed-rank objective functions. It is demonstrated that the method provides selection and estimation consistency in the presence of outliers and high-leverage points. Our simulation study considered both low and high-dimensional data. In both cases, it was shown that compared to penalized least squares, penalized rank-based estimators provided more accurate true negative and false negatives identification while providing higher efficiency in estimating true positives when the error distribution is heavy tailed or contaminated. The weighted versions of the rank-based estimators provided protection against high leverage points, even when the model is incorrectly specified for the high-leverage points as long as the proportion of high-leverage points is moderate.

While the results are encouraging, an interesting extension involves regression when the data are ultra-high dimensional; that is, the dimension of the predictor also goes to infinity. This is currently under consideration by the authors. Another interesting extension involves generalized linear and single index models or even functional data analysis. Variable selection remains a valid exercise in these cases, where the last case is usually dealt with using group-selection methods.

Acknowledgements We dedicate this work to Joseph W. McKean on the occasion of his 70th birthday. We are thankful for his mentorship and guidance over the years. We also thank the anonymous referee for suggestions that improved the presentation.

Appendix

This Appendix provides some lemmas and the proofs of the main results (Theorems 2.1 and 2.2). In the proofs we have taken $W = I$ to simplify notation. The general case follows by taking $W^{1/2}\boldsymbol{x}$ in place of $\boldsymbol{x}$ in the proofs.

Proofs

The following three lemmas, whose proofs follow from slight modifications of those given in Hössjer (1994) and Hettmansperger and McKean (2011), are key to deriving the proof of the main results.

Lemma 2.1. *Under assumptions* (I_1) *and* (I_2)*, we have* $\tilde{\boldsymbol{\beta}}_n \to \boldsymbol{\beta}_0$ *a.s.*

The proof of this lemma is given in Hössjer (1994) for $w \equiv 1$ and in Abebe et al. (2012) for any positive w, and a more general regression model. Also, as in Wu (1981), the proof of this lemma is obtained by showing that

$$\lim_{n\to\infty} \inf_{\boldsymbol{\beta}\in B^c} \big(D_n(\mathbf{v}_n, w, \boldsymbol{\beta}) - D_n(\mathbf{v}_n, w, \boldsymbol{\beta}_0)\big) > 0 \;\; a.s. \tag{2.11}$$

where B is an open subset of $\mathscr{B}$ and $\boldsymbol{\beta}_0 \in Int(B)$.

Lemma 2.2. *Putting* $U_n(\boldsymbol{\gamma}, \boldsymbol{\beta}) = \dfrac{\|S_n(\boldsymbol{\gamma}) - S_n(\boldsymbol{\beta}) - \boldsymbol{\xi}(\boldsymbol{\gamma}) + \boldsymbol{\xi}(\boldsymbol{\beta})\|_1}{n^{-1/2} + \|\boldsymbol{\xi}(\boldsymbol{\gamma})\|_1}$, *we have for small enough* $\delta > 0$ *that*

$$\sup_{\|\boldsymbol{\gamma}\|\le\delta} U_n(\boldsymbol{\gamma}, \boldsymbol{\beta}_0) \stackrel{a.s}{\longrightarrow} 0 \quad as \quad n \to \infty.$$

This lemma ensures that $n^{-1/2}S_n(\boldsymbol{\beta}_0)$ converges in distribution to a multivariate normal distribution with mean zero and covariance matrix $\gamma_{\varphi^+}\Sigma$. It also results in the following asymptotic linearity established in Hettmansperger and McKean (2011).

Lemma 2.3. *Under the assumption of the errors having a finite Fisher information, we have for all* $\epsilon > 0$ *and* $C > 0$

$$P\left[\sup_{\sqrt{n}\|\boldsymbol{\beta}-\boldsymbol{\beta}_0\|_1\le C} \|n^{-1/2}(S_n(\boldsymbol{\beta})-S_n(\boldsymbol{\beta}_0))+\zeta_{\varphi^+}\sqrt{n}(\boldsymbol{\beta}-\boldsymbol{\beta}_0)\|_1 \ge \epsilon\right] \to 0 \quad as\, n \to \infty.$$

From this asymptotic linearity follows that for all $\boldsymbol{\beta}$ such that $\|\boldsymbol{\beta} - \boldsymbol{\beta}_0\|_1 \le C/\sqrt{n}$, we have

$$n^{-1/2}S_n(\boldsymbol{\beta}) = n^{-1/2}S_n(\boldsymbol{\beta}_0) - \zeta_{\varphi^+}\sqrt{n}(\boldsymbol{\beta} - \boldsymbol{\beta}_0) + o(1) \tag{2.12}$$

Proof of Theorem 2.1. Set $B = \{\boldsymbol{\beta}_0 + n^{-1/2}\mathbf{u} : \;\; \|\mathbf{u}\|_1 < C\}$. Clearly B is an open neighborhood of $\boldsymbol{\beta}_0$ and therefore B^c is a closed subset of $\mathscr{B}$ not containing $\boldsymbol{\beta}_0$. To complete the proof, it is then sufficient to show that

$$\lim_{n\to\infty} \inf_{\boldsymbol{\beta}\in B^c} \big(Q(\boldsymbol{\beta}) - Q(\boldsymbol{\beta}_0)\big) > 0 \;\; a.s.$$

which from Lemma 1 of Wu (1981) will result in the $\sqrt{n}$-consistency of $\hat{\boldsymbol{\beta}}_n$. Indeed,

$$Q(\boldsymbol{\beta}) - Q(\boldsymbol{\beta}_0) = D_n(\mathbf{v}_n, w, \boldsymbol{\beta}) - D_n(\mathbf{v}_n, w, \boldsymbol{\beta}_0) + n\sum_{j=1}^{d}\big[P_{\lambda_j}(|\beta_j|) - P_{\lambda_j}(|\beta_{0j}|)\big]. \tag{2.13}$$

Now by the mean value theorem, assuming without loss of generality the $|\beta_{0j}| < |\beta_j|$, there exits $\alpha_j \in (|\beta_{0j}|, |\beta_j|)$ such that

$$P_{\lambda_j}(|\beta_j|) - P_{\lambda_j}(|\beta_{0j}|) = H_{\lambda_j}(|\alpha_j|)sgn(\alpha_j)(|\beta_j| - |\beta_{0j}|),$$

and therefore

$$|P_{\lambda_j}(|\beta_j|) - P_{\lambda_j}(|\beta_{0j}|)| \leq H_{\lambda_j}(|\alpha_j|)|\beta_j - \beta_{0j}|.$$

This together with Eq. (2.13) imply that

$$\begin{aligned} Q(\boldsymbol{\beta}) - Q(\boldsymbol{\beta}_0) &= D_n(\mathbf{v}_n, w, \boldsymbol{\beta}) - D_n(\mathbf{v}_n, w, \boldsymbol{\beta}_0) + n\sum_{j=1}^{d} H_{\lambda_j}(|\alpha_j|)sgn(\alpha_j)(|\beta_j| - |\beta_{0j}|) \\ &\geq D_n(\mathbf{v}_n, w, \boldsymbol{\beta}) - D_n(\mathbf{v}_n, w, \boldsymbol{\beta}_0) - \sqrt{n}a_n \sum_{j=1}^{p_0} |u_j|, \qquad (2.14) \end{aligned}$$

as $\boldsymbol{\beta} \in B^c$ implies that $\boldsymbol{\beta}$ can be written as $\boldsymbol{\beta} = \boldsymbol{\beta}_0 + n^{-1/2}\mathbf{u}$ with $\|\mathbf{u}\|_1 \geq C$. Being a closed subset of a compact space, B^c is compact, and hence, is closed and bounded. Then, there exists a constant M such that $C \leq \|\mathbf{u}\|_1 \leq M$. From the last term of equation (2.14), note that $\sum_{j=1}^{p_0} |u_j| \leq \|\mathbf{u}\|_1 \leq M$ from which, we have $-\sqrt{n}a_n \sum_{j=1}^{p_0} |u_j| \geq -\sqrt{n}a_n M$. Thus,

$$Q(\boldsymbol{\beta}) - Q(\boldsymbol{\beta}_0) \geq D_n(\mathbf{v}_n, w, \boldsymbol{\beta}) - D_n(\mathbf{v}_n, w, \boldsymbol{\beta}_0) - \sqrt{n}a_n M,$$

and so,

$$\lim_{n\to\infty} \inf_{\boldsymbol{\beta}\in B^c} \left(Q(\boldsymbol{\beta}) - Q(\boldsymbol{\beta}_0)\right) \geq \lim_{n\to\infty} \inf_{\boldsymbol{\beta}\in B^c} \left(D_n(\mathbf{v}_n, w, \boldsymbol{\beta}) - D_n(\mathbf{v}_n, w, \boldsymbol{\beta}_0)\right) - \lim_{n\to\infty} \left[\sqrt{n}a_n M\right].$$

By assumption (I_3), $\lim_{n\to\infty} \left[\sqrt{n}a_n M\right] = 0$, and by Lemma 2.1, we have

$$\lim_{n\to\infty} \inf_{\boldsymbol{\beta}\in B^c} \left(Q(\boldsymbol{\beta}) - Q(\boldsymbol{\beta}_0)\right) > 0 \;\; a.s.$$

Proof of Theorem 2.2. From the proof of Theorem 2.1 to obtain the oracle property, it is sufficient to show that for any $\boldsymbol{\beta}^*$ satisfying $\|\boldsymbol{\beta}_a^* - \boldsymbol{\beta}_{0a}\|_1 = O_p(n^{-1/2})$ and $|\beta_j^*| < Cn^{-1/2}$ for $j = p_0 + 1, \ldots, d$, $\left.\dfrac{\partial Q(\boldsymbol{\beta})}{\partial \beta_j}\right|_{\boldsymbol{\beta}=\boldsymbol{\beta}^*}$ and β_j^* have the same sign. Indeed,

$$\begin{aligned} n^{-1/2}\left.\frac{\partial Q(\boldsymbol{\beta})}{\partial \beta_j}\right|_{\boldsymbol{\beta}=\boldsymbol{\beta}^*} &= -n^{-1/2}S_n^j(\boldsymbol{\beta}_0) + \zeta_{\varphi} + \sqrt{n}(\boldsymbol{\beta}^* - \boldsymbol{\beta}_0) + \sqrt{n}H_{\lambda_j}(|\beta_j^*|)\mathrm{sgn}(\beta_j^*) + o(1) \\ &= O_P(1) + \sqrt{n}H_{\lambda_j}(|\beta_j^*|)\mathrm{sgn}(\beta_j^*) \;\; \text{for } j = p_0 + 1, \ldots, d, \end{aligned}$$

where $S_n^j(\boldsymbol{\beta}_0)$ is the jth component of $S_n(\boldsymbol{\beta}_0)$. Note that by assumption (I_3), $\sqrt{n}H_{\lambda_j}(|\beta_j^*|) \geq \sqrt{n}b_n \to \infty$ as $n \to \infty$, and thus the sign of $\left.\frac{\partial Q(\boldsymbol{\beta})}{\partial \beta_j}\right|_{\boldsymbol{\beta}=\boldsymbol{\beta}^*}$ is fully determined by that of β_j^* for n large enough. This together with Theorem 2.1 implies that $\lim_{n\to\infty} P(\hat{\boldsymbol{\beta}}_{nb} = \mathbf{0}) = 1$.

Moreover, by definition of $\hat{\boldsymbol{\beta}}_n$, it is obtained in a straightforward manner that $\left.\frac{\partial Q(\boldsymbol{\beta})}{\partial \boldsymbol{\beta}_a}\right|_{\boldsymbol{\beta}=(\hat{\boldsymbol{\beta}}_a,0)} = o_P(1)$. From this, partitioning $S_n(\boldsymbol{\beta}_0)$ as $(S_{n,a}(\boldsymbol{\beta}_0), S_{n,b}(\boldsymbol{\beta}_0))$, it follows from Eq. (2.12) that

$$o_P(1) = n^{-1/2}S_{n,a}(\boldsymbol{\beta}_0) - \zeta_{\varphi+}\sqrt{n}(\hat{\boldsymbol{\beta}}_{na} - \boldsymbol{\beta}_{0a}) + \sqrt{n}\sum_{j=1}^{p_0} H_{\lambda_j}(|\hat{\beta}_{na,j}|)\text{sgn}(\hat{\beta}_{na,j}),$$

and $|\sqrt{n}\sum_{j=1}^{p_0} H_{\lambda_j}(|\hat{\beta}_{na,j}|)\text{sgn}(\hat{\beta}_{na,j})| \leq p_0\sqrt{n}a_n \to 0$ as $n \to \infty$ by assumption (I_3). Hence,

$$\sqrt{n}(\hat{\boldsymbol{\beta}}_{na} - \boldsymbol{\beta}_{0a}) = \zeta_{\varphi+}^{-1}n^{-1/2}S_{n,a}(\boldsymbol{\beta}_0) + o_P(1).$$

As $n^{-1/2}S_{n,a}(\boldsymbol{\beta}_0) \xrightarrow{\mathscr{D}} N\big(0,\ \gamma_{\varphi+}\Sigma_a\big)$, we have

$$\sqrt{n}\big(\hat{\boldsymbol{\beta}}_{na} - \boldsymbol{\beta}_{0a}\big) \xrightarrow{\mathscr{D}} N\big(0,\ \zeta_{\varphi+}^{-2}\gamma_{\varphi+}\Sigma_a\big).$$

References

Abebe, A., McKean, J. W., & Bindele, H. F. (2012). On the consistency of a class of nonlinear regression estimators. *Pakistan Journal of Statistics and Operation Research, 8*(3), 543–555.

Arslan, O. (2012). Weighted LAD-LASSO method for robust parameter estimation and variable selection in regression. *Computational Statistics & Data Analysis, 56*(6), 1952–1965.

Bindele, H. F., & Abebe, A. (2012). Bounded influence nonlinear signed-rank regression. *Canadian Journal of Statistics, 40*(1), 172–189.

Fan, J., & Li, R. (2001). Variable selection via nonconcave penalized likelihood and its oracle properties. *Journal of the American Statistical Association, 96*(456), 1348–1360.

Hettmansperger, T. P., & McKean, J. W. (2011). Robust nonparametric statistical methods. In *Monographs on statistics and applied probability* (Vol. 119, 2nd ed.). Boca Raton, FL: CRC Press.

Hössjer, O. (1994). Rank-based estimates in the linear model with high breakdown point. *Journal of the American Statistical Association, 89*(425), 149–158.

Johnson, B. A. (2009). Rank-based estimation in the ℓ_1-regularized partly linear model for censored outcomes with application to integrated analyses of clinical predictors and gene expression data. *Biostatistics, 10*(4), 659–666.

Johnson, B. A., Lin, D., & Zeng, D. (2008). Penalized estimating functions and variable selection in semiparametric regression models. *Journal of the American Statistical Association, 103*(482), 672–680.

Johnson, B. A., & Peng, L. (2008). Rank-based variable selection. *Journal of Nonparametric Statistics, 20*(3), 241–252.

Leng, C. (2010). Variable selection and coefficient estimation via regularized rank regression. *Statistica Sinica, 20*(1), 167.

Rousseeuw, P. J. (1984). Least median of squares regression. *Journal of the American Statistical Association, 79*(388), 871–880.

Tibshirani, R. (1996). Regression shrinkage and selection via the lasso. *Journal of the Royal Statistical Society, Series B, 58*, 267–288.

Wang, H., & Leng, C. (2008). A note on adaptive group lasso. *Computational Statistics & Data Analysis, 52*(12), 5277–5286.

Wang, H., Li, G., & Jiang, G. (2007). Robust regression shrinkage and consistent variable selection through the lad-lasso. *Journal of Business & Economic Statistics, 25*(3), 347–355.

Wang, L., & Li, R. (2009). Weighted Wilcoxon-type smoothly clipped absolute deviation method. *Biometrics, 65*(2), 564–571.

Wu, C. F. (1981). Asymptotic theory of nonlinear least squares estimation. *Annals of Statistics, 9*(3), 501–513.

Xu, J., Leng, C., & Ying, Z. (2010). Rank-based variable selection with censored data. *Statistics and Computing, 20*(2), 165–176.

Zou, H. (206). The adaptive lasso and its oracle properties. *Journal of the American Statistical Association, 101*(476), 1418–1429.

Zou, H., & Hastie, T. (2005). Regularization and variable selection via the elastic net. *Journal of the Royal Statistical Society: Series B (Statistical Methodology), 67*(2), 301–320.

Chapter 3
Generalized Rank-Based Estimates for Linear Models with Cluster Correlated Data

John Kloke

Abstract This paper focuses on rank-based (R) estimation of parameters in a linear model with cluster correlated errors. The clusters are assumed to be independent, however, within a cluster the responses are allowed to be dependent. The method is applicable to general within cluster error structure. Application of a model which assumes the within cluster errors which follow an AR(1) process is developed. Discussion of an estimate of the AR(1) parameter is included. The algorithm first estimates the correlation structure by obtaining a robust rank-based estimate of the AR(1) parameter. The responses are then transformed to working independence and the model parameters are fit using ordinary rank regression. Estimates of standard errors—which utilize a sandwich estimate—are provided. An example and simulation results are discussed.

Keywords AR(1) • JR estimator • Robust • Wilcoxon

3.1 Introduction

Rank-based (R) estimation for linear models with independent and identically distributed (iid) errors was first developed in the 70s by Jureǎková (1971) and Jaeckel (1972). In the subsequent three to four decades a complete inference—including estimates of standard errors, tests of hypothesis, confidence intervals, and diagnostic procedures—was developed; in addition geometry and robustness properties were established. Extensions to multivariate regression models, nonlinear models, timeseries, and cluster correlated error data have been established. A summary of these published works is provided in the monograph by Hettmansperger and McKean (2011). R (R Development Core Team 2010) software implementation is discussed in Kloke and McKean (2015).

J. Kloke (✉)
Department of Biostatistics, University of Wisconsin School of Medicine and Public Health, Madison, WI 53726, USA
e-mail: kloke@biostat.wisc.edu

© Springer International Publishing Switzerland 2016
R.Y. Liu, J.W. McKean (eds.), *Robust Rank-Based and Nonparametric Methods*, Springer Proceedings in Mathematics & Statistics 168,
DOI 10.1007/978-3-319-39065-9_3

This paper develops a robust R procedure for estimation of parameters in a linear model with cluster correlated errors. The individual clusters are assumed to be independent and the structure of the within cluster errors is assumed to be known. A sandwich estimator is used in estimating the standard errors of the model estimates. Given a robust estimate of the within cluster correlation observations are transformed to working independence. Estimates of the linear model parameters are obtained via rank-estimation where the ranking is over the entire set of responses. Focus is on the case where the within subject errors follow an AR(1) process.

Previous work has been done on R estimation for linear models with cluster correlated errors. Methods which work with the correlated (non-transformed) observations include Kloke et al. (2009) and Rashid et al. (2012). Kloke et al. (2009) developed a joint rankings (JR) estimator which utilizes the rankings of the entire set of responses. Rashid et al. (2012) developed a many rankings (MR) estimator which ignores the within cluster effect. Kloke and McKean (2011) considered a transformation approach for exchangeable (compound symmetric) within block errors. A comparison between MR and JR is given in Rashid et al. (2013).

The data are assumed to be collected in clusters or blocks and the observations within each cluster are assumed to be correlated. The design is general and can include baseline covariates, time varying covariates, or variables based on the design (e.g. sets of indicator variables denoting treatment). In this new approach we transform the responses, in a manner similar to generalized least squares (GLS), to uncorrelated and then apply ordinary rank regression, or more generally, JR estimation to the problem. In this paper we call these GJR estimates. A similar approach was implemented in Bilgic et al. (2015) for nested designs. The focus of this paper is on a repeated measures problem which assumes the errors within an experimental unit follow an AR(1) process. Estimates of the AR(1) parameter are based on the work of Terpstra et al. (2000, 2001).

In Sect. 3.2 we provide a review or rank-based estimation and joint rankings (JR) estimation. In Sect. 3.3 we develop a new method for estimation of parameters in a linear model with cluster correlated errors based on a transformation of the model to working independence and discuss estimation the AR(1) parameter. In Sect. 3.4 we discuss the results of a small simulation study and in Sect. 3.5 we present the a real data example. Section 3.6 is a brief summary of the work.

3.2 R and JR Estimation

In this section we briefly review rank-based (R) estimation for linear models with independent and identically distributed (iid) errors as well as joint-rankings (JR) estimation for linear models with cluster correlated error.

3.2.1 Rank-Based Estimator

R estimates for linear models were first developed by Jureăková (1971) and Jaeckel (1972). A complete inference has been developed since that time and is summarized in the monograph by Hettmansperger and McKean (2011). R estimates and associated inference can be computed using the R (R Development Core Team 2010) package `Rfit` (Kloke and McKean 2012). `Rfit` computes R estimates, standard errors, and diagnostics for a general linear model.

In this section, we consider the following linear model

$$y_i = \alpha + \boldsymbol{x}_i^T \boldsymbol{\beta} + e_i \text{ for } i = 1, \ldots n \tag{3.1}$$

where y_i is a continuous response variable, $\boldsymbol{x}_i$ is the vector of explanatory variables, α is the intercept parameter, $\boldsymbol{\beta}$ is the vector of regression coefficients, and e_i is the error term. The errors are assumed to be independent and identically distributed with continuous pdf f. Further assume f has finite Fisher information. Additionally, there are design assumptions that are similar to the least squares analysis [see Sect. 3.4 of Hettmansperger and McKean (2011)].

Rewrite (3.1) in matrix notation as follows

$$\boldsymbol{y} = \alpha\mathbf{1} + \mathbf{X}\boldsymbol{\beta} + \mathbf{e}$$

where $\boldsymbol{y} = [y_1, \ldots, y_n]^T$ is a $n \times 1$ vector of responses, $\mathbf{X} = [\boldsymbol{x}_1, \ldots, \boldsymbol{x}_n]^T$ is an $n \times p$ design matrix, and $\mathbf{e} = [e_1, \ldots, e_n]^T$ is an $n \times 1$ vector of error terms. Recall that the least squares estimator is the minimizor of Euclidean distance between $\boldsymbol{y}$ and $\hat{\boldsymbol{y}}_{LS} = \mathbf{X}\hat{\boldsymbol{\beta}}_{LS}$. To obtain the R estimator of $\boldsymbol{\beta}$ one uses a different measure of distance which is referred to as Jaeckel's (1972) dispersion function. Jaeckel's dispersion function is defined as

$$D(\boldsymbol{\beta}) = \|\boldsymbol{y} - \mathbf{X}\boldsymbol{\beta}\|_\varphi \tag{3.2}$$

where $\|\cdot\|_\varphi$ is a pseudo-norm defined as

$$\|\mathbf{u}\|_\varphi = \sum_{i=1}^{n} a(R(u_i))u_i \text{ for } \mathbf{u} \in R^n,$$

$a(t) = \varphi\left(\frac{t}{n+1}\right)$, and φ is a **score function**. Assume φ is nondecreasing on $(0, 1)$ and WLOG standardized so that $\int_0^1 \varphi(u)\ du = 0$ and $\int_0^1 \varphi(u)^2\ du = 1$. Under these assumptions (3.2) is a convex function of $\boldsymbol{\beta}$ and provides a robust measure of distance between $\boldsymbol{y}$ and $\mathbf{X}\boldsymbol{\beta}$. Commonly used score functions are Wilcoxon (linear) scores $\varphi_W(u) = \sqrt{12}\left(u - \frac{1}{2}\right)$ and the sign scores (L1) $\varphi_S(u) = \text{sign}\left(u - \frac{1}{2}\right)$. A theoretical result shows that for distribution f, the optimal scores are $\varphi(u) = -\frac{f'(F^{-1}(u))}{f(F^{-1}(u))}$; these scores are optimal in the sense that they are asymptotically efficient relative to the maximum likelihood estimate. Wilcoxon scores are commonly used as they robust to outliers[1] and have 95 % efficiency relative to least squares at the normal distribution. Further, the efficiency of the Wilcoxon relative to least squares can be much greater at distributions with heavier tails than the normal. Wilcoxon scores are the default is scores in `Rfit`.

The R estimator of $\boldsymbol{\beta}$ is defined as

$$\hat{\boldsymbol{\beta}}_\varphi = \text{Argmin}\|\boldsymbol{y} - \mathbf{X}\boldsymbol{\beta}\|_\varphi. \tag{3.3}$$

Note that closed form solutions exist for least squares, however, this is not the case for rank estimation. The R estimates are obtained by minimizing a convex optimization problem. The intercept parameter, α, is estimated separately using a robust rank-based estimate of location; the median of the residuals is typically used. It can be shown, see for example Hettmansperger and McKean (2011), that the solution to (3.3) is consistent and asymptotically normal. Symbolically we write

$$\hat{\boldsymbol{\beta}}_\varphi \dot{\sim} N\left(\boldsymbol{\beta}, \tau_\varphi^2(\mathbf{X}^T\mathbf{X})^{-1}\right)$$

where τ_φ is a scale parameter which depends on f and the score function φ

$$\tau_\varphi^{-1} = \int_0^1 \varphi(u)\ \varphi_f(u)\ du.$$

Koul et al. (1987) provide a consistent estimate of τ_φ which is implemented in `Rfit`. So that the estimated variance covariance matrix is

$$\hat{\tau}_\varphi^2(\mathbf{X}^T\mathbf{X})^{-1}.$$

Wald type tests and confidence intervals are then easily calculated. Hettmansperger and McKean (2011) discuss tests based on a reduction in dispersion as well as a gradient (scores) test.

3.2.2 *Joint Rankings Estimator*

Kloke et al. (2009) showed that rank-based analysis can be extended to cluster correlated data. In this section we summarize these methods. An experimental R software package for computing these joint rankings (JR) estimates is `jrfit` which is available at https://github.com/kloke/jrfit.

[1]Having bounded influence function and 29 % breakdown point.

Assume an experiment is done over m blocks or clusters. Note that we use the terms block and cluster interchangeably. Let n_k denote the number of measurements taken within the kth block. Let Y_{ki} denote the response variable for the ith experimental unit within the kth block; let $\boldsymbol{x}_{ki}$ denote the corresponding vector of covariates. Note that the design is general in that $\boldsymbol{x}_{ki}$ may contain, for example, covariates, baseline values, or treatment indicators. The response variable is then modeled as

$$y_{ki} = \alpha + \boldsymbol{x}_{ki}^T\boldsymbol{\beta} + e_{ki} \text{ for } k = 1, \ldots, m, i = 1, \ldots, n_k, \tag{3.4}$$

where α is the intercept parameter, $\boldsymbol{\beta}$ is a $p \times 1$ vector of unknown parameters, and e_{ki} is an error term. We assume that the errors within a block are correlated (i.e. e_{ki} and $e_{ki'}$) but the errors between blocks are independent (i.e. e_{ki} and $e_{k'j}$). Further, we assume that e_{ki} has pdf and cdf $f(x)$ and $F(x)$, respectively. Now write model (3.4) in block vector notation as

$$\boldsymbol{y}_k = \alpha \mathbf{1}_{n_k} + \mathbf{X}_k\boldsymbol{\beta} + \mathbf{e}_k.$$

where $\mathbf{1}_{n_k}$ is an $n_k \times 1$ vector of ones and $\mathbf{X}_k = [\boldsymbol{x}_{k1} \ldots \boldsymbol{x}_{kn_k}]^T$ is a $n_k \times p$ design matrix and $\mathbf{e}_k = [e_{k1}, \ldots e_{kn_k}]^T$ is a $n_k \times 1$ vector of error terms. Let $N = \sum_{k=1}^m n_k$ denote the total sample size. Let $\boldsymbol{y} = (\boldsymbol{y}_1^T, \ldots, \boldsymbol{y}_m^T)^T$ be the $N \times 1$ vector of all measurements (responses) and consider the matrix formulation of the model as

$$\boldsymbol{y} = \alpha \mathbf{1}_N + \mathbf{X}\boldsymbol{\beta} + \mathbf{e}$$

where $\mathbf{1}_N$ is an $N \times 1$ vector of ones and $\mathbf{X} = [\mathbf{X}_1^T \ldots \mathbf{X}_m^T]^T$ is a $N \times p$ design matrix and $\mathbf{e} = [\mathbf{e}_1^T, \ldots \mathbf{e}_m^T]^T$ is a $N \times 1$ vector of error terms. Since there is an intercept in the model, we may assume (WLOG) that $\mathbf{X}$ is centered.

The rank-based estimator of $\boldsymbol{\beta}$ is given by

$$\hat{\boldsymbol{\beta}}_\varphi = \text{Argmin}\|\boldsymbol{y} - \mathbf{X}\boldsymbol{\beta}\|_\varphi \text{ where } \|\mathbf{u}\|_\varphi = \sum_{t=1}^N a(R(u_t))u_t, \quad \mathbf{u} \in R^N, \tag{3.5}$$

is Jaeckel's dispersion function.

For formal inference, Kloke et al. (2009) develop the asymptotic distribution of the $\hat{\boldsymbol{\beta}}_\varphi$ under the assumption that the marginal distribution functions of the random vector $\mathbf{e}_k$ are the same. This includes two commonly assumed error structures: exchangeable within block errors as well as the components of $\mathbf{e}_k$ following a stationary time series, such as autoregressive of general order. This asymptotic distribution of $\hat{\boldsymbol{\beta}}$ is given by

$$\hat{\boldsymbol{\beta}}_\varphi \dot{\sim} N_p\left(\boldsymbol{\beta}_0, \tau_\varphi^2(\mathbf{X}^T\mathbf{X})^{-1}\left(\sum_{k=1}^m \mathbf{X}_{c_k}\boldsymbol{\Sigma}_{\varphi_k}\mathbf{X}_{c_k}\right)(\mathbf{X}^T\mathbf{X})^{-1}\right)$$

where $\boldsymbol{\Sigma}_{\varphi_k} = \text{var}(\varphi(F(\mathbf{e}_k)))$ and $F(\mathbf{e}_k) = [F(e_{k1}), \ldots, F(e_{kn_k})]^T$. As with the iid case, the scale parameter τ_φ can be estimated by the Koul et al. (1987) method. A sandwich estimator is recommended to estimate $\boldsymbol{\Sigma}_{\varphi_k}$ which is given by

$$\frac{m}{m-p}\sum_{k=1}^{m}\mathbf{X}_k^T a(R(\hat{\mathbf{e}}_k))a(R(\hat{\mathbf{e}}_k))^T\mathbf{X}_k.$$

As they will be useful in the development of the GJR estimator in the next section, we present the asymptotic representation of the parameters. The asymptotic representation for $\hat{\boldsymbol{\beta}}$ is given by

$$\sqrt{N}(\hat{\boldsymbol{\beta}} - \boldsymbol{\beta}_0) = \tau_\varphi\sqrt{N}(\mathbf{X}^T\mathbf{X})^{-1}\mathbf{X}^T\varphi(F(\mathbf{e})) + o_p(1) \tag{3.6}$$

and the asymptotic representation for $\hat{\alpha}$ is given by

$$\sqrt{N}(\hat{\alpha} - \alpha_0) = \frac{\tau_s}{\sqrt{N}}\mathbf{1}_N^T\text{sgn}(\mathbf{e}) + o_p(1). \tag{3.7}$$

3.3 GJR Estimator

In this section we describe a generalized rank-based estimator. This estimator is akin to the generalized least squares estimator in that the working covariance matrix is applied to the vector of responses in each cluster to transform them to working independence; following that the usual estimation procedure is followed. For the rank-based estimator we present, we utilized JR estimation and the covariance is estimated using a sandwich estimator as was discussed in the previous section. Thus we call the estimator a generalized joint rankings (GJR) estimator.

3.3.1 *Model and Notation*

As with the JR estimator of the previous section, assume an experiment done over m block or clusters. Note we use the terms block and cluster interchangeably. The results presented in this paper are general and can be utilized in a variety of settings. For example, blocks may denote centers in a multi-center clinical trial or subjects in a single site trial.

We now establish notation which is similar to the last section. Let n_k denote the number of measurements observed for the kth block. Let y_{ki} denote the ith measurement observed for the kth block and let $\boldsymbol{x}_{ki}$ denote the corresponding vector of covariates. We model the response as

$$y_{ki} = \boldsymbol{x}_{ki}^T\boldsymbol{\beta} + e_{ki} \text{ for } k = 1, \ldots, m, i = 1, \ldots, n_k \tag{3.8}$$

where $\boldsymbol{\beta}$ is a $p\times 1$ vector of unknown parameters and e_{ki} is an error term. We assume that the errors within a block are correlated (i.e. e_{ki} and $e_{ki'}$) but the errors between subjects are independent (i.e. e_{ki} and $e_{k'j}$). Further we assume that $e_{ki} \sim F,f$. Now write model (3.8) in block vector notation as

$$y_k = \mathbf{X}_k\boldsymbol{\beta} + \mathbf{e}_k.$$

where $\boldsymbol{y}_k$ is an $n_k \times 1$ vector of responses, $\mathbf{X}_k$ is an $n_k \times 1$ matrix, and $\mathbf{e}_k$ is a vector of random errors. Note we need not estimate the intercept separately as is typically the case in R estimation. We assume that $\text{var}(\mathbf{e}_k) = \boldsymbol{\Sigma}_k$.[2]

3.3.2 GJR Estimation Procedure

Assume the dependence structure between the responses is known and is given by $\boldsymbol{\Sigma}_k$. In practice we will estimate $\boldsymbol{\Sigma}_k$, one such estimate is the AR(1) parameter for when the within subject errors follow an AR(1) process.

We first transform the responses to working independence:

$$\mathbf{y}_k^* = \boldsymbol{\Sigma}_k^{-1/2}\boldsymbol{y}_k \text{ for } k = 1,\ldots,m$$

along with the corresponding design matrices

$$\mathbf{X}_k^* = \boldsymbol{\Sigma}_k^{-1/2}\mathbf{X}_k \text{ for } k = 1,\ldots,m.$$

So the transformed model is

$$y_k^* = \mathbf{X}_k^*\boldsymbol{\beta} + \mathbf{e}_k^* \text{ for } k = 1,\ldots,m$$

where $\mathbf{e}_k^* = \boldsymbol{\Sigma}_k^{-1/2}\mathbf{e}_k$, $\text{var}(\mathbf{e}_k^*) = \sigma^2\mathbf{I}_{n_k}$, and $\mathbf{I}_{n_k}$ is an $n_k\times n_k$ identity matrix. Next we form one linear model by stacking the transformed response vectors, corresponding design matrices and error vectors :

$$\mathbf{y}^* = \mathbf{X}^*\boldsymbol{\beta} + \mathbf{e}^* \tag{3.9}$$

so that $\boldsymbol{y}^{*T} = [y_1^{*T},\ldots,y_m^{*T}]^T$, $\mathbf{X}^{*T} = [\mathbf{X}_1^{*T},\ldots,\mathbf{X}_m^{*T}]^T$, and $\mathbf{e}^{*T} = [\mathbf{e}_1^{*T},\ldots\mathbf{e}_m^{*T}]^T$. Except in special cases the design matrix $\mathbf{X}^*$ does not have $\mathbf{1}_N$ in the column space. So, as discussed in Dixon and McKean (1996), Model (3.9) cannot be estimated

[2] We do not need the variance to exist. Only that a linear transformation exists which has the goal of reducing the dependence in the response variables. The variance notation is adopted for convenience.

directly and we must first fit a model with an intercept term and then transform back. That is we fit

$$\mathbf{y}^* = \alpha_1 \mathbf{1}_N + \mathbf{X}_c^* \boldsymbol{\beta}_1 + \mathbf{e}^*, \tag{3.10}$$

where $\mathbf{X}_c^*$ is the centered version of $\mathbf{X}^*$; i.e. $\mathbf{X}_c^* = (\mathbf{I}_N - \mathbf{H}_{\mathbf{1}_N})\mathbf{X}^*$ where $\mathbf{H}_{\mathbf{1}_N}$ is a projection matrix onto the space spanned by the $N \times 1$ vector $\mathbf{1}_N$. The estimate of $\boldsymbol{\beta}$ is the one which minimizes Jaeckel's dispersion function (3.2). For inference we utilize the results for JR estimation (Kloke et al. 2009) and the extension of Bilgic et al. (2015). To obtain a first set of intermediate fitted values

$$\hat{\mathbf{y}}_1^* = \hat{\alpha}_1 \mathbf{1}_N + \mathbf{X}_c^* \hat{\boldsymbol{\beta}}_1.$$

From which we can project these first intermediate fitted values to the correct space [colSpace($\mathbf{X}^*$)] to obtain the second intermediate fitted values

$$\hat{\mathbf{y}}^* = \mathbf{H}_{X^*} \hat{y}_1^*.$$

Now we re-transform these second intermediate fitted values to obtain the model fits.

$$\hat{y}_k = \boldsymbol{\Sigma}^{1/2} \hat{\mathbf{y}}_k^* \text{ for } k = 1, \ldots m.$$

The estimate of $\boldsymbol{\beta}$ is given by solving the generalized least squares problem $\mathbf{X}\hat{\boldsymbol{\beta}} = \hat{y}$ where $\hat{y} = [\hat{y}_1, \ldots, \hat{y}_m]^T$. That is

$$\hat{\boldsymbol{\beta}} = (\mathbf{X}^T\mathbf{X})^{-1}\mathbf{X}^T\hat{y}.$$

The estimate of $\boldsymbol{\beta}$ is a linear combination of the JR estimates of Model (3.10), i.e.

$$\hat{\boldsymbol{\beta}} = \left(\mathbf{X}^{*T}\mathbf{X}^*\right)^{-1} \mathbf{X}^{*T}[\hat{\alpha}_1 \mathbf{1}_N + \mathbf{X}_c^* \hat{\boldsymbol{\beta}}_1]$$

To determine the asymptotic distribution of $\hat{\boldsymbol{\beta}}$ we utilize the asymptotic representation of $(\hat{\boldsymbol{\beta}}_1)$ and $(\hat{\alpha}_1)$ given in (3.6) and (3.7) along with the theory developed in Kloke et al. (2009) and Bilgic et al. (2015). The asymptotic distribution of $\hat{\boldsymbol{\beta}}$ is normal with mean vector $\boldsymbol{\beta}$ and variance covariance matrix

$$\begin{aligned} \text{var}(\hat{\boldsymbol{\beta}}) = {} & (\mathbf{X}^{*T}\mathbf{X}^*)^{-1}\mathbf{X}^{*T}[\sigma_1^2(0)\tau_s^2\mathbf{H}_1 + \\ & \tau_\varphi^2 \mathbf{X}_c^* (\mathbf{X}_c^{*T}\mathbf{X}_c^*)^{-1} \left(\sum_{k=1}^m \mathbf{X}_{ck}^{*T} \boldsymbol{\Sigma}_{\varphi k} \mathbf{X}_{ck}^* \right) (\mathbf{X}_c^{*T}\mathbf{X}_c^*)^{-1}\mathbf{X}_c^{*T}]\mathbf{X}^*(\mathbf{X}^{*T}\mathbf{X}^*)^{-1} \end{aligned} \tag{3.11}$$

where $\sigma_1^2(0) = \sum_{k=1}^m \sum_{i=1}^{n_k} \sum_{j=1}^{n_k} \text{cov}(\text{sgn}(e_{ki})\text{sgn}(e_{kj}))$.

3.3.3 *Estimation of the AR(1) Parameter*

In this section we consider an R analysis of repeated measures taken on m subjects. It is common to assume that the within subject errors follow an AR(1) process and that is the approach we take. That is we assume $\text{var}(e_{ki}, e_{kj}) = \sigma^2\rho^{|i-j|}$. The estimate of ρ is based on the one proposed by Terpstra et al. (2000, 2001).

The algorithm is as follows.

1. Fit ordinary rank regression (ORR) to the model

$$y = \alpha\mathbf{1}_N + \mathbf{X}\boldsymbol{\beta} + \mathbf{e}.$$

and form the residuals: $\hat{\mathbf{e}} = y - \mathbf{X}\hat{\boldsymbol{\beta}}$. Kloke et al. (2009) ensures that the estimator of $\boldsymbol{\beta}$ is consistent.

2. Using the residuals from Step 1 estimate the AR(1) parameter, again using ORR, though based on regression through the origin (Dixon and McKean 1996)[3]:

$$\hat{e}_{ki} = \rho\hat{e}_{k,i-1} + u_{ki}.$$

3. (GJR step) Let $\hat{\mathbf{C}}_k$ be the estimated AR(1) correlation structure of size n_k. Transform the response vectors and design matrices for each of the m subjects:

$$\mathbf{y}_k^* = \hat{\mathbf{C}}_k^{-1/2}\mathbf{y}_k \text{ and } \mathbf{X}_k^* = \hat{\mathbf{C}}_k^{-1/2}\mathbf{X}_k.$$

Now one may use GJR as discussed in the previous section to obtain estimates and inference for $\boldsymbol{\beta}$.

3.4 Simulation Study

We conducted a small simulation study to investigate the properties of the proposed method. The model we looked at was a simple analysis of covariance problem:

$$y_k = \alpha\mathbf{1}_n + \boldsymbol{x}_k\beta + w_k\mathbf{1}_n\Delta + \mathbf{e}_k$$

where $\boldsymbol{x}_k \sim \text{N}_n(\mathbf{0}, \mathbf{I})$ and $w_k \sim \text{bin}(1, 0.5)$, for $k = 1, \ldots, m$ and n is the number of within subject measurements. We looked at sample sizes of $m = 20, 50,$ and 100 and the number of measures per cluster was $n = 2, 4,$ or 8. Three error distributions were considered: a standard normal distribution (N(0,1)), a contaminated normal distribution with 10 % contamination and a standard deviation of 3 (CN(0.1,3)), and a contaminated normal distribution with 25 % contamination and a standard

[3] As the intercept was fit in the previous step, the residuals should have location zero.

Table 3.1 Empirical confidence levels for joint rank (JR) and generalized joint rank (GJR) estimators

			$\rho = 0$				$\rho = 0.4$			
			GJR		JR		GJR		JR	
	m	n	Δ	β	Δ	β	Δ	β	Δ	β
N(0,1)	20	2	0.942*	0.939*	0.952	0.953	0.944*	0.939*	0.964*	0.974*
	20	4	0.940*	0.940*	0.952	0.951	0.916*	0.942*	0.977*	0.975*
	20	8	0.940*	0.943*	0.951	0.955*	0.942*	0.940*	0.970*	0.974*
	50	2	0.947	0.946	0.952	0.950	0.952	0.942*	0.958*	0.959*
	50	4	0.948	0.944*	0.953	0.947	0.945*	0.951	0.956*	0.964*
	50	8	0.946	0.943*	0.951	0.948	0.943*	0.948	0.956*	0.961*
	100	2	0.954	0.947	0.956*	0.951	0.948	0.948	0.953	0.958*
	100	4	0.953	0.944*	0.955*	0.946	0.945*	0.951	0.950	0.959*
	100	8	0.947	0.947	0.949	0.949	0.946	0.948	0.954	0.957*
CN(0.1,3)	20	2	0.948	0.935*	0.957*	0.942*	0.949	0.944*	0.972*	0.977*
	20	4	0.936*	0.942*	0.949	0.949	0.932*	0.942*	0.973*	0.977*
	20	8	0.944*	0.941*	0.957*	0.948	0.940*	0.945*	0.972*	0.974*
	50	2	0.951	0.950	0.956*	0.950	0.950	0.944*	0.957*	0.962*
	50	4	0.948	0.948	0.952	0.949	0.945*	0.944*	0.954*	0.956*
	50	8	0.949	0.942*	0.954	0.945*	0.947	0.945*	0.959*	0.962*
	100	2	0.953	0.944*	0.954	0.940*	0.950	0.948	0.953	0.958*
	100	4	0.954	0.948	0.956*	0.949	0.952	0.946	0.953	0.953
	100	8	0.949	0.948	0.950	0.948	0.952	0.947	0.955*	0.955*
CN(0.25,5)	20	2	0.957*	0.951	0.961*	0.946	0.944*	0.952	0.978*	0.980*
	20	4	0.950	0.948	0.960*	0.950	0.938*	0.952	0.976*	0.978*
	20	8	0.939*	0.944*	0.955*	0.950	0.892*	0.946	0.990*	0.978*
	50	2	0.952	0.954	0.957*	0.945*	0.958*	0.952	0.963*	0.966*
	50	4	0.952	0.950	0.957*	0.949	0.949	0.952	0.957*	0.960*
	50	8	0.950	0.952	0.955*	0.952	0.947	0.950	0.960*	0.963*
	100	2	0.954	0.956*	0.956*	0.950	0.950	0.945*	0.954	0.952
	100	4	0.955*	0.953	0.959*	0.954*	0.950	0.954	0.955*	0.956*
	100	8	0.948	0.950	0.951	0.950	0.952	0.950	0.956*	0.954

A * denotes the value is outside of the interval $0.95 \pm 2\sqrt{\frac{0.95 \cdot 0.05}{10{,}000}}$

deviation of 5 (CN(0.25, 5)). Two AR(1) parameters were considered: $\rho = 0.0, 0.4$. The simulation size was 10,000. Two rank-based estimates were computed (JR and GJR) and for relative efficiency comparisons the REML estimates were computed as well. We use the RE = $\text{MSE}_{\text{REML}}/\text{MSE}_{\text{(G)JR}}$ and the 95 % confidence interval coverage as performance metrics. Results are presented in Tables 3.1 and 3.2.

Looking at the empirical confidence levels in Table 3.1 we see that the GJR estimate can be slightly liberal when the sample size is small ($m = 20$), however,

Table 3.2 Empirical relative efficiencies (to REML) for joint rank (JR) and generalized joint rank (GJR) estimators

			$\rho = 0$				$\rho = 0.4$			
			GJR		JR		GJR		JR	
	m	n	Δ	β	Δ	β	Δ	β	Δ	β
N(0,1)	20	2	0.933	0.946	0.937	1.000	0.939	0.930	0.953	0.822
	20	4	0.944	0.958	0.951	0.979	0.952	0.957	0.948	0.767
	20	8	0.952	0.962	0.954	0.973	0.953	0.949	0.951	0.735
	50	2	0.947	0.947	0.946	0.966	0.945	0.942	0.956	0.805
	50	4	0.944	0.961	0.947	0.972	0.955	0.960	0.951	0.752
	50	8	0.951	0.955	0.951	0.961	0.955	0.952	0.956	0.726
	100	2	0.954	0.948	0.954	0.958	0.948	0.947	0.961	0.803
	100	4	0.950	0.961	0.952	0.966	0.957	0.959	0.950	0.748
	100	8	0.954	0.955	0.955	0.958	0.952	0.949	0.940	0.718
CN(0.1,3)	20	2	1.313	1.323	1.325	1.392	1.231	1.213	1.228	1.064
	20	4	1.344	1.376	1.358	1.413	1.269	1.277	1.205	0.998
	20	8	1.369	1.362	1.379	1.377	1.300	1.285	1.218	0.939
	50	2	1.363	1.378	1.368	1.403	1.269	1.256	1.256	1.060
	50	4	1.350	1.368	1.356	1.381	1.296	1.297	1.230	1.006
	50	8	1.361	1.363	1.363	1.371	1.319	1.330	1.246	0.974
	100	2	1.367	1.378	1.371	1.387	1.255	1.300	1.241	1.093
	100	4	1.363	1.370	1.365	1.374	1.292	1.314	1.213	0.993
	100	8	1.347	1.377	1.349	1.382	1.308	1.320	1.210	0.957
CN(0.25,5)	20	2	2.574	2.543	2.704	2.769	1.788	1.878	1.700	1.573
	20	4	2.909	2.850	3.002	2.942	2.036	2.114	1.740	1.547
	20	8	3.032	2.922	3.081	2.975	2.168	2.298	1.759	1.543
	50	2	2.900	2.903	2.935	2.958	2.078	2.110	1.945	1.688
	50	4	3.044	3.003	3.082	3.039	2.172	2.262	1.844	1.615
	50	8	3.023	3.073	3.040	3.094	2.349	2.268	1.865	1.529
	100	2	3.010	3.066	3.026	3.092	2.063	2.197	1.929	1.719
	100	4	3.039	3.091	3.051	3.114	2.221	2.326	1.884	1.640
	100	8	3.036	3.051	3.045	3.062	2.328	2.397	1.850	1.605

when the sample size increases ($m = 50, 100$) the coverage is closer the nominal 95 % level. The JR estimate is slightly conservative when the within cluster correlation has a autocorrelation structure.

At the normal distribution (N(0,1)) the relative efficiency is in the neighborhood of 95.5 %. One notable exception is the JR estimate of β when $\rho = 0.4$, which looses efficiency. When there is no correlation the performance of the JR and GJR estimates are practically the same for estimating Δ (numerically the JR has a slight advantage), however, when there is autocorrelation ($\rho = 0.4$) the GJR has an advantage.

3.5 Example

This example is based on the `Orthodont` data from Pinheiro and Bates (2006). This is a longitudinal study on 27 young subjects with data collected at 8, 10, 12, and 14 years. The outcome variable is the distance from the pituitary to the pterygomaxillary fissure (mm). Explanatory variables include age and sex of the subject. Though the authors of that text conclude a compound symmetric covariance structure may be suitable; we utilize this example to illustrate the robustness of the method proposed in this paper. Namely, we model an AR(1) correlation structure for subjects and estimate the linear increase in distance over time; a indicator variable is included in the model for females as well as an interaction term.

A spaghetti plot is presented in Fig. 3.1. There are one or more mild outliers in the data. For example, one subject (`M09`) has a pattern much different than the other children The outcome measures decrease from 23 mm at age 8 to 20.5 mm at age 10 then increase to 31 mm at age 12 and then decrease to 26 mm at age 14.

The results of the GJR and GLS fits are displayed in Table 3.3 (in the left most columns, under Original Data). The estimates and standard errors are similar for the two methods. The conclusions that both sex and age are important in estimating distance are the same for the two methods.

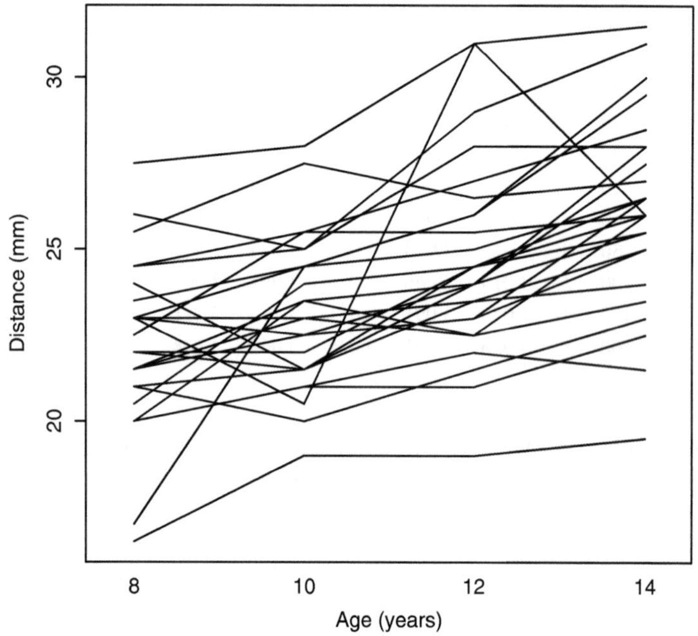

Fig. 3.1 Spaghetti plot of 27 young subjects in a longitudinal study to examine distance [from pituitary to the pterygomaxillary fissure (in mm)] over time

Table 3.3 Estimates (Est) and standard errors (SE) of GJR and GLS estimates for orthodont data

	Original data				Error introduced			
	GJR		GLS		GJR		GLS	
	Est	SE	Est	SE	Est	SE	Est	SE
Intercept	25.29	0.5	25.06	0.44	25.41	0.7	41.2	12.6
Female	−2.39	0.86	−2.42	0.69	−2.72	1.02	−18.56	19.75
Age-11	0.69	0.1	0.77	0.12	0.69	0.1	0.19	0.4
Female × (Age-11)	−0.21	0.14	−0.29	0.18	−0.21	0.14	0.29	0.63
rho	0.75		0.63		0.84		0.99	

To examine the robustness properties of the proposed method we introduced an outlier by reordering the first subject's values and multiplying by 10 (i.e. the data were entered in the wrong order in the wrong units/with a missing decimal point). The results of the GJR and GLS fits are displayed in Table 3.3 (in the right most columns, under Error Introduced).

The estimates and standard errors change somewhat for the GJR approach, though the conclusions would remain the same. For the GLS method, the estimates and standard errors change substantially and the conclusions would now be then neither sex nor age is an important explanatory variable.

This example demonstrates that the procedure is robust to gross outliers and gives estimates similar to those of traditional methods when there are only mild outliers in the data.

In practice, it is prudent to select an appropriate correlation structure as doing so may affect the conclusions; we are working on model selection procedures to aide in this selection process.

3.6 Summary and Future Work

In this paper we developed generalized rank-based estimates for cluster correlated data when the within cluster correlation structure is AR(1). A robust estimate of the AR(1) parameter was discussed as well as estimates of standard errors. A small simulation study was conducted showing the validity (especially when the number of clusters is 100 or more) as well as the gains in efficiency to REML methods when the data contain contamination. An example demonstrates the robustness of the method to severe outliers.

In future work, we plan to extend these analyses to other correlation structures (e.g. compound symmetric) as well as develop Studentized residuals. We are also working to develop model selection, including choice of correlation structure.

Software for the method is in development, though an experimental release is available at https://github.com/kloke/gjrfit.

References

Bilgic, Y. K., McKean, J. W., Kloke, J. D., & Abebe, A. (2015, submitted). Iteratively reweighted generalized rank-based methods for hierarchical mixed models.

Dixon, S. L., & McKean, J. W. (1996). *Journal of the American Statistical Association, 91*, 699 (1996).

Hettmansperger, T. P., & McKean, J. W. (2011). *Robust nonparametric statistical methods* (2nd ed.). Boca Raton, FL: Chapman & Hall/CRC Press.

Jaeckel, L. A. (1972). *Estimating regression coefficients by minimizing the dispersion of residuals, 43*, 1449.

Jureǎková, J. (1971). *Nonparametric estimate of regression coefficients, 42*, 1328.

Kloke, J. D., & McKean, J. W. (2011). In D. R. Hunter, D. P. Richards, & J .L. Rosenberger (Eds.), *Nonparametric statistics and mixture models: A festschrift in honor of Thomas P. Hettmansperger* (pp. 183–203). Hackensack, NJ: World Scientific Publishing Co. Pte. Ltd.

Kloke, J. D., & McKean, J. W. (2012). *The R Journal, 4*(2), 57.

Kloke, J. D., McKean, J. W., & Rashid, M. (2009). *Journal of the American Statistical Association, 104*, 384.

Kloke, J., & McKean, J. W. (2015). *Nonparametric statistical methods using R*. Boca Raton, FL: Chapman & Hall/CRC Press.

Koul, H. L., Sievers, G. L., & McKean, J. W. (1987). *Scandinavian Journal of Statistics, 14*, 131.

Pinheiro, J., & Bates, D. (2006). *Mixed-effects models in S and S-PLUS*. Springer Science & Business Media, New York, NY: Springer

R Development Core Team (2010). *R: A Language and Environment for Statistical Computing*. Vienna: R Foundation for Statistical Computing. http://www.R-project.org. ISBN:3-900051-07-0.

Rashid, M. M., McKean, J. W., & Kloke, J. D. (2012). *Statistics in Biopharmaceutical Research , 4*(1), 37.

Rashid, M. M., McKean, J. W., & Kloke, J. D. (2013). *Journal of Biopharmaceutical Statistics, 23*(6), 1207.

Terpstra, J., McKean, J. W., & Naranjo, J. D. (2000). *Statistics, 35*, 45.

Terpstra, J., McKean, J. W., & Naranjo, J. D. (2001). *Statistics and Probability Letters, 51*, 165.

Chapter 4
Iterated Reweighted Rank-Based Estimates for GEE Models

Asheber Abebe, Joseph W. McKean, John D. Kloke, and Yusuf K. Bilgic

Abstract Repeated measurement designs occur in many areas of statistical research. In 1986, Liang and Zeger offered an elegant analysis of these problems based on a set of generalized estimating equations (GEEs) for regression parameters, that specify only the relationship between the marginal mean of the response variable and covariates. Their solution is based on iterated reweighted least squares fitting. In this paper, we propose a rank-based fitting procedure that only involves substituting a norm based on a score function for the Euclidean norm used by Liang and Zeger. Our subsequent fitting, while also an iterated reweighted least squares solution to GEEs, is robust to outliers in response space and the weights can easily be adapted for robustness in factor space. As with the fitting of Liang and Zeger, our rank-based fitting utilizes a working covariance matrix. We prove that our estimators of the regression coefficients are asymptotically normal. The results of a simulation study show that the our proposed estimators are empirically efficient and valid. We illustrate our analysis on a real data set drawn from a hierarchical (three-way nested) design.

Keywords Asymptotic theory • Hierarchical models • Longitudinal models • Nonlinear Models • Nonparametric methods • Rank scores • Robust estimation

---

A. Abebe
Department of Mathematics and Statistics, Auburn University, 221 Parker Hall, Auburn, AL 36849, USA
e-mail: ash@auburn.edu

J.W. McKean (✉)
Department of Statistics, Western Michigan University, Kalamazoo, MI 49008, USA
e-mail: joseph.mckean@wmich.edu

J.D. Kloke
Department of Biostatistics, University of Wisconsin School of Medicine and Public Health, Madison, WI 53726, USA
e-mail: kloke@biostat.wisc.edu

Y.K. Bilgic
Department of Mathematics, SUNY-Geneseo, Geneseo, NY 14454, USA
e-mail: bilgic@geneseo.edu

© Springer International Publishing Switzerland 2016

R.Y. Liu, J.W. McKean (eds.), *Robust Rank-Based and Nonparametric Methods*, Springer Proceedings in Mathematics & Statistics 168,
DOI 10.1007/978-3-319-39065-9_4

4.1 Introduction

Repeated measurement designs are often used by researchers to study patterns of response over time or to simply use available subjects more efficiently. The analysis of data resulting from such designs often becomes complicated due to the non-zero within-subject correlation. An elegant solution was given by Liang and Zeger (1986) who proposed the generalized estimating equations (GEEs) for regression parameters, that specify only the relationship between the marginal mean of the response variable and covariates. Within-subject correlation is then accounted for through a 'working' correlation matrix.

The GEE method of Liang and Zeger (1986) gives consistent estimators of the regression parameter. The method, however, is not robust against outliers since it is based on score equations from the maximum likelihood method of estimation. A solution proposed by Qaqish and Preisser (1999) is to use M-type estimation by involving downweighting schemes. Another solution is one given by Jung and Ying (2003) who proposed an adaptation of the Wilcoxon–Mann–Whitney method of estimating linear regression parameters for use in longitudinal data analysis under the working independence model. They used joint ranking (JR) of all observations in their development. Wang and Zhu (2006) consider the same model as Jung and Ying (2003) but they use separate between-subject and within-subject ranks to specify their Wilcoxon–Mann–Whitney estimating equations.

The purpose of the current paper is to provide a direct rank-based estimation analogue of the GEE model of Liang and Zeger (1986) obtained via the minimization of rank dispersion functions of Jaeckel (1972). This extends the work of Jung and Ying (2003) in three directions: (1) specifying a generalized linear model where the responses are nonlinear as a function of the regression coefficients, (2) allowing user specified 'working' correlation matrices, and (3) allowing general score functions including the one that gives rise to the Wilcoxon–Mann–Whitney method. Our development uses the iterated reweighted least squares (IRLS) formulation of the rank dispersion function given in Sievers and Abebe (2004). This makes our approach closer in spirit to that of Qaqish and Preisser (1999); however, while they require the specification of a separate weight function, our weights are a result of the score function used in the rank dispersion function.

Kloke et al. (2009) studied rank estimation for linear models with cluster correlated errors using JR along with general score functions. The asymptotic variance of their estimators includes terms based on the underlying dependency and they provided techniques for their estimation. The model specified in the current paper includes the model studied in Kloke et al. (2009); however, the approaches of estimation are different. The current paper uses marginal models along with a 'working' correlation matrix to account for within-subject correlations. Kloke et al. (2009) specify the joint distribution under the assumption that the within-subject marginal univariate distributions are the same. So our approach here is much more general. Earlier versions of our procedure were discussed in a technical report by Abebe et al. (2010); see, also, Sect. 5.5 of Hettmansperger and McKean (2011).

The paper is organized as follows. The model and rank estimation of its parameters using the IRLS estimation technique are given in Sect. 4.2. In Sect. 4.3, we state the asymptotic distribution of our estimators, giving a detailed proof of this theory in Appendix. Our estimator is highly efficient and robust in response space. We also develop a complement procedure which offers protection to outliers in factor space in Sect. 4.3.2. In Sect. 4.4, we present the results of a Monte Carlo study over two- and three-way nested designs comparing our procedures with the traditional REML procedure. In Sect. 4.5, we compare these analyses in their handling of a real data set drawn from a hierarchical model. The Monte Carlo study and example offers empirical evidence confirming the robustness and validity of our procedures.

Packages written in R are available for the computation of our procedures. Bilgic and Susmann (2013) developed the R package `rlme`, downloadable at CRAN. In Sect. 8.4 of Kloke and McKean (2014), the R package `rbgee` for fitting these robust procedures is discussed.

4.2 Rank-Based Estimator

Our notation follows that of Liang and Zeger (1986). Consider a longitudinal set of observations over K subjects. Let y_{it} denote the tth response for ith subject for $t = 1, 2, \ldots, n_i$ and $i = 1, 2, \ldots, K$. Assume that $\boldsymbol{x}_{it}$ is a $p \times 1$ vector of corresponding covariates. Let $N = \sum_{i=1}^{K} n_i$ denote the total sample size. Assume that the marginal distribution of y_{it} is of the exponential class of distributions and is given by

$$f(y_{it}) = \exp\{[y_{it}\theta_{it} - a(\theta_{it}) + b(y_{it})]\phi\} , \tag{4.1}$$

where $\phi > 0$, $\theta_{it} = h(\eta_{it})$ and $\eta_{it} = \boldsymbol{x}_{it}^T\boldsymbol{\beta}$. Thus the mean and variance of y_{it} are given by

$$E(y_{it}) = a'(\theta_{it}) \text{ and } \mathrm{Var}(y_{it}) = a''(\theta_{it})/\phi . \tag{4.2}$$

In this notation, the link function is $h^{-1} \circ (a')^{-1}$. More assumptions are stated later for the theory.

Let $\mathbf{Y}_i = (y_{i1}, \ldots, y_{in_i})^T$ and $\mathbf{X}_i = (\mathbf{x}_{i1}, \ldots, \mathbf{x}_{in_i})^T$ denote the $n_i \times 1$ vector of responses and the $n_i \times p$ matrix of covariates, respectively, for the ith individual. We consider the general case where the components of the vector of responses for the ith subject, $\mathbf{Y}_i$, are dependent. Let $\boldsymbol{\theta}_i = (\theta_{i1}, \theta_{i2}, \ldots, \theta_{in_i})^T$, so that $E(\mathbf{Y}_i) = \mathbf{a}'(\boldsymbol{\theta}_i) = (a'(\theta_{i1}), \ldots, a'(\theta_{in_i}))^T$. For a $s \times 1$ vector of unknown parameters $\boldsymbol{\alpha}$, let $\mathbf{R}_i = \mathbf{R}_i(\boldsymbol{\alpha})$ denote a $n_i \times n_i$ correlation matrix. Define the matrix

$$\mathbf{V}_i = \mathbf{A}_i^{1/2}\mathbf{R}_i(\boldsymbol{\alpha})\mathbf{A}_i^{1/2}/\phi , \tag{4.3}$$

where $\mathbf{A}_i = \text{diag}\{a''(\theta_{i1}), \ldots, a''(\theta_{in_i})\}$. The matrix $\mathbf{V}_i$ may or may not be the covariance matrix of $\mathbf{Y}_i$. In any case, we refer to $\mathbf{R}_i$ as the working correlation matrix. For estimation, let $\hat{\mathbf{V}}_i$ be an estimate of $\mathbf{V}_i$. This, in general, requires estimation of $\boldsymbol{\alpha}$ and often an initial estimate of $\boldsymbol{\beta}$. In general, we will denote the estimator of $\boldsymbol{\alpha}$ by $\hat{\boldsymbol{\alpha}}(\boldsymbol{\beta}, \phi)$ to reflect its dependence on $\boldsymbol{\beta}$ and ϕ.

Liang and Zeger (1986) defined their estimate in terms of general estimating equations (GEE). Define the $n_i \times p$ Hessian matrix,

$$\mathbf{D}_i = \frac{\partial \mathbf{a}'(\boldsymbol{\theta}_i)}{\partial \boldsymbol{\beta}}, \quad i = 1, \ldots, K. \tag{4.4}$$

Then their GEE estimator $\hat{\boldsymbol{\beta}}_{LS}$ is the solution to the equations

$$\sum_{i=1}^{K} \mathbf{D}_i^T \hat{\mathbf{V}}_i^{-1} [\mathbf{Y}_i - \mathbf{a}'(\boldsymbol{\theta}_i)] = 0. \tag{4.5}$$

To motivate our estimator, it is convenient to write this in terms of the Euclidean norm. Define the dispersion function,

$$\begin{aligned} D_{LS}(\boldsymbol{\beta}) &= \sum_{i=1}^{K} [\mathbf{Y}_i - \mathbf{a}'(\boldsymbol{\theta}_i)]^T \hat{\mathbf{V}}_i^{-1} [\mathbf{Y}_i - \mathbf{a}'(\boldsymbol{\theta}_i)] \\ &= \sum_{i=1}^{K} [\hat{\mathbf{V}}_i^{-1/2} \mathbf{Y}_i - \hat{\mathbf{V}}_i^{-1/2} \mathbf{a}'(\boldsymbol{\theta}_i)]^T [\hat{\mathbf{V}}_i^{-1/2} \mathbf{Y}_i - \mathbf{V}_i^{-1/2} \mathbf{a}'(\boldsymbol{\theta}_i)] \\ &= \sum_{i=1}^{K} \sum_{t=1}^{n_i} [y_{it}^* - d_{it}(\boldsymbol{\beta})]^2, \end{aligned} \tag{4.6}$$

where $\mathbf{Y}_i^* = \hat{\mathbf{V}}_i^{-1/2} \mathbf{Y}_i = (y_{i1}^*, \ldots, y_{in_i}^*)^T$, $d_{it}(\boldsymbol{\beta}) = \mathbf{c}_t^T \mathbf{a}'(\boldsymbol{\theta}_i)$, and $\mathbf{c}_t^T$ is the *tth* row of $\hat{\mathbf{V}}_i^{-1/2}$. The gradient of $D_{LS}(\boldsymbol{\beta})$ is

$$\triangledown D_{LS}(\boldsymbol{\beta}) = -\sum_{i=1}^{K} \mathbf{D}_i^T \hat{\mathbf{V}}_i^{-1} [\mathbf{Y}_i - \mathbf{a}'(\boldsymbol{\theta})]. \tag{4.7}$$

Thus the solution to the GEE equations (4.5) also can be expressed as

$$\hat{\boldsymbol{\beta}}_{LS} = \text{Argmin } D_{LS}(\boldsymbol{\beta}). \tag{4.8}$$

From this point of view, $\hat{\boldsymbol{\beta}}_{LS}$ is a nonlinear least squares (LS) estimator. We will refer to it as GEEWL2 estimator.

Nonlinear LS methods are an extension of linear LS procedures. Their geometry is similar in that both linear and nonlinear estimators minimize the Euclidean norm of the residuals. Abebe and McKean (2007) developed a class of nonlinear robust estimators. Similar to nonlinear LS estimators, these estimators minimize a norm of the residuals where, for a vector $\mathbf{v} \in R^n$, the norm is defined by

$$\|v\| = \sum_{i=1}^{n} \varphi[R(v_i)/(n+1)]v_i \,, \tag{4.9}$$

where $R(v_i)$ denotes the rank of v_i among $v_1, \ldots, v_n$ and the score function $\varphi(u)$ is a nondecreasing, square-integrable function defined on the interval $(0, 1)$. Without loss of generality, we standardized φ so that

$$\int \varphi(u)\, du = 0 \text{ and } \int \varphi^2(u)\, du = 1. \tag{4.10}$$

Two commonly used score functions are the Wilcoxon score function, $\varphi(u) = \sqrt{12}[u - (1/2)]$, and the sign score function, $\varphi(u) = \text{sgn}[u - (1/2)]$. Ranked-based robust methods for linear models are based on this norm; see Chap. 3 of Hettmansperger and McKean (2011) or Chap. 9 of Hollander and Wolfe (1999).

Next consider the general model defined by expressions (4.1) and (4.2). As in the LS development, let $\mathbf{Y}_i^* = \hat{\mathbf{V}}_i^{-1/2}\mathbf{Y}_i = (y_{i1}^*, \ldots, y_{in_i}^*)^T$, $g_{it}(\boldsymbol{\beta}) = \mathbf{c}_t^T \mathbf{a}'(\boldsymbol{\theta}_i)$, where $\mathbf{c}_t^T$ is the *tth* row of $\hat{\mathbf{V}}_i^{-1/2}$, and let $\mathbf{G}_i^* = [g_{it}]$. The rank-based dispersion function is given by

$$D_R(\boldsymbol{\beta}) = \sum_{i=1}^{K} \sum_{t=1}^{n_i} \varphi[R(y_{it}^* - g_{it}(\boldsymbol{\beta}))/(n+1)][y_{it}^* - g_{it}(\boldsymbol{\beta})] \,. \tag{4.11}$$

We next write the R estimator as weighted LS estimator. Depending on the score function, there are two cases for the weights. For Case (1), assume that the score function $\varphi(u)$ is odd about $1/2$, i.e.,

$$\varphi(1-u) = -\varphi(u). \tag{4.12}$$

As discussed on page 101 of Hettmansperger and McKean (2011), such scores are appropriate for symmetric error distributions. Let $e_{it}(\boldsymbol{\beta}) = y_{it}^* - g_{it}(\boldsymbol{\beta})$ denote the $(i, t)th$ residual and let $m(\boldsymbol{\beta}) = \text{med}_{(i,t)}\{e_{it}(\boldsymbol{\beta})\}$ denote the median of all the residuals. Then because the scores sum to 0 we have the identity,

$$\begin{aligned} D_R(\boldsymbol{\beta}) &= \sum_{i=1}^{K} \sum_{t=1}^{n_i} \varphi[R(e_{it}(\boldsymbol{\beta}))/(n+1)][e_{it}(\boldsymbol{\beta}) - m(\boldsymbol{\beta})] \\ &= \sum_{i=1}^{K} \sum_{t=1}^{n_i} \frac{\varphi[R(e_{it}(\boldsymbol{\beta}))/(n+1)]}{e_{it}(\boldsymbol{\beta}) - m(\boldsymbol{\beta})}[e_{it}(\boldsymbol{\beta}) - m(\boldsymbol{\beta})]^2 \\ &= \sum_{i=1}^{K} \sum_{t=1}^{n_i} w_{it}(\boldsymbol{\beta})[e_{it}(\boldsymbol{\beta}) - m(\boldsymbol{\beta})]^2 \,, \end{aligned} \tag{4.13}$$

where $w_{it}(\boldsymbol{\beta}) = \varphi[R(e_{it}(\boldsymbol{\beta}))/(n+1)]/[e_{it}(\boldsymbol{\beta})-m(\boldsymbol{\beta})]$ is a weight function. If $e_{it}(\boldsymbol{\beta})-m(\boldsymbol{\beta}) = 0$, we set its weight to the maximum of the weights. Note that by using the median of the residuals in conjunction with the fact that the score function is odd about $1/2$, (4.12), ensures that the weights are positive.

For Case (2), consider score functions in general which may not satisfy (4.12). Because the scores sum to 0, there is an i_0 such that $a(i_0) \leq 0$ and $a(j) > 0$, for all $j > i_0$. Take $m(\boldsymbol{\beta})$ to be the i_0th quantile of the residuals. As with Case (1), the resulting weights will be nonnegative. An example of this situation is discussed in Sect. 5.5.3 of Hettmansperger and McKean (2011).

Now let $\hat{\boldsymbol{\beta}}_R^{(0)}$ denote an initial estimator of $\boldsymbol{\beta}$. As estimates of the weights, we use $\hat{w}_{it}\left(\hat{\boldsymbol{\beta}}_R^{(0)}\right)$; i.e., the weight function evaluated at $\hat{\boldsymbol{\beta}}^{(0)}$. Expression (4.13) leads to the dispersion function

$$D_R^*\left(\boldsymbol{\beta}|\hat{\boldsymbol{\beta}}_R^{(0)}\right) = \sum_{i=1}^{K}\sum_{t=1}^{n_i} \hat{w}_{it}\left(\hat{\boldsymbol{\beta}}_R^{(0)}\right)\left[e_{it}(\boldsymbol{\beta}) - m\left(\hat{\boldsymbol{\beta}}_R^{(0)}\right)\right]^2$$
$$= \sum_{i=1}^{K}\sum_{t=1}^{n_i}\left[\sqrt{\hat{w}_{it}\left(\hat{\boldsymbol{\beta}}_R^{(0)}\right)}e_{it}(\boldsymbol{\beta}) - \sqrt{\hat{w}_{it}\left(\hat{\boldsymbol{\beta}}_R^{(0)}\right)}m\left(\hat{\boldsymbol{\beta}}_R^{(0)}\right)\right]^2 . \quad (4.14)$$

Let

$$\hat{\boldsymbol{\beta}}_R^{(1)} = \text{Argmin}\, D^*\left(\boldsymbol{\beta}|\hat{\boldsymbol{\beta}}_R^{(0)}\right) . \quad (4.15)$$

This establishes a sequence of IRLS estimates, $\left\{\hat{\boldsymbol{\beta}}_R^{(k)}\right\}$, $k = 1, 2, \ldots$.

After some algebraic simplification, we obtain the gradient

$$\nabla D_R^*\left(\boldsymbol{\beta}|\hat{\boldsymbol{\beta}}_R^{(k)}\right) = -2\sum_{i=1}^{K}\mathbf{D}_i^T\hat{\mathbf{V}}_i^{-1/2}\hat{\mathbf{W}}_i\hat{\mathbf{V}}_i^{-1/2}\left[\mathbf{Y}_i - \mathbf{a}'(\boldsymbol{\theta}) - \mathbf{M}^*\left(\hat{\boldsymbol{\beta}}_R^{(k)}\right)\right] , \quad (4.16)$$

where $\mathbf{M}^*\left(\hat{\boldsymbol{\beta}}_R^{(k)}\right) = \hat{\mathbf{V}}_i^{1/2} m\left(\hat{\boldsymbol{\beta}}_R^{(k)}\right)\mathbf{1}$, $\mathbf{1}$ denotes a $n_i \times 1$ vector all of whose elements are 1, and $\hat{\mathbf{W}}_i = \text{diag}\{\hat{w}_{i1}, \ldots, \hat{w}_{in_i}\}$ is the diagonal matrix of weights for the ith subject. Hence, $\hat{\boldsymbol{\beta}}_R^{(k+1)}$ satisfies the general estimating equations (GEE) given by,

$$\sum_{i=1}^{K}\mathbf{D}_i^T\hat{\mathbf{V}}_i^{-1/2}\hat{\mathbf{W}}_i\hat{\mathbf{V}}_i^{-1/2}\left[\mathbf{Y}_i - \mathbf{a}'(\boldsymbol{\theta}) - \mathbf{M}^*\left(\hat{\boldsymbol{\beta}}_R^{(k)}\right)\right] = \mathbf{0} . \quad (4.17)$$

We will refer to this weighted, general estimation equations estimator as the GEERB estimator.

4.2.1 Computation

There are several software packages for computing these rank-based GEE estimates. The CRAN package `rlme` uses an iterated reweighted least squares algorithm based on expression (4.14); see Bilgic and Susmann (2013). As discussed in Sect. 8.6 of Kloke and McKean (2014), the estimating equations (4.17) naturally lead to a Gauss-Newton type algorithm. The R function `geerfit` uses this approach for computation of GEERB estimates; see Sect. 8.6 of Kloke and McKean (2014) for illustrations.

4.3 Asymptotic Theory

Recall that both the GEEWL2 and GEERB estimators were defined in terms of the univariate variables y^*_{it}. These of course are transformations of the original observations by the estimates of the covariance matrix $\mathbf{V}_i$ and the weight matrix $\mathbf{W}_i$. For the theory, we need to consider similar transformed variables using the matrices $\mathbf{V}_i$ and $\mathbf{W}_i$, where this notation means that $\mathbf{V}_i$ and $\mathbf{W}_i$ are evaluated at the true parameters. For $i = 1, \ldots, K$ and $t = 1, \ldots, n_i$, let

$$\begin{aligned} \mathbf{Y}^\dagger_i &= \mathbf{V}_i^{-1/2}\mathbf{Y}_i = (y^\dagger_{i1}, \ldots, y^\dagger_{in_i})^T \\ \mathbf{G}^\dagger_i(\boldsymbol{\beta}) &= \mathbf{V}_i^{-1/2}\mathbf{a}'_i(\boldsymbol{\theta}) = [g^\dagger_{it}] \\ e^\dagger_{it} &= y^\dagger_{it} - g^\dagger_{it}(\boldsymbol{\beta}). \end{aligned} \tag{4.18}$$

To obtain asymptotic distribution theory for a GEE procedure, assumptions concerning the random errors $e^\dagger_{it}$ must be made. The assumptions for the GEEWL2 estimates are discussed in Liang and Zeger (1986). These include $E(e^\dagger_{it}) = 0$, $\text{Var}(e^\dagger_{it}) < \infty$, the regularity condition cited under expression (4.31), and Assumptions A.1–A.4 listed below. For the GEERB estimates, Assumptions A.1–A.6 are needed.

A.1 $\sqrt{K}[\hat{\phi}(\boldsymbol{\beta}) - \phi] = O_p(1)$, as $K \to \infty$, when $\boldsymbol{\beta}$ is known.

A.2 $\sqrt{K}[\hat{\boldsymbol{\alpha}}(\boldsymbol{\beta}, \phi) - \boldsymbol{\alpha}] = O_p(1)$ when $\boldsymbol{\beta}$ and ϕ are known.

A.3 $|\partial\hat{\boldsymbol{\alpha}}(\boldsymbol{\beta}, \phi)/\partial\phi| \leq H(\mathbf{Y}, \boldsymbol{\beta})$ which is $O_p(1)$.

A.4 (Lindeberg-Feller Conditions): For $i = 1, \ldots, K$, let $\mathbf{u}_i = \mathbf{V}_i^{-1/2}\mathbf{D}_i$ and $\mathbf{u}_N = [\mathbf{u}_1^T \; \mathbf{u}_2^T \; \cdots \; \mathbf{u}_K^T]^T$. Denote the (l, j)th entry of $\mathbf{u}_N$ by u_{lj}, $l = 1, 2, \ldots, N; j = 1, 2, \ldots p$. Then

$$\max_{1 \leq l \leq N} \frac{u_{lj}^2}{\sum_{m=1}^N u_{mj}^2} \to 0, \quad \text{for all } j = 1, \ldots, p,$$

and

$$\lim_{N \to \infty} \tfrac{1}{N}\mathbf{u}_N^T\mathbf{u}_N \text{ exists and is positive definite.}$$

A.5 The score function $\varphi(u)$ is bounded and satisfies the standardizing conditions (4.10).
A.6 The marginal pdf of $e_{it}^{\dagger}$ is continuous and variance-covariance matrix given in (4.19) is positive definite.

Theorem 4.1. *Assume that the initial estimate satisfies* $\sqrt{K}(\hat{\boldsymbol{\beta}}_R^{(0)} - \boldsymbol{\beta}) = O_p(1)$. *Then under the above assumptions, for* $k \geq 1$, $\sqrt{K}(\hat{\boldsymbol{\beta}}_R^{(k)} - \boldsymbol{\beta})$ *has an asymptotic normal distribution with mean* $\mathbf{0}$ *and covariance matrix,*

$$\lim_{K\to\infty} K\left\{\sum_{i=1}^{K} \mathbf{D}_i^T \mathbf{V}_i^{-1/2} \mathbf{W}_i \mathbf{V}_i^{-1/2} \mathbf{D}_i\right\}^{-1} \left\{\sum_{i=1}^{K} \mathbf{D}_i^T \mathbf{V}_i^{-1/2} Var(\boldsymbol{\varphi}_i^{\dagger}) \mathbf{V}_i^{-1/2} \mathbf{D}_i\right\}$$
$$\times \left\{\sum_{i=1}^{K} \mathbf{D}_i^T \mathbf{V}_i^{-1/2} \mathbf{W}_i \mathbf{V}_i^{-1/2} \mathbf{D}_i\right\}^{-1}, \tag{4.19}$$

where $\boldsymbol{\varphi}_i^{\dagger}$ *denotes the* $n_i \times 1$ *vector* $(\varphi[R(e_{i1}^{\dagger})/(N+1)], \ldots, \varphi[R(e_{in_i}^{\dagger})/(N+1)])^T$.

The proof is sketched in Appendix. It involves a Taylor series expansion, as in Liang and Zeger's (1986) proof, and the rank-based theory found in Brunner and Denker (1994) for dependent observations.

Provided the score function $\varphi(u)$ is bounded, it follows from the asymptotic representation of $\hat{\boldsymbol{\beta}}_R^{(k)}$ given in expression (4.35) of Appendix that the estimator is resistant to outliers in response (y) space. This is verified by the results of the simulation study presented in the next section. For resistance to outliers in factor space, the usual high breakdown weights used in linear models can easily be incorporated into the weight function $w(\boldsymbol{\beta})$ as discussed in Sect. 4.3.2.

4.3.1 Implementation

For practical use of the GEERB estimate, the asymptotic covariance matrix (4.19) requires estimation. This is true even in the case where percentile bootstrap confidence intervals are employed for inference, because appropriate standardized bootstrap estimates are generally used.

The covariance structure suggests a simple moment estimator. Let $\hat{\boldsymbol{\beta}}^{(k)}$ and (for the ith subject) $\hat{\mathbf{V}}_i^{(k)}$ denote the final estimates of $\boldsymbol{\beta}$ and $\mathbf{V}_i$, respectively. Then the residuals which estimate $\mathbf{e}_i^{\dagger} \equiv (e_{i1}^{\dagger}, \ldots, e_{in_i}^{\dagger})^T$ are given by

$$\hat{\mathbf{e}}_i^{\dagger} = \left[\hat{\mathbf{V}}_i^{(k)}\right]^{-1/2} \mathbf{Y}_i - \hat{\mathbf{G}}_i^{(k)}(\hat{\boldsymbol{\beta}}^{(k)}), \quad i = 1, \ldots, K, \tag{4.20}$$

where $\hat{\mathbf{G}}_i^{(k)} = \left[\hat{\mathbf{V}}_i^{(k)}\right]^{-1/2} \mathbf{a}'\left(\hat{\boldsymbol{\theta}}^{(k)}\right)$ and $\hat{\boldsymbol{\theta}}_{it}^{(k)} = h\left(\mathbf{x}_{it}^T\hat{\boldsymbol{\beta}}^{(k)}\right)$. Let $R(\hat{e}_{it}^\dagger)$ denote the rank of $\hat{e}_{it}^\dagger$ among $\{\hat{e}_{i't'}^\dagger\}$, $t = 1,\ldots,n_i; i = 1,\ldots,K$. Let $\hat{\boldsymbol{\varphi}}_i^\dagger = (\varphi[R(\hat{e}_{i1}^\dagger)/(N+1)],\ldots,\varphi[R(\hat{e}_{in_i}^\dagger)/(N+1)])^T$. Let $\hat{\mathbf{S}}_i = \hat{\boldsymbol{\varphi}}_i^\dagger - \overline{\hat{\varphi}_i^\dagger}\mathbf{1}_{n_i}$ where $\overline{\hat{\varphi}_i^\dagger} = \sum_{j=1}^{n_i}\varphi[R(\hat{e}_{ij}^\dagger)/(N+1)]$.

Then a moment estimator of the covariance matrix (4.19) is that expression with $\mathrm{Var}(\boldsymbol{\varphi}_i^\dagger)$ estimated by

$$\widehat{\mathrm{Var}}(\boldsymbol{\varphi}_i^\dagger) = \hat{\mathbf{S}}_i\hat{\mathbf{S}}_i^T, \tag{4.21}$$

and, of course, final estimates of $\mathbf{D}_i$ and $\mathbf{V}_i$. Although, this is a simple nonparametric estimate of the covariance structure, the simulation study discussed in Sect. 5.5.2 of Hettmansperger and McKean (2011) shows that this estimate may lead to a liberal inference. Werner and Brunner (2007) discovered this in a corresponding rank testing problem.

The form of the weights, though, suggests a simple approximation, which is based on certain ideal conditions. Suppose the model is correct. Assume that the true transformed errors are independent. Then, because the scores have been standardized, asymptotically $\mathrm{Var}(\boldsymbol{\varphi}_i^\dagger)$ converges to $\mathbf{I}_{n_i}$, so replace it with $\mathbf{I}_{n_i}$. This is the first part of the approximation.

Next consider the weights. The functional for the weights is of the form $\varphi[F(e)]/e$. Assuming that $F(0) = 1/2$, a simple application of the Mean Value Theorem gives the approximation $\varphi[F(e)]/e = \varphi'[F(e)]f(e)$. The expected value of this approximation can be expressed as

$$\tau^{-1} = \int_{-\infty}^{\infty} \varphi'[F(t)]f^2(t)\,dt = \int_0^1 \varphi(u)\left\{-\frac{f'[F^{-1}(u)]}{f[F^{-1}(u)]}\right\} du, \tag{4.22}$$

where the second integral is derived from the first using integration by parts followed by a substitution. The parameter τ is the scale parameter for the usual R estimates in the linear model for both the independent errors case and for clustered correlated errors; see Hettmansperger and McKean (2011) and Kloke et al. (2009), respectively. A consistent estimate for τ is given in Koul et al. (1987). The second part of the approximation is to replace the weight matrix by $(1/\hat{\tau})\mathbf{I}$. The results of a simulation study discussed in Sect. 5.5.2 of Hettmansperger and McKean (2011) show that the above approximation leads to a valid inference over the situations of the study. The validity results of the GEERB estimators in the simulation study in Sect. 4.4 offer further confirmation.

4.3.2 GEEhbr Estimate

As discussed above in Sect. 4.3, if the score function is bounded then the rank-based GEE estimator offers protection to outliers in response space. To additionally obtain protection in factor space we can modify the weights.

As in Sect. 4.2, let $e_{it}(\boldsymbol{\beta}) = y_{it}^* - g_{it}(\boldsymbol{\beta})$ denote the $(i,t)th$ residual. Then, under Wilcoxon scores, the objective function (4.11) can be expressed as

$$D_R(\boldsymbol{\beta}) = \frac{2(N+1)}{\sqrt{3}} \sum_{i=1}^{K} \sum_{t=1}^{n_i} \sum_{i'=1}^{K} \sum_{t'=1}^{n_{i'}} |e_{it}(\boldsymbol{\beta}) - e_{i't'}(\boldsymbol{\beta})|;$$

see, for example, page 82 of Hettmansperger and McKean (2011). This easily lends itself to a weighted versions; i.e.,

$$D_{W,R}(\boldsymbol{\beta}) = \frac{2(N+1)}{\sqrt{3}} \sum_{i=1}^{K} \sum_{t=1}^{n_i} \sum_{i'=1}^{K} \sum_{t'=1}^{n_{i'}} w_{it} w_{i't'} |e_{it}(\boldsymbol{\beta}) - e_{i't'}(\boldsymbol{\beta})|;$$

for a set of nonnegative weights $\{w_{it}\}$. This is a nonlinear objective function and Abebe and McKean (2013) showed the existence of a minimizing value. They further discussed several sets of weights including the high breakdown weights discussed in Chang et al. (1999) for the linear model case. These weights adjust for both outliers in response space as well as in factor space. In the linear model case they achieve 50 % breakdown. Their high breakdown argument holds for our GEE model, if the objective function is convex. We use this estimator in our example and Monte Carlo study under the label GEEhbr. It is computed by the `rlme` package.

4.4 Monte Carlo Study

In this section, we present the results of a Monte Carlo study of the small sample properties of the GEERB procedure. Besides the GEERB fit, we also obtained the results of the REML fit, the restricted maximum-likelihood method. We compare these two procedures in terms of the empirical validity of their 95 % confidence intervals and the empirical relative efficiencies [ratios of empirical mean square errors (MSE)]. The REML procedure is computed by the R package *nlme* developed by Pinheiro et al. (2016), while the GEERB estimates were computed by the package `rmle`; see Bilgic and Susmann (2013).

Our models of interest are the two- and three-level nested random effects models. The model for the two-way nested design is given by

$$y_{ij} = \alpha + x_{ij}^T \beta + a_i + \epsilon_{j(i)}, \quad i = 1, \ldots, I, j = 1, \ldots n_i, \tag{4.23}$$

where a_i is the random effect for cluster i, I is the number of clusters, n_i is the size of the ith cluster, and x_{ij} is the vector of predictor variables. For the simulation, we used $I = 7$ clusters. The total sample size is 185, so the design is slightly unbalanced. The three-way nested design simulated is given by

$$y_{ijk} = \alpha + x_{ijk}^T\beta + a_i + w_{j(i)} + \epsilon_{k(ij)},\ i = 1, \ldots, I, j = 1, \ldots, J_i, k = 1, \ldots, n_{ij}, \tag{4.24}$$

where a_i is the random effect for cluster i, $w_{j(i)}$ is the random effect for the jth subcluster of cluster i, J_i is the number of subclusters in cluster i, n_{ij} is the size of jth subcluster in cluster i, and x_{ijk} is the vector of predictors. Again $I = 7$. For each of the 7 clusters, there were 2 or 3 subclusters. The sample sizes in these clusters range from 6 to 20, with median about 9. As with the two-way, the total sample size is 185. For both designs, the random errors are uncorrelated and independent. This nested analog could be adopted for any repeated measure design, randomized block design, cluster correlated, and other hierarchical models. We chose two predictors for these nested models. One predictor is binary and it is labeled as Treatment; while the other is a continuous predictor and it is labeled as Covariate. Hence, for each design, the predictor vector is of length 2.

Both normal (N) and contaminated normal (CN), (25 % contamination and variance ratio of 100), were used for error distributions. The random effects for the normal models were also normal. Likewise, the random effects for the contaminated normal models were also contaminated normal with the same parameters as for the errors. The various variance component situations are given by the vector in Tables 4.1 and 4.2 under the column Error Dists. For a two-level design, the vector is $\boldsymbol{\theta} = (\sigma_a, \sigma_\epsilon)^T$, while for a three-level design it is $\boldsymbol{\theta} = (\sigma_a, \sigma_w, \sigma_\epsilon)^T$. For example, $\boldsymbol{\theta} = (1, 3)^T$ indicates that intra-class correlation is formed with the parameter of $1^2/(1^2 + 3^2) = 0.10$.

Procedures include the REML estimates, the GEERB estimates with Wilcoxon scores, and the GEEhbr estimates as discussed in Sect. 4.3.2. For the rank-based fits estimates of the matrices $\mathbf{V}_i$, $i = 1, \ldots, K$, (4.3), are needed. As discussed in the theory section, $\mathbf{V}_i$ need not be the variance-covariance of the response $\mathbf{Y}_i$. For the simulation study, though, we did use robust estimates of the variance components as proposed by Kloke et al. (2009); see Sect. 8.4 of Kloke and McKean (2014), also. We describe it briefly for the three-level nested design. Let $\hat{\beta}^{(l)}$ denote the current estimate of the fixed effects β. Fixing i and j, we write Model 4.24 as

$$y_{ijk} - \boldsymbol{x}_{ijk}^T\boldsymbol{\beta} = a_i + w_{j(i)} + \epsilon_{k(ij)} \tag{4.25}$$

Let $\boldsymbol{r}_{ij}$ denote the $n_{ij} \times 1$ vector of the residuals $r_{ijk} = y_{ijk} - \boldsymbol{x}_{ijk}^T\hat{\boldsymbol{\beta}}^{(l)}$. For these residuals, the model is

$$r_{ijk} \doteq a_i + w_{j(i)} + \epsilon_{k(ij)} \tag{4.26}$$

With i and j fixed, this is a simple location model with location $a_i + w_{j(i)}$. Let $\hat{u}_{ij} = \text{med}(\boldsymbol{r}_{ij})$. We call this estimate the predictor of $a_i + w_{j(i)}$. To separate a_i and $w_{j(i)}$, let

$$\hat{u}_{i.} = \text{med}\,\{\hat{u}_{i1}, \hat{u}_{i2}, \cdots, \hat{u}_{iJ_i}\}.$$

For severely unbalanced data, the $\hat{u}_{it}$'s could be weighted. Then, for $j = 1, \ldots, J_i$, the difference $\hat{u}_{ij} - \hat{u}_{i.}$ is free of a_i and, hence, is a predictor of $w_{j(i)}$. That is, the prediction of $w_{j(i)}$ is

$$\hat{w}_{j(i)} = \hat{u}_{ij} - \hat{u}_{i.}.$$

Finally, move this prediction to the left side of Eq. (4.26) to obtain the model

$$r_{ijk} - \hat{w}_{j(i)} \doteq a_i + \epsilon_{k(ij)}.$$

This simple location model yields as the prediction of a_i,

$$\hat{a}_i = \text{med}_{jk}\{r_{ijk} - \hat{w}_{j(i)}\}.$$

Proceeding over all sections, we obtain the predictions of the random effects. Since we used the median as the location predictor, we adjust these with their common median; i.e., $\hat{a}_i \leftarrow \hat{a}_i - \text{med}_s\{\hat{a}_s\}$ and $\hat{w}_{j(i)} \leftarrow \hat{w}_{j(i)} - \text{med}_{s,t}\{\hat{w}_{t(s)}\}$.

Define the random effects as $\hat{\boldsymbol{a}} = (\hat{a}_1, \cdots, \hat{a}_I)$, and $\hat{\boldsymbol{w}} = (\hat{w}_{1(1)}, \cdots, \hat{w}_{J_I(I)})$. Our robust estimators of the variance components σ_ϵ^2, σ_a^2 and σ_w^2 are, respectively, $\text{MAD}_w^2(\hat{\boldsymbol{\epsilon}})$, $\text{MAD}_w^2(\hat{\boldsymbol{a}})$, and $\text{MAD}_w^2(\hat{\boldsymbol{w}})$. These estimators are then substituted for the variance components in in the true variance covariance of $\mathbf{Y}_i$.

For each situation 1000 simulations were run. We are interested in the estimates of fixed effects parameters under normal errors (N) and contaminated normal errors (CN). The results are displayed in Tables 4.1 and 4.2. The entries in the column ARE are the ratios of the mean-squared errors of the rank-based estimates to REML estimates. We also considered confidence intervals of the form Est$\pm 1.96\cdot$SE, where SE's are the asymptotic standard errors of the estimates computed as discussed in Sect. 4.3.1. Thus the nominal α-level of the confidence interval is 0.05 and the empirical results for α are recorded in the column called level.

For the two-level design, for the normal situations, the GEERB's empirical efficiency relative to the REML's were close to the linear model (independent case) efficiency of 0.95. For the contaminated situations, the GEERB estimates were much more efficient than their REML counterparts. The validity (level) results for the GEERB estimates were close to 0.05 or conservative in all cases except the normal $(3, 1)$ situation. These CN situations do not have outliers in factor space, so while the GEEhbr estimates are more efficient than the REML estimates it is not surprising that they are less efficient than the Wilcoxon estimates.

The results for the three-level design were similar. Note that in four of the normal (N) cases, the GEERB estimates are slightly more efficient than the REML estimates. As in the two-level situations, the GEERB were much more efficient than the REML estimates for the contaminated normal (CN) situations.

Table 4.1 Two-level nested design results

		Treatment		Covariate	
Error dist.	Methods	Level	ARE	Level	ARE
(0,1), N	REML	0.056	1	0.06	1
	GEERB	0.059	0.9362	0.055	0.9433
(1,3), N	REML	0.047	1	0.047	1
	GEERB	0.056	0.9342	0.043	0.9693
(2,3), N	REML	0.06	1	0.047	1
	GEERB	0.048	0.943	0.052	0.9177
(1,1), N	REML	0.044	1	0.044	1
	GEERB	0.049	0.9245	0.038	0.9237
(3,1), N	REML	0.042	1	0.073	1
	GEERB	0.045	0.903	0.071	0.862
(1,3), CN	REML	0.057	1	0.058	1
	GEERB	0.031	6.1068	0.039	5.5319
	GEERB (hbr)	0.008	2.1108	0.02	1.1302
(1,1), CN	REML	0.059	1	0.056	1
	GEERB	0.028	4.5263	0.035	4.2173
	GEERB (hbr)	0.02	1.7524	0.047	1.3536
(3,1), CN	REML	0.067	1	0.052	1
	GEERB	0.047	3.3953	0.031	3.2894
	GEERB (hbr)	0.042	1.8252	0.078	1.4715

4.5 Example of a Hierarchical Model

For an example, we consider a data set drawn from an educational study discussed in the book by West et al. (2006). This data is originally from The Study of Instructional Improvement, called SII. The study concerns the math achievement scores of first-grade students in randomly selected classrooms within randomly selected schools from a national U.S. sample of elementary schools. There are 1081 observations in 105 schools with the total of 285 classrooms after deleting some missing observations. The dependent variable is mathgain score; it measures the change in a student's math achievement score from the spring of kindergarten to the spring of first grade. As covariates, we used the variables mathkind (continuous), gender (female–male), minority (no–yes), and ses (continuous). The study design has a three-level nested structure, in which students are nested within classrooms, and classrooms are nested within schools; see expression (4.24) for the model formulation. The package `rlme` contains the data set under `instruction`.

We fit the data with the REML, GEERB, and GEEhbr procedures. These procedures were computed as they were for the Monte Carlo study of Sect. 4.4. The results for the data analysis are shown in Table 4.3. The table indicates that for the REML and GEERB procedures, at the 5 % level, all of the fixed effects are

Table 4.2 Three-level nested design results

		Treatment		Covariate	
Error dist	Methods	Level	ARE	Level	ARE
(0,0,1), N	REML	0.043	1	0.054	1
	GEERB	0.047	0.9171	0.063	0.8894
(1,0,1), N	REML	0.048	1	0.05	1
	GEERB	0.05	0.9347	0.045	0.9656
(1,1,3), N	REML	0.06	1	0.047	1
	GEERB	0.045	1.0598	0.051	0.9212
(2,1,3), N	REML	0.058	1	0.058	1
	GEERB	0.045	1.0057	0.041	1.0813
(1,1,1), N	REML	0.052	1	0.059	1
	GEERB	0.065	0.871	0.045	1.0583
(3,2,1), N	REML	0.049	1	0.054	1
	GEERB	0.047	0.9547	0.051	0.9483
(1,1,3), CN	REML	0.065	1	0.051	1
	GEERB	0.033	5.6231	0.044	4.81
	GEERB (hbr)	0.004	1.7636	0.052	1.0415
(1,1,1), CN	REML	0.045	1	0.036	1
	GEERB	0.032	3.5458	0.043	3.3896
	GEERB (hbr)	0.011	1.5708	0.048	1.3638
(3,2,1), CN	REML	0.041	1	0.051	1
	GEERB	0.024	2.8718	0.04	3.1456
	GEERB (hbr)	0.033	1.4897	0.057	1.439

Table 4.3 SII analysis: original data

		Intercept	Mathkind	Gender	Minority	ses
REML	*est*	57.49	−0.48	−1.36	−8.51	5.40
	se	1.33	0.02	1.72	2.37	1.27
	p-value	0.00	0.0000	0.4290	0.0003	0.0000
GEERB	*est*	55.64	−0.47	−2.03	−8.55	4.78
	se	2.37	0.02	1.60	2.54	1.21
	p-value	0.00	0.0000	0.2033	0.0007	0.0001
GEERB (hbr)	*est*	53.90	−0.49	−2.85	−1.93	3.81
	se	2.50	0.02	1.61	2.57	1.23
	p-value	0.00	0.0000	0.0774	0.4541	0.0019

significant, except for the female effect. Thus, REML and GEERB yielded similar analyses. The GEEhbr analysis differs some in that it did not find the minority effect significant.

For a sensitivity analysis, we corrupted one observation of the response variable and its continuous mathkind variable replacing the original values with points at 10 standard deviations from their centers. Hence, the corrupted observation is a

Table 4.4 SII Analysis with a bad point of high leverage

		Intercept	Mathkind	Gender	Minority	ses
REML	*est*	58.80	0.12	−2.21	−0.05	−7.52
	se	1.62	0.01	2.26	3.01	1.55
	p-value	0.00	0.0000	0.3276	0.9875	0.0000
GEERB	*est*	55.27	−0.26	−2.03	−6.84	1.29
	se	2.46	0.01	1.66	2.62	1.18
	p-value	0.00	0.0000	0.2194	0.0091	0.2747
GEERB (hbr)	*est*	53.30	−0.48	−2.69	−2.10	3.90
	se	2.45	0.01	1.61	2.56	1.15
	p-value	0.00	0.0000	0.0957	0.4107	0.0007

bad point of high leverage. The corresponding results for parameter estimates and inference dramatically changed in the REML analysis. As shown in Table 4.4, the estimates for the mathkind and ses variables not only reversed sign, they also are significant in the reversed sign sense. However, the GEERB and GEEhbr analyses are insensitive to the corrupted case. The GEEhbr analysis, estimates and inferences, changed only slightly, whereas the GEERB analysis changed only with respect to the ses variable, (all signs of their regression coefficients remain the same).

McKean et al. (1996) developed a set of diagnostics to differentiate between difference fits; see Sect. 3.13 of Hettmansperger and McKean (2011) for discussion. The overall difference in fits (TDBETAS) is measured by a quadratic form in the difference of the regression coefficients, where the standardizing matrix is the inverse of the variance-covariance matrix for the GEERB estimates. If this exceeds a benchmark, then casewise changes in fits (CFITS) are obtained. In the original SII data, TDBETAS between REML and GEERB is 1.55 which exceeds the benchmark 0.08, $(4(p + 1)^2/N)$. Boxplots of the residuals, (not shown), show numerous outliers in the data. The effect of outliers is also apparent in the table of regression coefficients, Table 4.3. For example, the difference in the estimates of the gender predictor is close to one-half of a standard error.

For the corrupted data, TDBETAS between REML and GEEhbr is 63.95, which far exceeds the suggested benchmark 0.08. For the corrupted case, CFITS between REML and GEEhbr is −7.85 which far exceeds in the absolute value the suggested benchmark of 0.14. Hence, for this corrupted data, using these diagnostics, the experimenter is alerted to "bad" data, knowing the location of at least one bad case.

4.6 Conclusion

In this paper we proposed a rank-based analogue of the GEE method of Liang and Zeger (1986) for estimation of parameters of a generalized linear model where responses are observed longitudinally. This procedure (GEERB) is based on minimization of the general rank dispersion function of Jaeckel (1972). Our procedure allows for general scores functions for symmetric and asymmetric distributions. As in the linear model case, the analysis can be optimized by careful selection of the score function. We express the dispersion function in such a way that its minimizer can be obtained via iterated reweighted least squares estimation. Using similar assumptions as Liang and Zeger (1986), we established the asymptotic normality of the rank-based estimator. As we cite in Sect. 4.1, there are computational R packages available for our estimators.

We confirm the robustness and validity of our procedures via a simulation study. Over the normal situations simulated in this study, the GEERB estimates with Wilcoxon scores had close to linear model (independent errors) efficiency of 0.955 relative the traditional REML estimates. For contaminated situations, the GEERB estimates were much more efficient than the REML estimates. Moreover, applying the method to the real data set of Sect. 4.5 confirms that the rank method gives reasonable robust estimates in the existence of outliers. For data with outliers in factor (covariate) space, we also develop a procedure (GEEhbr) which incorporates high breakdown weights and gives resistance to these outliers.

Appendix

Proof of Theorem 4.1. Let $\boldsymbol{\alpha}^*(\boldsymbol{\beta}) = \hat{\boldsymbol{\alpha}}(\boldsymbol{\beta}, \hat{\phi}(\boldsymbol{\beta}))$. Let $k \geq 1$ be arbitrary but fixed. For $i = 1, \ldots, K$, let

$$\begin{aligned} \mathbf{Z}_i(\boldsymbol{\beta}, \boldsymbol{\alpha}^*(\boldsymbol{\beta})) &= \mathbf{D}_i^T \hat{\mathbf{V}}_i^{-1/2} \mathbf{W}_i \hat{\mathbf{V}}_i^{-1/2} \left[\mathbf{Y}_i - \mathbf{a}'(\boldsymbol{\theta}) - \mathbf{M}^*(\boldsymbol{\beta})\right] \\ &= \mathbf{D}_i^T \hat{\mathbf{V}}_i^{-1/2} \mathbf{W}_i [\mathbf{Y}_i^* - \mathbf{G}_i^*(\boldsymbol{\beta}) - \mathbf{M}\mathbf{1}]. \end{aligned} \tag{4.27}$$

We then write the GEERB estimating equations (4.17) in the compact form

$$\sum_{i=1}^{K} \mathbf{Z}_i(\boldsymbol{\beta}, \boldsymbol{\alpha}^*(\boldsymbol{\beta})) = \mathbf{0}\,. \tag{4.28}$$

The GEERB estimator $\hat{\boldsymbol{\beta}}_R^{(k)}$ solves this equation.

Similar to Liang and Zeger (1986), we first expand $K^{-1/2}\sum_{i=1}^{K} \mathbf{Z}_i(\boldsymbol{\beta}, \boldsymbol{\alpha}^*(\boldsymbol{\beta}))$ in a Taylor series about the true parameter $\boldsymbol{\beta}$ and evaluated at $\hat{\boldsymbol{\beta}}_R^{(k)}$. By the chain rule, the gradient in this expansion is given by,

$$\begin{aligned} \boldsymbol{\nabla}_i &= \frac{\partial \mathbf{Z}_i(\boldsymbol{\beta}, \boldsymbol{\alpha}^*(\boldsymbol{\beta}))}{\partial \boldsymbol{\beta}} + \frac{\partial \mathbf{Z}_i(\boldsymbol{\beta}, \boldsymbol{\alpha}^*(\boldsymbol{\beta}))}{\partial \boldsymbol{\alpha}} \frac{\partial \boldsymbol{\alpha}}{\partial \boldsymbol{\beta}} \\ &= \mathbf{A}_i + \mathbf{B}_i \mathbf{C}\,. \end{aligned} \tag{4.29}$$

Because $\hat{\boldsymbol{\beta}}_R^{(k)}$ solves Eq. (4.28), the Taylor expansion evaluated at $\hat{\boldsymbol{\beta}}_R^{(k)}$ is

$$\mathbf{0} = \sum_{i=1}^{K} \mathbf{Z}_i(\boldsymbol{\beta}, \boldsymbol{\alpha}^*(\boldsymbol{\beta})) + \sum_{i=1}^{K} \nabla_i(\hat{\boldsymbol{\beta}}_R^{(k)} - \boldsymbol{\beta}) .$$

Solving for $\sqrt{K}(\hat{\boldsymbol{\beta}}_R^{(k)} - \boldsymbol{\beta})$, we obtain

$$\sqrt{K}(\hat{\boldsymbol{\beta}}_R^{(k)} - \boldsymbol{\beta}) = \left\{ \frac{1}{K} \sum_{i=1}^{K} \nabla_i \right\}^{-1} \left[\frac{1}{\sqrt{K}} \sum_{i=1}^{K} \mathbf{Z}_i(\boldsymbol{\beta}, \boldsymbol{\alpha}^*(\boldsymbol{\beta})) \right] . \tag{4.30}$$

Secondly, we fix $\boldsymbol{\beta}$ and expand $K^{-1/2} \sum_{i=1}^{K} \mathbf{Z}_i(\boldsymbol{\beta}, \boldsymbol{\alpha}^*(\boldsymbol{\beta}))$ about the true parameter $\boldsymbol{\alpha}$ and evaluated at $\boldsymbol{\alpha}^*$ to get

$$\begin{aligned} \frac{1}{\sqrt{K}} \sum_{i=1}^{K} \mathbf{Z}_i(\boldsymbol{\beta}, \boldsymbol{\alpha}^*(\boldsymbol{\beta})) &= \frac{1}{\sqrt{K}} \sum_{i=1}^{K} \mathbf{Z}_i(\boldsymbol{\beta}, \boldsymbol{\alpha}) + \frac{1}{K} \sum_{i=1}^{K} \frac{\partial}{\boldsymbol{\alpha}} \mathbf{Z}_i(\boldsymbol{\beta}, \boldsymbol{\alpha}) \sqrt{K}(\boldsymbol{\alpha}^* - \boldsymbol{\alpha}) + o_p(1) \\ &= \frac{1}{\sqrt{K}} \sum_{i=1}^{K} Z_i(\boldsymbol{\beta}, \boldsymbol{\alpha}) + \mathbf{B}^* \mathbf{C}^* + o_p(1) , \end{aligned} \tag{4.31}$$

where the $o_p(1)$ term is due to regularity conditions which imply that the remainder term is $\frac{1}{K} O_p(1)$. Note that the weights are evaluated at the true parameters in this expansion too.

Because $\mathbf{Z}_i(\boldsymbol{\beta}, \boldsymbol{\alpha})$ is evaluated at the true parameters, we use the notation given in (4.18). Letting $\mathbf{h}_{it}^T$ be the *tth* row the product $\mathbf{D}_i^T \mathbf{V}_i^{-1/2}$, we then have

$$\begin{aligned} \frac{1}{\sqrt{K}} \sum_{i=1}^{K} Z_i(\boldsymbol{\beta}, \boldsymbol{\alpha}) &= \frac{1}{\sqrt{K}} \sum_{i=1}^{K} \sum_{t=1}^{n_i} \mathbf{h}_{it}^T w_{it} [y_{it}^\dagger - g_{it}^\dagger(\boldsymbol{\beta}) - m(\boldsymbol{\beta})] \\ &= \frac{1}{\sqrt{K}} \sum_{i=1}^{K} \sum_{t=1}^{n_i} \mathbf{h}_{it}^T a [R(y_{it}^\dagger - g_{it}^\dagger(\boldsymbol{\beta}))]. \\ &= \frac{1}{\sqrt{K}} \sum_{i=1}^{K} \mathbf{D}_i^T \mathbf{V}_i^{-1/2} \mathbf{a} [R(\mathbf{Y}_i^\dagger - \mathbf{G}_i^\dagger(\boldsymbol{\beta}))]. \end{aligned} \tag{4.32}$$

The second equality holds because the weights are evaluated at the true parameters.

By Assumptions [A.5] and [A.6], it follows from Theorem 3.1 of Brunner and Denker (1994) and the usual Cramer-Wold device that $\frac{1}{\sqrt{K}} \sum_{i=1}^{K} Z_i(\boldsymbol{\beta}, \boldsymbol{\alpha})$ is asymptotically normal with mean $\mathbf{0}$ and variance-covariance matrix

$$\mathbf{M} = \frac{1}{K} \sum_{i=1}^{K} \mathbf{D}_i^T \mathbf{V}_i^{-1/2} \mathrm{Var}(\boldsymbol{\varphi}_i^\dagger) \mathbf{V}_i^{-1/2} \mathbf{D}_i . \tag{4.33}$$

Note that the form of the variance-covariance matrix follows from (4.32) and the independence between subjects.

Returning to expression (4.31), from the assumptions we have $\mathbf{C}^* = O_p(1)$. Because the scores can change value at only a finite number of points [e.g., Sect. 3.2.1 of Hettmansperger and McKean (2011)], we can write $\mathbf{B}^*$ as

$$\mathbf{B}^* = \frac{1}{K}\sum_{i=1}^{K}\left\{\frac{\partial}{\partial\boldsymbol{\alpha}}\mathbf{D}_i^T\mathbf{V}_i^{-1/2}\right\}\mathbf{a}[R(\mathbf{Y}_i^\dagger - \mathbf{G}_i^\dagger(\boldsymbol{\beta}))]. \tag{4.34}$$

Assuming Lindeberg-Feller conditions for the quantity in braces, as in (A.4), it follows, similar to (4.32) that $\sqrt{K}\mathbf{B}^* = O_p(1)$, and, hence, $\mathbf{B}^* = o_p(1)$.

To finish the proof, we need to consider the terms in expression (4.29). A simple derivation shows that

$$\mathbf{A}_i = -\mathbf{D}_i^T\mathbf{V}_i^{-1/2}\mathbf{W}_i\mathbf{V}_i^{-1/2}\mathbf{D}_i^T.$$

By assumption, $\mathbf{C} = O_p(1)$. Since,

$$\mathbf{B}_i = \frac{\partial\mathbf{Z}_i(\boldsymbol{\beta}, \boldsymbol{\alpha}^*(\boldsymbol{\beta}))}{\partial\boldsymbol{\alpha}}$$

arguments similar to those above show that $K^{-1}\sum_{i=1}^{K}\mathbf{B}_i = o_p(1)$. Hence,

$$\frac{1}{K}\nabla_i = \frac{1}{K}\sum_{i=1}^{K}\mathbf{D}_i^T\mathbf{V}_i^{-1/2}\mathbf{V}_i^{-1/2}\mathbf{D}_i + o_p(1).$$

This and the discussion around expressions (4.32) and (4.33) finish the proof of Theorem 4.1.

As a final note, the asymptotic representation of the estimator is

$$\sqrt{K}(\hat{\boldsymbol{\beta}}_R^{(k)} - \boldsymbol{\beta}) = \left\{\frac{1}{K}\sum_{i=1}^{K}\mathbf{A}_i\right\}^{-1}\left[\frac{1}{\sqrt{K}}\sum_{i=1}^{K}\mathbf{D}_i^T\mathbf{V}_i^{-1/2}\mathbf{a}[R(\mathbf{Y}_i^\dagger - \mathbf{G}_i^\dagger(\boldsymbol{\beta}))]\right] + o_p(1). \tag{4.35}$$

References

Abebe, A., & McKean, J. W. (2007). Highly efficient nonlinear regression based on the Wilcoxon norm. In D. Umbach (Ed.), *Festschrift in honor of Mir Masoom Ali on the occasion of his retirement* (pp. 340–357).

Abebe, A., & McKean, J. W. (2013). Weighted Wilcoxon estimators in nonlinear regression. *Australian & New Zealand Journal of Statistics, 55*, 401–420.

Abebe, A., McKean, J. W., & Kloke, J. D. (2010). *Iterated Reweighted Rank-Based Estimates for GEE Models*. Technical report, Western Michigan University.

Bilgic, Y. K., & Susmann, H. (2013). *rlme:* an R package for rank-based estimation and prediction in random effects nested models. *The R Journal, 5*, 71–79.

Brunner, E., & Denker, M. (1994). Rank statistics under dependent observations and applications to factorial designs. *Journal of Statistical Inference and Planning, 42*, 353–378.

Chang, W., McKean, J. W., Naranjo, J. D., & Sheather, S. J. (1999). High breakdown rank-based regression. *Journal of the American Statistical Association, 94*, 205–219.

Hettmansperger, T. P., & McKean, J. W. (2011). *Robust nonparametric statistical methods* (2nd ed.). Boca Raton, FL: Chapman-Hall.

Hollander, M., & Wolfe, D. A. (1999). *Nonparametric statistical methods* (2nd ed.). New York: Wiley.

Jaeckel, L. A. (1972). Estimating regression coefficients by minimizing the dispersion of the residuals. *Annals of Mathematical Statistics, 43*, 1449–1458.

Jung, S.-H., & Ying, Z. (2003). Rank-based regression with repeated measurements data. *Biometrika, 90*, 732–740.

Kloke, J., & McKean, J. W. (2014). *Nonparametric statistical methods using R*. Boca raton, FL: Chapman-Hall.

Kloke, J., McKean, J. W., & Rashid, M. (2009). Rank-based estimation and associated inferences for linear models with cluster correlated errors. *Journal of the American Statistical Association, 104*, 384–390.

Koul, H. L., Sievers, G. L., & McKean, J. W. (1987). An estimator of the scale parameter for the rank analysis of linear models under general score functions. *Scandinavian Journal of Statistics, 14*, 131–141.

Liang, K.-Y., & Zeger, S. L. (1986). Longitudinal data analysis using generalized linear models. *Biometrika, 73*, 13–22.

McKean, J. W., Naranjo, J. D., & Sheather, S. J. (1996). Diagnostics to detect differences in robust fits of linear models. *Computational Statistics, 11*, 223–243.

Pinheiro, J., Bates, D., DebRoy, S., Sarkar, D., & R Core Team (2016). *nlme: Linearand Nonlinear Mixed Effects Models*. R package version 3.1-128. http://CRAN.R-project.org/package=nlme

Qaqish, B. F., & Preisser, J. S. (1999). Resistant fits for regression with correlated outcomes an estimating equations approach. *Journal of Statistical Planning and Inference, 75*, 415–431.

Sievers, G. L., & Abebe, A. (2004). Rank estimation of regression coefficients using iterated reweighted least squares, *Journal of Statistical Computation and Simulation, 74*, 821–831.

Wang, Y. -G., & Zhu, M. (2006). Rank-based regression for analysis of repeated measures. *Biometrika, 93*, 459–464.

Werner, C., & Brunner, E. (2007). Rank methods for the analysis of clustered data. *Computational Statistics and Data Analysis, 51*, 6041–5054.

West, B. T., Welch, K. B., & Gatecki, A. T. (2006). *Linear mixed models: a practicalguide using statistical software*. Boca Raton, FL: CRC Press

Chapter 5
On the Asymptotic Distribution of a Weighted Least Absolute Deviation Estimate for a Bifurcating Autoregressive Process

Jeff T. Terpstra

Abstract This paper introduces a new class of estimates for estimating the parameter vector of a first-order bifurcating autoregressive model. The estimates minimize a sum of weighted absolute deviations where the weights are of the Schweppe variety. Asymptotic linearity properties are derived for the so called WL1-estimate. Based on these properties, the WL1-estimate is shown to be asymptotically normal at rate $n^{1/2}$. The results hinge on two new law of large numbers theorems for bifurcating processes. As an application of the theory, some asymptotic relative efficiency comparisons are made.

Keywords Asymptotic normality • Bifurcating autoregressive model • L1-estimates • Median • Schweppe weights

5.1 Introduction

Bifurcating processes are generally utilized to model tree structured data. For instance, these processes can be used to study cell lineage data where cell characteristics (e.g. diameters, lifetimes, and/or volumes) are of interest. Briefly, let $Z_1, Z_2, \ldots, Z_n$ denote the random variates from a perfectly observed binary tree with g generations. Here, Z_1 corresponds to the initial node (i.e. generation 0) while observations $Z_{2^i}, Z_{2^i+1}, \ldots, Z_{2^{i+1}-1}$ correspond to the 2^i observations from generation i, $i = 1, 2, \ldots, g$. Thus, the total number of observations in terms of g is given by $n = 2^{g+1} - 1$. Furthermore, the indexing is such that observations Z_{2t} and

J.T. Terpstra (✉)
Department of Statistics, Western Michigan University, Kalamazoo, MI 49008, USA
e-mail: jeffrey.terpstra@wmich.edu

© Springer International Publishing Switzerland 2016

R.Y. Liu, J.W. McKean (eds.), *Robust Rank-Based and Nonparametric Methods*, Springer Proceedings in Mathematics & Statistics 168,
DOI 10.1007/978-3-319-39065-9_5

Z_{2t+1} are offspring of observation Z_t, $t = 1, 2, \ldots, 2^g - 1$.Hence, the process (i.e. $\{Z_t\}$) is referred to as a bifurcating process. Adopting the notation of Terpstra and Elbayoumi (2012) let

$$[x] = \begin{cases} \lfloor x \rfloor & ; x \geq 1 \\ \lfloor \log_2(x) \rfloor + 1 & ; 0 < x < 1, \end{cases} \tag{5.1}$$

where $\lfloor x \rfloor$ denotes the largest integer less than or equal to x. In this way, for a given observation Z_t, the ancestors can be written as $Z_{[t/2^1]}, Z_{[t/2^2]}, Z_{[t/2^3]}, \ldots$, where Z_0, $Z_{-1}, Z_{-2}, \ldots$ denote the unobserved predecessors of Z_1.

A popular model for tree structured data is the (first-order) bifurcating autoregressive model, denoted here and after as the BAR(1) model. Using the notation in (5.1), the model can be written as

$$Z_t = \phi_0 + \phi_1 Z_{[t/2]} + \varepsilon_t \stackrel{\text{def}}{=} \boldsymbol{X}_{[t/2]}^{\mathsf{T}} \boldsymbol{\phi} + \varepsilon_t, \tag{5.2}$$

where $\boldsymbol{X}_{[t/2]} = (1, Z_{[t/2]})^{\mathsf{T}}$ and $\boldsymbol{\phi} = (\phi_0, \phi_1)^{\mathsf{T}}$. Furthermore, ε_t is an element of one of the vectors given in $\{(\varepsilon_{2t}, \varepsilon_{2t+1})^{\mathsf{T}}\}_{t=-\infty}^{\infty}$, which is assumed to be an independently and identically distributed (iid) sequence of bivariate exchangeable random variables with continuous joint distribution F_J and (common) marginal distribution F. Note that Z_t can be written as $\sum_{i=0}^{\infty} \phi_1^i (\varepsilon_{[t/2^i]} + \phi_0)$. Thus, Z_t is well-defined as an almost sure limit provided $|\phi_1| < 1$ and $E[|\varepsilon_1|] < \infty$, which we will assume throughout the paper. More importantly, since $\{\varepsilon_{[t/2^i]}\}_{i=0}^{\infty}$ is an iid sequence for all t it follows that the distribution of Z_t is the same for all t. In particular, $E[h(Z_t)] = E[h(Z_1)]$ and $E\left[h(Z_{[t/2]}, \varepsilon_t)\right] = E[h(Z_1, \varepsilon_2)]$ for all t and h (assuming the expectations exists).

A number of papers have been written on the robust estimation of the parameters in (5.2). See, for example, Huggins and Marschner (1991), Marschner (1992), Huggins (1996), Bui and Huggins (1998) and Basawa and Zhou (2004). However, many of these methods are rooted in the use of Mallows-type weighting schemes. That is, weighting schemes that only incorporate information on the design point (i.e. $Z_{[t/2]}$ in the present context). This can lead to reduced efficiency. Incorporating information from an initial fit into the weights (via the corresponding residuals) can regain some of this lost efficiency. This is the premise for considering Schweppe-type weighting schemes. In fact, in the autoregressive time series context, Terpstra et al. (2000, 2001) have shown (via Monte Carlo) that estimates based on Schweppe-type weighting schemes can be both robust and efficient simultaneously. Since the BAR model implies that each line in the tree follows an autoregressive process one can expect similar properties to hold in the BAR context.

More specifically, the proposed estimate of $\boldsymbol{\phi}$, say $\hat{\boldsymbol{\phi}}_n = (\hat{\boldsymbol{\phi}}_{n0}, \hat{\boldsymbol{\phi}}_{n1})^{\mathsf{T}}$, is a value of $\boldsymbol{\phi}$ that minimizes

$$D(\boldsymbol{\phi}) = \sum_{t=2}^{n} b(Z_{[t/2]}, \hat{\varepsilon}_t) \, |\varepsilon_t(\boldsymbol{\phi})| \, , \tag{5.3}$$

where $b(Z_{[t/2]}, \hat{\varepsilon}_t)$ denotes a Schweppe-type weight to be used for the tth term, $\hat{\varepsilon}_t$ is a residual based on some initial fit, and $\varepsilon_t(\boldsymbol{\phi}) = Z_t - \boldsymbol{X}_{[t/2]}^{\mathsf{T}}\boldsymbol{\phi}$. Note that $b \equiv 1$ yields the least absolute deviation estimate. Thus, we will refer to the estimate obtained by minimizing (5.3) as a weighted L1-estimate (WL1).

As long as b is positive, it can be shown that $D(\boldsymbol{\phi})$ is non-negative, piecewise linear and convex. Hence, a minimum of $D(\boldsymbol{\phi})$ is guaranteed. Although this minimum is not necessarily unique, it turns out that the diameter of the set of solutions is $o_p\left(n^{-1/2}\right)$. Alternatively, the estimate of $\boldsymbol{\phi}$ can be viewed as an approximate solution of the equation $\boldsymbol{S}(\boldsymbol{\phi}) = -\nabla D(\boldsymbol{\phi}) = \boldsymbol{0}$ where

$$\boldsymbol{S}(\boldsymbol{\phi}) = \sum_{t=2}^{n} b(Z_{[t/2]}, \hat{\varepsilon}_t)\boldsymbol{X}_{[t/2]}\operatorname{sgn}(\varepsilon_t(\boldsymbol{\phi})) .$$

This paper derives the asymptotic distribution of the WL1-estimate in the context of the BAR(1) model. Specifically, Sect. 5.2 presents and discusses two law of large numbers theorems. These theorems play a critical role in the theoretical development of the estimate. Section 5.3 presents the main results of the paper, namely, asymptotic uniform linearity, asymptotic uniform quadraticity, and the asymptotic normality of the WL1-estimate. Proofs for all of the results are given in Sect. 5.4 and some applications of the theory are given in Sect. 5.5. Some concluding remarks are given in Sect. 5.6.

5.2 Some Law of Large Numbers Results

In this section we present two law of large numbers theorems which play critical roles in the theoretical development of this paper. The first theorem is basically an extension of the law of large numbers result given in Terpstra and Elbayoumi (2012). For instance, letting $a_1 = a_2 = a - 1$ and $w_{a-1,0} = w_{0,a-1} = 1$ ($w_{kl} = 0$ o.w.) in (5.4) yields the generalized Lipschitz condition used by Terpstra and Elbayoumi (2012). Moreover, (5.4) also satisfies the condition needed in the proof of Corollary 3.2 of that paper. The theorem is as follows; see Sect. 5.4 for a proof.

Theorem 5.1. *Let $\{Z_t\}_{t=1}^{n}$ denote the n random variables from a perfect binary tree. Then, $\overline{h}_n = n^{-1}\sum_{t=1}^{n} h(Z_t) = E\left[h(Z_1)\right] + o_p(1)$ provided the following four conditions are satisfied:*

A1. $\{(\varepsilon_{2t}, \varepsilon_{2t+1})^{\mathsf{T}}\}_{t=-\infty}^{\infty}$ is an iid sequence of bivariate exchangeable random variables,
A2. $Z_t = \sum_{i=0}^{\infty} \psi_i \varepsilon_{[t/2^i]}$, where $|\psi_i| = O(\beta^i)$ for some $\beta \in (0, 1)$,
A3. h is a real-valued function that satisfies the following inequality for some $C > 0$:

$$|h(x) - h(y)| \leq C|x - y| \left(\sum_{k=0}^{a_1} \sum_{l=0}^{a_2} w_{kl}|x|^k|y|^l \right), \tag{5.4}$$

where w_{kl} is either 1 (indicates presence of $|x|^k|y|^l$) or 0 (indicates absence of $|x|^k|y|^l$) and
A4. $E[|\varepsilon_1|^\gamma] < \infty$, where

$$\gamma = 2\max\{1, \max_{(k,l)}\{w_{kl}(k+l)\}, \max_{(k,l)}\{w_{kl}(k+1)\}, \max_{(k,l)}\{w_{kl}l\}\}. \tag{5.5}$$

As an application of Theorem 5.1 let $h(x) = w(x)p_d(x)$, where $w(x)$ is a bounded Lipschitz function and $p_d(x)$ is an arbitrary d-degree polynomial in x. Adding in and subtracting out $w(x)p_d(y)$ leads to the following inequality

$$|h(x) - h(y)| \le |w(x)||p_d(x) - p_d(y)| + |p_d(y)||w(x) - w(y)|. \tag{5.6}$$

It follows from the Mean Value Theorem that

$$|p_d(x) - p_d(y)| = |p_d'(\lambda x + (1-\lambda)y)||x - y|$$

where $\lambda \in (0,1)$. Moreover, it now follows from the c_r-inequality [e.g. page 44 of Jurečková and Sen (1996)] that $|p_d'(\lambda x + (1-\lambda)y)| \le C\sum_{k=0}^{d-1}\sum_{l=0}^{d-1} w_{kl}|x|^k|y|^l$ where C only depends on the coefficients of the polynomial and the constants from the c_r-inequality and $w_{k,0} = w_{0,l} = 1$ ($w_{kl} = 0$ o.w.). Hence, $p_d(x)$ satisfies (5.4) and (5.5) with $a_1 = a_2 = d-1$ and $\gamma = 2d$. This fact, combined with the bound on $w(x)$ and it's Lipschitz property, imply that $h(x)$ also satisfies (5.4) and (5.5) with $a_1 = d-1$, $a_2 = d$, $w_{k,0} = w_{0,l} = 1$ ($w_{kl} = 0$ o.w.) and $\gamma = 2d$. We can now apply Theorem 5.1 to show, for example, that

$$\frac{1}{n}\sum_{t=1}^{n} w(Z_t)(Z_t - \overline{Z})^2 = E\left[w(Z_1)(Z_1 - E[Z_1])^2\right] + o_p(1);$$

which corresponds to a weighted variance result.

The second theorem is a new law of large numbers result for a bifurcating process. It can be viewed as a generalization of Theorem 5.1 from the perspective that the summands are now allowed to depend on $(\varepsilon_{2t}, \varepsilon_{2t+1})^\mathsf{T}$. The need for such a theorem stems from the fact that the weights being considered in this paper are of the Schweppe variety. Once more, the proof of this theorem is given in Sect. 5.4.

Theorem 5.2. *Let $\{\boldsymbol{v}_t\}_{t=1}^m$, where $\boldsymbol{v}_t = (Z_t, \varepsilon_{2t}, \varepsilon_{2t+1})^\mathsf{T}$, denote the $m = (n-1)/2$ tri-vectors from a perfect binary tree. Then, $\overline{h}_m = m^{-1}\sum_{t=1}^m h(\boldsymbol{v}_t) = E[h(\boldsymbol{v}_1)] + o_p(1)$ provided A1 and A2 of Theorem 5.1 are satisfied in addition to the following three conditions:*

A1. $E\left[|h(\boldsymbol{v}_1)|^{1+\delta}\right] < \infty$ for some $\delta > 0$,
A2. $|h(x,\varepsilon_2,\varepsilon_3) - h(y,\varepsilon_2,\varepsilon_3)| \le C(\varepsilon_2,\varepsilon_3)|x-y|\left(\sum_{k=0}^{a_1}\sum_{l=0}^{a_2} w_{kl}|x|^k|y|^l\right)$ for some $C(\varepsilon_2,\varepsilon_3) > 0$ such that $E[C(\varepsilon_2,\varepsilon_3)] < \infty$ and
A3. $E[|\varepsilon_1|^\gamma] < \infty$, where γ is defined in (5.5), but now uses the w_{kl} given in A2 of this theorem.

As a special case of Theorem 5.2 let $h(\boldsymbol{v}_t) = h_1(Z_t, \varepsilon_{2t}) + h_1(Z_t, \varepsilon_{2t+1})$. Furthermore, suppose h_1 satisfies A1 of the theorem and a condition analogous to A2 of the theorem. Then, it follows from Theorem 5.2 that

$$\frac{1}{n}\sum_{t=2}^{n} h_1(Z_{[t/2]}, \varepsilon_t) = \left(\frac{2m}{n}\right)\left(\frac{1}{m}\sum_{t=1}^{m} \frac{h_1(Z_t, \varepsilon_{2t}) + h_1(Z_t, \varepsilon_{2t+1})}{2}\right)$$
$$= E\left[h_1(Z_1, \varepsilon_2)\right] + o_p(1). \tag{5.7}$$

5.3 Asymptotic Distribution Theory

We begin this section with a list of assumptions that are sufficient for the asymptotic normality of the estimate. Assumptions labeled with an "F" pertain to the (marginal) error distribution (F) while assumptions labeled with a "B" correspond to the weight function (b).

F1. $f(\varepsilon) = F'(\varepsilon)$ is continuous.
F2. $f(\varepsilon)$ is bounded.
F3. $f(0) > 0$.
F4. $E\left[\varepsilon_1^4\right] < \infty$.
B1. $b(z, \varepsilon)$ is bounded.
B2. $|b(z_1, \varepsilon_1) - b(z_2, \varepsilon_2)| \le C\|(z_1, \varepsilon_1)^{\mathsf{T}} - (z_2, \varepsilon_2)^{\mathsf{T}}\|$ for some $C > 0$.
B3. $E\left[b(Z_1, \varepsilon_2)\mathrm{sgn}(\varepsilon_2) \mid Z_1\right] = 0$.

With regard to B3, we note that it is a priori satisfied provided f is symmetric about zero and $b(z, -\varepsilon) = b(z, \varepsilon)$.

Moreover, recall from Sect. 5.1 that the weights depend on the residuals from some initial fit. For instance, in practice, b is typically defined as $b(Z_{[t/2]}, \hat{\varepsilon}_t) = b^*(m_{n1}(Z_{[t/2]}), m_{n2}(\hat{\varepsilon}_t))$ where m_{n1} and m_{n2} denote robust versions of Mahalanobis distance and b^* is a corresponding weight function. Thus, the weights implicitly depend on an initial estimate of $\boldsymbol{\phi}$, measures of location and dispersion and possible stochastic tuning constants. In such cases it is more appropriate to denote the weights by $b(Z_{[t/2]}, \varepsilon_t; \hat{\boldsymbol{\theta}})$ where $\hat{\boldsymbol{\theta}}$ denotes a vector of estimated nuisance parameters. However, we assume throughout this paper that $\hat{\boldsymbol{\theta}}$ can be replaced by its non-stochastic counterpart without affecting the asymptotic results. Lemma 5.4 of Appendix gives a set of sufficient conditions that justifies such an assumption. Consequently, for all theoretical results, we write the weight function as $b(Z_{[t/2]}, \varepsilon_t)$ and emphasize that the only random elements for theoretical purposes are $Z_{[t/2]}$ and ε_t.

Next, recall that the estimate, say $\hat{\boldsymbol{\phi}}_n$, is such that $D\left(\hat{\boldsymbol{\phi}}_n\right) = \min_{\boldsymbol{\phi}} D(\boldsymbol{\phi})$. Ultimately, we wish to show that this estimate is asymptotically normal. To this end, the true parameter vector will be denoted by $\boldsymbol{\phi}_0$. Furthermore, let $\boldsymbol{\Delta} \in \Re^2$, and define the following functions of $\boldsymbol{\Delta}$

$$D_n(\boldsymbol{\Delta}) = D\left(\boldsymbol{\phi}_0 + \boldsymbol{\Delta} n^{-1/2}\right),$$

$$\boldsymbol{S}_n(\boldsymbol{\Delta}) = -\frac{\partial}{\partial \boldsymbol{\Delta}} D_n(\boldsymbol{\Delta}) \;=\; n^{-1/2} \boldsymbol{S}\left(\boldsymbol{\phi}_0 + \boldsymbol{\Delta} n^{-1/2}\right) \quad \text{and}$$

$$Q_n(\boldsymbol{\Delta}) = D_n(\mathbf{0}) - \boldsymbol{S}_n^{\mathsf{T}}(\mathbf{0})\,\boldsymbol{\Delta} + \boldsymbol{\Delta}^{\mathsf{T}}\left(\frac{1}{2\tau}\boldsymbol{C}\right)\boldsymbol{\Delta},$$

where $\tau = (2f(0))^{-1}$ and $\boldsymbol{C} = E\left[b(Z_1, 0)\boldsymbol{X}_1\boldsymbol{X}_1^{\mathsf{T}}\right]$.

Two key results needed to prove the fundamental result are asymptotic uniform linearity (AUL) and asymptotic uniform quadraticity (AUQ). Briefly, the AUL and AUQ results refer to the following two propositions, respectively: for all $c > 0$

$$\sup_{\|\boldsymbol{\Delta}\| \le c} \left\|\boldsymbol{S}_n(\boldsymbol{\Delta}) - \boldsymbol{S}_n(\mathbf{0}) + \tau^{-1}\boldsymbol{C}\boldsymbol{\Delta}\right\| = o_p(1) \text{ and } \sup_{\|\boldsymbol{\Delta}\| \le c} |D_n(\boldsymbol{\Delta}) - Q_n(\boldsymbol{\Delta})| = o_p(1).$$

For future reference, we note that asymptotic linearity (AL) is the same as AUL, less the supremum. Theorem 5.3 establishes the above results; see Sect. 5.4 for a proof.

Theorem 5.3. *AL, AUL and AUQ hold under model assumption (5.2), F1–F4 and B1–B2.*

With AUL and AUQ established, the next step is to derive the asymptotic distribution of $\boldsymbol{S}_n(\mathbf{0})$. This result is given in Theorem 5.4 and its proof can be found in Sect. 5.4.

Theorem 5.4. *If model assumption (5.2), F4 and B1–B3 hold then* $\boldsymbol{S}_n(\mathbf{0}) \xrightarrow{D} N(\mathbf{0}, \boldsymbol{\Sigma})$ *where* $\boldsymbol{\Sigma} = E[(\boldsymbol{X}_1\nu_1)(\boldsymbol{X}_1\nu_1)^{\mathsf{T}}]$ *and* $\nu_1 = 2^{-1/2}(b(Z_1, \varepsilon_2)sgn(\varepsilon_2) + b(Z_1, \varepsilon_3)sgn(\varepsilon_3))$.

Finally, to obtain the asymptotic distribution of $\hat{\boldsymbol{\phi}}_n$ let $\boldsymbol{\Delta} = n^{1/2}(\boldsymbol{\phi} - \boldsymbol{\phi}_0)$ and let $\tilde{\boldsymbol{\Delta}}_n$ denote the value that minimizes $Q_n(\boldsymbol{\Delta})$. Taking the derivative of this function with respect to $\boldsymbol{\Delta}$ and equating it to zero yields the following representation for $\tilde{\boldsymbol{\phi}}_n$

$$\tilde{\boldsymbol{\Delta}}_n \;=\; \sqrt{n}\left(\tilde{\boldsymbol{\phi}}_n - \boldsymbol{\phi}_0\right) = \tau\boldsymbol{C}^{-1}\boldsymbol{S}_n(\mathbf{0}). \tag{5.8}$$

Note that because $\tilde{\boldsymbol{\phi}}_n$ depends on the true value of the process, it is not a statistic. That said, its asymptotic distribution can still be derived. For instance, upon combining Theorem 5.4 with (5.8) we obtain

$$\sqrt{n}\left(\tilde{\boldsymbol{\phi}}_n - \boldsymbol{\phi}_0\right) \xrightarrow{D} N\left(\mathbf{0}, \tau^2\boldsymbol{C}^{-1}\boldsymbol{\Sigma}\boldsymbol{C}^{-1}\right). \tag{5.9}$$

Now consider the quantity $\hat{\boldsymbol{\Delta}}_n = n^{1/2}\left(\hat{\boldsymbol{\phi}}_n - \boldsymbol{\phi}_0\right)$ where $\hat{\boldsymbol{\phi}}_n$ denotes the estimate obtained from minimizing the objective function given in (5.3). Jaeckel's (1972) convexity argument along with the AUQ result can then be used to show that $n^{1/2}\left(\hat{\boldsymbol{\phi}}_n - \tilde{\boldsymbol{\phi}}_n\right) = \hat{\boldsymbol{\Delta}}_n - \tilde{\boldsymbol{\Delta}}_n = \boldsymbol{o}_p(1)$. The main result of this paper, which we state as Theorem 5.5, now follows from this fact and (5.9).

Theorem 5.5. *Under model assumption (5.2), F1–F4 and B1–B3 we have the following*

$$\sqrt{n}\left(\hat{\boldsymbol{\phi}}_n - \boldsymbol{\phi}_0\right) \xrightarrow{D} N\left(\mathbf{0}, \tau^2 \boldsymbol{C}^{-1}\boldsymbol{\Sigma}\boldsymbol{C}^{-1}\right).$$

5.4 Technical Details

5.4.1 *Proof of Theorem 5.1*

Note that (5.4) yields a condition that is similar to that of A3 in Terpstra and Elbayoumi (2012). In fact, it is easily shown that this assumption is also sufficient for Lemma 3.1 of Terpstra and Elbayoumi (2012). For instance, simply replace the inequality in (1) of that paper with the following inequality

$$|h(Z_t) - h(Q_{t,j})| \leq \beta^j K_a \left(\sum_{k=0}^{a_1} \sum_{l=0}^{a_2} w_{kl}(|T_{t,j}||Q_{t,j}|^{k+l} + |T_{t,j}|^{k+1}|Q_{t,j}|^l) \right)$$

and then continue as in that paper. The only additional adjustment is to replace Eq. (7) of that paper with the following inequality

$$V\left[h(Z_1) - h(c)\right] \leq C_a^2 E\left[(|Z_1|^2 + |c|^2) \left(\sum_{k=0}^{a_1} \sum_{l=0}^{a_2} w_{kl}|Z_1|^{2k} \right) \right],$$

where c is any constant such that $h(c) < \infty$. Hence, upon making these minor modifications, Theorem 3.1 of Terpstra and Elbayoumi (2012) remains valid. That is, $\overline{h}_n = E\left[h(Z_1)\right] + o_p(1)$. □

5.4.2 *Proof of Theorem 5.2*

To begin, note that $\overline{h}_m = \overline{h}_{m1} + \overline{h}_{m2}$ where

$$\overline{h}_{m1} = \frac{1}{m}\sum_{t=1}^{m} E\left[h(\boldsymbol{v}_t) \mid Z_t\right] \text{ and } \overline{h}_{m2} = \frac{1}{m}\sum_{t=1}^{m} \left(h(\boldsymbol{v}_t) - E\left[h(\boldsymbol{v}_t) \mid Z_t\right]\right).$$

Consider $\overline{h}_{m1}$ first. In doing so, let $H(x) = E\left[h(\boldsymbol{v}_t) \mid Z_t = x\right]$ and note that the independence of Z_t and $(\varepsilon_{2t}, \varepsilon_{2t+1})^{\mathsf{T}}$ along with A2 of the theorem imply that

$$\begin{aligned} |H(x) - H(y)| &\leq E\left[|h(x, \varepsilon_2, \varepsilon_3) - h(y, \varepsilon_2, \varepsilon_3)|\right] \\ &\leq E\left[C(\varepsilon_2, \varepsilon_3)\right] |x - y| \left(\sum_{k=0}^{a_1} \sum_{l=0}^{a_2} w_{kl}|x|^k|y|^l \right). \end{aligned} \tag{5.10}$$

Thus, (5.10) satisfies (5.4) so Theorem 5.1 yields $\overline{h}_{m1} = E[h(\boldsymbol{v}_1)] + o_p(1)$. Finally, let $U_t = h(\boldsymbol{v}_t) - E[h(\boldsymbol{v}_t) \mid Z_t]$ so that $\overline{h}_{m2} = m^{-1}\sum_{t=1}^{m} U_t$. It follows from A1 of the theorem and the definition given on page 190 of Hamilton (1994) that $\{U_t\}_{t=1}^{\infty}$ is an L^1-mixingale with respect to $\Omega_t = \sigma\text{-}field\{Z_1, \varepsilon_2, \varepsilon_3, \ldots, \varepsilon_{2t}, \varepsilon_{2t+1}\}$. For example, let $c_t = E[|U_t|]$ and $\xi_t = I(t = 0)$, where $I(\cdot)$ denotes the indicator function. Moreover, since $E[U_t] = 0$, it readily follows from Proposition 7.6 of Hamilton (1994) that $\overline{h}_{m2} = o_p(1)$. Hence, $\overline{h}_m = E[h(\boldsymbol{v}_1)] + o_p(1)$. □

5.4.3 Proof of Theorem 5.3

Heiler and Willers (1988) have shown that AL, AUL and AUQ are equivalent in the context of linear regression. The only two features of the regression model and objective function that are utilized in their proof are linearity and convexity, respectively. Since the BAR(1) model is linear in its parameters and our objective function is convex, the same proof can essentially be used to establish the equivalence of AL, AUL and AUQ in the current context. That is, AUL and AUQ follow from the AL result, which we subsequently prove.

To begin, let $U_n = \boldsymbol{\lambda}^{\mathsf{T}}(S_n(\boldsymbol{\Delta}) - S_n(\mathbf{0}))$ where $\boldsymbol{\lambda} \in \Re^2$ and $\boldsymbol{\Delta} \in \Re^2$ are arbitrary but fixed. Since vector convergence holds if and only if component-wise convergence holds (obtained via $\boldsymbol{\lambda}$), it suffices to show that $U_n = -\tau^{-1}\boldsymbol{\lambda}^{\mathsf{T}}\boldsymbol{C}\boldsymbol{\Delta} + o_p(1)$ in order to prove the theorem. Of course, this will follow if we can show that $E[U_n] = -\tau^{-1}\boldsymbol{\lambda}^{\mathsf{T}}\boldsymbol{C}\boldsymbol{\Delta} + o(1)$ and $V[U_n] = o(1)$. Consider $E[U_n]$ first. In doing so, let $w_t = w(Z_{[t/2]}, \varepsilon_t) = b(Z_{[t/2]}, \varepsilon_t)\boldsymbol{\lambda}^{\mathsf{T}}\boldsymbol{X}_{[t/2]}$ and note that

$$U_n = -2n^{-1/2}\sum_{t=2}^{n} w(Z_{[t/2]}, \varepsilon_t)\left(I\left(\varepsilon_t \le \boldsymbol{X}_{[t/2]}^{\mathsf{T}}\boldsymbol{\Delta}n^{-1/2}\right) - I(\varepsilon_t \le 0)\right) \quad a.e.$$

$$\stackrel{\text{def}}{=} -2n^{-1/2}\sum_{t=2}^{n} w_t d_{nt}.$$

However, since the distribution of $(Z_{[t/2]}, \varepsilon_t)^{\mathsf{T}}$ and $(Z_1, \varepsilon_2)^{\mathsf{T}}$ are the same for all t it follows that

$$E[U_n] = -2(n-1)n^{-1/2}E\left[w(Z_1, \varepsilon_2)\left(I\left(\varepsilon_2 \le \boldsymbol{X}_1^{\mathsf{T}}\boldsymbol{\Delta}n^{-1/2}\right) - I(\varepsilon_2 \le 0)\right)\right].$$

It now follows from Lemma 5.1 that $E[U_n] = -\tau^{-1}\boldsymbol{\lambda}^{\mathsf{T}}\boldsymbol{C}\boldsymbol{\Delta} + o(1)$.

Next, let us consider $V[U_n]$. Again, since the distribution of $(Z_{[t/2]}, \varepsilon_t)^{\mathsf{T}}$ and $(Z_1, \varepsilon_2)^{\mathsf{T}}$ are the same for all t it follows that

$$V[U_n] = 4\left(\frac{n-1}{n}\right)V[w_2 d_{n2}] + \frac{8}{n}\sum_{s<t} COV[w_s d_{ns}, w_t d_{nt}] \stackrel{\text{def}}{=} V_{n1} + V_{n2}.$$

With regard to V_{n1}, note that $V[w_2 d_{n2}] \leq E\left[(w_2 d_{n2})^2\right]$ and $E\left[(w_2 d_{n2})^2\right] = O(n^{-1/2})$ by Lemma 5.2. Hence, $V_{n1} = o(1)$.

Consider V_{n2} next. For $s < t$, there are four cases to consider. That is, s and t are such that:

1. ε_s and ε_t are sisters,
2. ε_s and ε_t belong to the same line,
3. an ancestor of ε_t is a sister of ε_s and
4. ε_s and ε_t belong to different lines, but are not sisters.

Dividing up the sum according to these cases yields $V_{n2} \stackrel{\text{def}}{=} \sum_{i=1}^{4} V_{n2i}$. Furthermore, it can be shown that the number of terms for each of these sums are $O(n)$, $O(n\log_2(n))$, $O(n\log_2(n))$ and $O(n^2)$, respectively. Since $V[w_2 d_{n2}] = O(n^{-1/2})$ (by Lemma 5.2) and $COV[X, Y] \leq \sqrt{V[X]\,V[Y]}$ it follows that $V_{n2i} = o(1)$ for $i = 1, 2, 3$. Therefore, we only need to show that $V_{n24} = o(1)$ to complete the proof. However, this follows from Lemma 5.3. The details are similar to those following the discussion of Eq. (9) in Terpstra and Elbayoumi (2012). □

5.4.4 *Proof of Theorem 5.4*

Since $\boldsymbol{S}_n(\boldsymbol{0})$ is a vector we will use the Cramer-Wold device and show $\boldsymbol{\lambda}^{\mathsf{T}}\boldsymbol{S}_n(\boldsymbol{0})$ is asymptotically normal where $\boldsymbol{\lambda} \in \Re^2$ is arbitrary but fixed. To begin, recall from Theorem 5.2 that there are $m = (n-1)/2$ tri-vectors of the form $(Z_t, \varepsilon_{2t}, \varepsilon_{2t+1})^{\mathsf{T}}$. Since ε_{2t} and ε_{2t+1} are dependent, we rewrite $\boldsymbol{\lambda}^{\mathsf{T}}\boldsymbol{S}_n(\boldsymbol{0})$ as follows

$$\begin{aligned}
\boldsymbol{\lambda}^{\mathsf{T}}\boldsymbol{S}_n(\boldsymbol{0}) &= \sqrt{\frac{2m}{n}}\sum_{t=1}^{m}\frac{1}{\sqrt{m}}(\boldsymbol{\lambda}^{\mathsf{T}}\boldsymbol{X}_t)\left(\frac{b(Z_t,\varepsilon_{2t})\mathrm{sgn}(\varepsilon_{2t}) + b(Z_t,\varepsilon_{2t+1})\mathrm{sgn}(\varepsilon_{2t+1})}{\sqrt{2}}\right) \\
&\stackrel{\text{def}}{=} \sqrt{\frac{2m}{n}}\sum_{t=1}^{m}\frac{1}{\sqrt{m}}(\boldsymbol{\lambda}^{\mathsf{T}}\boldsymbol{X}_t)\nu_t \\
&\stackrel{\text{def}}{=} \sqrt{\frac{2m}{n}}S_m(0).
\end{aligned}$$

Since $\sqrt{2m/n} = 1 + o(1)$ we need only consider $S_m(0)$. To that end, define

$$Y_{m,t} = \frac{1}{\sqrt{m}}(\boldsymbol{\lambda}^{\mathsf{T}}\boldsymbol{X}_t)\nu_t \text{ and } S^{*}_{m,j} = \sum_{t=1}^{j} Y_{m,t}.$$

Furthermore, let $\Omega_{m,t} = \sigma\text{-}\mathit{field}\{Z_1, \varepsilon_2, \varepsilon_3, \ldots, \varepsilon_{2t}, \varepsilon_{2t+1}\}$. Under F4 and B1 it is straight forward to show that

$$E\left[Y_{m,t}^2\right] = \frac{1}{m}E\left[(\boldsymbol{\lambda}^{\mathsf{T}}\boldsymbol{X}_t\nu_t)^2\right] = \frac{1}{m}\boldsymbol{\lambda}^{\mathsf{T}}\boldsymbol{\Sigma}\boldsymbol{\lambda} \leq \boldsymbol{\lambda}^{\mathsf{T}}\boldsymbol{\Sigma}\boldsymbol{\lambda} < \infty. \quad (5.11)$$

Next, it follows from B3 that

$$E\left[Y_{m,t} \mid \Omega_{m,t-1}\right] = E\left[\frac{1}{\sqrt{m}}\boldsymbol{\lambda}^{\mathsf{T}}\boldsymbol{X}_t\nu_t \mid \Omega_{m,t-1}\right] = \frac{1}{\sqrt{m}}(\boldsymbol{\lambda}^{\mathsf{T}}\boldsymbol{X}_t)E\left[\nu_t \mid \Omega_{m,t-1}\right] = 0. \tag{5.12}$$

It now follows from (5.11) and (5.12) that $\{S^*_{m,j}, \Omega_{m,j}\}$ is a zero-mean square-integrable martingale array with differences $Y_{m,j}$. Thus, we set out to show that the assumptions used in this paper imply the four conditions of the Martingale Central Limit Theorem (MCLT) stated as Corollary 3.1 in Hall and Heyde (1980). Consider the first condition of the MCLT. Using B1 to bound ν_t, we obtain

$$\max_{1\le t\le m} |Y_{m,t}| = \max_{1\le t\le m}\left|\frac{1}{\sqrt{m}}(\boldsymbol{\lambda}^{\mathsf{T}}\boldsymbol{X}_t)\nu_t\right| \le K\frac{1}{\sqrt{m}}\max_{1\le t\le m}\|\boldsymbol{X}_t\|,$$

where K is a constant depending only on the bound in B1 and $\boldsymbol{\lambda}$. However, since

$$P\left[\frac{1}{\sqrt{m}}\max_{1\le t\le m}\|\boldsymbol{X}_t\| > \varepsilon\right] \le \frac{1}{\varepsilon^2}E\left[\|\boldsymbol{X}_t\|^2 I\left(\|\boldsymbol{X}_t\|^2 > m\varepsilon^2\right)\right]$$

it follows that $m^{-1/2}\max_{1\le t\le m}\|\boldsymbol{X}_t\| = o_p(1)$ provided $E\left[\|\boldsymbol{X}_t\|^2\right] < \infty$ (which is implied by F4). Hence, $\max_{1\le t\le m}|Y_{m,t}| = o_p(1)$ and the first condition of the MCLT is satisfied. Consider the second condition of the MCLT next. In doing so, let $h(Z_t, \varepsilon_{2t}, \varepsilon_{2t+1}) = (\boldsymbol{\lambda}^{\mathsf{T}}\boldsymbol{X}_t\nu_t)^2$. It follows from F4 and B1 that A1 of Theorem 5.2 is satisfied (e.g. $\delta = 1$). Moreover, it follows from B1 and B2 that A2 and A3 of Theorem 5.2 are satisfied with $a_1 = 1$, $a_2 = 2$, $w_{0,0} = w_{1,0} = w_{0,1} = w_{0,2} = 1$ ($w_{kl} = 0$ o.w.) and $\gamma = 4$. Here, $C(\varepsilon_2, \varepsilon_3)$ turns out to be constant so the finite expectation assumption is trivial. The details are similar to those that follow (5.6). Hence, it follows from Theorem 5.2 that

$$\sum_{t=1}^{m} Y_{m,t}^2 = \frac{1}{m}\sum_{t=1}^{m}(\boldsymbol{\lambda}^{\mathsf{T}}\boldsymbol{X}_t\nu_t)^2 = \boldsymbol{\lambda}^{\mathsf{T}}E\left[\boldsymbol{X}_t\boldsymbol{X}_t^{\mathsf{T}}\nu_t^2\right]\boldsymbol{\lambda} + o_p(1).$$

This verifies the second condition of the MCLT. Now, the above derivations also imply the following

$$E\left[\max_{1\le t\le m}\left|Y_{m,t}^2\right|\right] \le E\left[\sum_{t=1}^{m} Y_{m,t}^2\right] = E\left[(\boldsymbol{\lambda}^{\mathsf{T}}\boldsymbol{X}_t\nu_t)^2\right] = O(1).$$

Hence, the third condition of the MCLT is satisfied. For verification of the fourth condition one is referred to the "Remarks" paragraph on page 59 of Hall and Heyde (1980). Thus, since all four conditions of the MCLT have been verified, we have proven that

$$S_m(0) \;=\; S^*_{m,m} \xrightarrow{D} N(0, \boldsymbol{\lambda}^{\mathsf{T}} \boldsymbol{\Sigma} \boldsymbol{\lambda})$$

which, by the Cramer-Wold device, completes the proof. □

5.5 Applications of Theory

Recall that when the weights only incorporate information on the design point (i.e. $b(z, \varepsilon) = b(z)$) they are referred to as Mallows weights. The corresponding estimates can be viewed as a special case of the theory presented in this paper. For example, the result given in Theorem 5.5 remains valid with $\boldsymbol{C} = E\left[b(Z_1)\boldsymbol{X}_1\boldsymbol{X}_1^{\mathsf{T}}\right]$ and $\boldsymbol{\Sigma} = (1+\rho_s)E\left[b^2(Z_1)\boldsymbol{X}_1\boldsymbol{X}_1^{\mathsf{T}}\right]$, where $\rho_s = COR(\text{sgn}(\varepsilon_2), \text{sgn}(\varepsilon_3))$. In particular, when $b(z) \equiv 1$ we obtain the result for the L1-estimate.

We note that the result for the least squares estimate [see e.g. Zhou and Basawa (2005)] is very similar to that of the L1-estimate. For instance, simply replace ρ_s with $\rho = COR(\varepsilon_2, \varepsilon_3)$ and τ^2 with $\sigma^2 = V[\varepsilon_2]$. Upon doing so, the asymptotic relative efficiency (ARE) of the L1-estimate of ϕ_1 relative to the least squares estimate of ϕ_1 is given by $[\sigma^2(1+\rho)]/[\tau^2(1+\rho_s)]$. Note that σ^2/τ^2 represents the corresponding ARE for the location model and $(1+\rho)/(1+\rho_s)$ is an adjustment that takes into account the dependence between ε_2 and ε_3. As an example, suppose F_J corresponds to the following bivariate contaminated normal distribution

$$(1-\gamma)\text{N}(\boldsymbol{0}, \boldsymbol{\Omega}) + \gamma\text{N}(\boldsymbol{0}, \omega^2\boldsymbol{\Omega}) \text{ where } \boldsymbol{\Omega} \;=\; \begin{bmatrix} 1 & \theta \\ \theta & 1 \end{bmatrix}. \tag{5.13}$$

While the values of σ^2, τ^2, and ρ can be calculated directly, a Monte Carlo approximation ($N = 100{,}000$) was used for the values of ρ_s. Table 5.1 presents some ARE values for different values of θ, γ, and ω^2. Unfortunately, the ARE calculations are not as straightforward when $b \not\equiv 1$. The difficulties primarily stem from the fact that the matrices (i.e. $\boldsymbol{C}$ and $\boldsymbol{\Sigma}$) which define the asymptotic variance-covariance matrix do not necessarily simplify when $b \not\equiv 1$. In addition, the exact distribution of Z_t is difficult to determine (for $\gamma > 0$) when the errors have the distribution given in (5.13). However, since $\boldsymbol{C}$ and $\boldsymbol{\Sigma}$ correspond to expectations, one can readily approximate these quantities via Monte Carlo.

Table 5.1 AREs of the L1-estimate of ϕ_1 Relative to the Least Squares Estimate

$(\gamma, \omega^2)^{\mathsf{T}}$	$\theta = 0$	$\theta = 0.1$	$\theta = 0.5$	$\theta = 0.9$
$(0, \text{NA})^{\mathsf{T}}$	0.637	0.661	0.718	0.707
$(0.1, 25)^{\mathsf{T}}$	1.831	1.895	2.071	2.035

As an example, let us approximate the ARE between two different WL1-estimates ($b \not\equiv 1$ for both estimates). Consider first a Mallows-type estimate where the weights are defined as

$$b_m(z) = \min\left\{1, \frac{\sigma_z^2 \chi^2_{0.95}(1)}{(z-\mu_z)^2}\right\},$$

where $\mu_z = E[Z_t]$, $\sigma_z^2 = V[Z_t]$, and $\chi^2_{0.95}(1) = 3.84$ corresponds to the 95th percentile of a Chi-square distribution with one degree of freedom. Here, *all* points of high leverage (i.e. large values of $Z_{[t/2]}$) will be downweighted.

On the other hand, consider the following Schweppe-type estimate where the weights are defined as

$$b_s(z,\varepsilon) = 1 - I\left(\varepsilon^2 > \sigma_\varepsilon^2 \chi^2_{0.95}(1)\right) I\left((z-\mu_z)^2 > \sigma_z^2 \chi^2_{0.95}(1)\right)(1 - b_m(z)),$$

where $I(\cdot)$ denotes the indicator function and $\sigma_\varepsilon^2 = V[\varepsilon_t]$. Note that these weights will only downweight points of high leverage when $\hat{\varepsilon}_{2t}$ and/or $\hat{\varepsilon}_{2t+1}$ are deemed to be large. Furthermore, the degree to which these so-called *bad* leverage points are downweighted is identical to that of the Mallows estimate.

As previously discussed, the values of $\boldsymbol{C}$ and $\boldsymbol{\Sigma}$ can be approximated via Monte Carlo ($N = 100,000$). In fact, only the more simple AR(1) model given by

$$Z_t = \phi_0 + \phi_1 Z_{t-1} + \varepsilon_t$$

is needed to generate the $\{Z_t\}$ process. That is, it is not necessary to generate a bifurcating tree to approximate these matrices. Moreover, this can be done independently from the generation of $(\varepsilon_{2t}, \varepsilon_{2t+1})^{\mathsf{T}}$ since the BAR(1) model implies that Z_t and $(\varepsilon_{2t}, \varepsilon_{2t+1})^{\mathsf{T}}$ are independent.

In what follows, the values of ϕ_1 and ϕ_0 are such that $\phi_1 \in \{0.1, 0.5, 0.9\}$ and $\phi_0 = 10(1-\phi_1)$. The distribution of $(\varepsilon_{2t}, \varepsilon_{2t+1})^{\mathsf{T}}$, F_J, is the same as that given in (5.13). The values used for ϕ_0 imply that the mean of the process (μ_z) is always equal to 10. Additionally, $\sigma_z^2 = \sigma_\varepsilon^2/(1-\phi_1^2)$ where $\sigma_\varepsilon^2 = 1 - \gamma + \gamma\omega^2$.

The ARE results for this example are given in Table 5.2. Note that when $\gamma = 0$ there is only a small gain in efficiency and this gain is relatively stable across the different values of ϕ_1 and θ. On the other hand, when $\gamma > 0$, there are some sizable

Table 5.2 AREs of the Mallows WL1-estimate of ϕ_1 Relative to the Schweppe WL1-estimate

$(\gamma,\omega^2)^{\mathsf{T}}$	ϕ_1	$\theta = 0$	$\theta = 0.1$	$\theta = 0.5$	$\theta = 0.9$
$(0, \text{NA})^{\mathsf{T}}$	0.1	1.041	1.043	1.045	1.044
$(0, \text{NA})^{\mathsf{T}}$	0.5	1.043	1.047	1.046	1.042
$(0, \text{NA})^{\mathsf{T}}$	0.9	1.045	1.048	1.046	1.045
$(0.1, 25)^{\mathsf{T}}$	0.1	1.581	1.579	1.618	1.595
$(0.1, 25)^{\mathsf{T}}$	0.5	1.378	1.405	1.406	1.384
$(0.1, 25)^{\mathsf{T}}$	0.9	1.104	1.103	1.116	1.122

gains in efficiency. These gains are relatively stable across the different values of θ, but appear to be a decreasing function of ϕ_1. More importantly, the gain in efficiency can be attributed to how the *good* leverage points are being weighted. That is, by $b_m(Z_t)$ (< 1) in the case of the Mallows WL1-estimate and not at all (e.g. $b_s(Z_t, \varepsilon_{2t}) = 1$) in the case of the Schweppe WL1-estimate. Once more, these results support the well-known fact that incorporating information from residuals into the weights can produce more efficient estimates.

5.6 Conclusion

We commence this section with a few observations regarding the theory presented in this paper. For instance, we note that there is an alternative representation of $\boldsymbol{\Sigma}$, the asymptotic variance-covariance matrix of $\boldsymbol{S}_n(\mathbf{0})$. More specifically, if we let

$$\begin{aligned}\rho_{bs}(Z_1) &= COR(b(Z_1, \varepsilon_2)\text{sgn}(\varepsilon_2), b(Z_1, \varepsilon_3)\text{sgn}(\varepsilon_3)|Z_1) \quad \text{and} \\ \sigma^2_{bs}(Z_1) &= V\left[b(Z_1, \varepsilon_2)\text{sgn}(\varepsilon_2)|Z_1\right]\end{aligned}$$

denote the conditional (on Z_1) correlation of $b(Z_1, \varepsilon_2)\text{sgn}(\varepsilon_2)$ and $b(Z_1, \varepsilon_3)\text{sgn}(\varepsilon_3)$ and the conditional variance of $b(Z_1, \varepsilon_2)\text{sgn}(\varepsilon_2)$, respectively, then it can be shown that

$$\boldsymbol{\Sigma} = E\left[\sigma^2_{bs}(Z_1)(1 + \rho_{bs}(Z_1))\boldsymbol{X}_1\boldsymbol{X}_1^{\mathsf{T}}\right].$$

Recall that ε_2 and ε_3 are, in general, dependent random variables. Thus, this alternative representation of $\boldsymbol{\Sigma}$ reflects how the asymptotic distribution of $\hat{\boldsymbol{\phi}}_n$ is affected by this dependence.

Next, we note from a robustness perspective (e.g. bounded influence function) that the weight function b is typically chosen so that $b(z, \varepsilon)z$ is uniformly bounded. Under this scenario, assume further that $b(z, \varepsilon)z$ satisfies the Lipschitz property (see B2). Then, a careful inspection of the proofs reveals that the finite fourth moment assumption given in F4 can be relaxed to a finite second moment assumption.

Finally, we note that the theory can be extended to the more general BAR(p) model. That is, a bifurcating autoregressive model of order p. The result given in Theorem 5.5 is essentially the same. The fundamental difference is that $\boldsymbol{X}_{[t/2]}$ needs to be redefined as $(1, Z_{[t/2^1]}, Z_{[t/2^2]}, \ldots, Z_{[t/2^p]})^{\mathsf{T}}$ in the definitions of $\boldsymbol{C}$ and $\boldsymbol{\Sigma}$. However, to prove this result Theorems 5.1 and 5.2 need to be generalized to the case where $b(z, \varepsilon)$ is a function from $\Re^p \times \Re$ to $\Re$.

In closing, the results in this paper establish the asymptotic distribution of the WL1-estimate for a bifurcating autoregressive model. Martingale and mixingale theorems combined with two new law of large numbers theorems are the underpinning tools needed to obtain the results. That said, Theorems 5.1 and 5.2 are applicable to other situations as well. For instance, in establishing consistency results for basic summary statistics such as (weighted) means and variances. Moreover, Theorems 5.3–5.5 can be used to derive tests of hypotheses for the model parameters based on Reduction in Dispersion, Aligned Rank, and/or Wald-type statistics. The interested reader is referred to Sect. 3.6 of Hettmansperger and McKean (2011) for details in the linear regression context.

Acknowledgements I would like to thank an anonymous referee for providing helpful comments regarding the initial version of this paper.

Appendix

Lemma 5.1. *Let* $w(Z_1, \varepsilon_2) = b(Z_1, \varepsilon_2)\boldsymbol{\lambda}^{\mathsf{T}}\boldsymbol{X}_1$*, where* $\boldsymbol{X}_1^{\mathsf{T}} = (1, Z_1)$*. Then, assuming F1, F2, F4, B1 and B2, we have the following:*

$$\begin{aligned}
&E\left[w(Z_1,\varepsilon_2)\left(I\left(\varepsilon_2 \le \boldsymbol{X}_1^{\mathsf{T}}\boldsymbol{\Delta} n^{-1/2}\right) - I\left(\varepsilon_2 \le 0\right)\right)\right] \\
&= n^{-1/2}E\left[b(Z_1,\xi_n)f(\xi_n)(\boldsymbol{\lambda}^{\mathsf{T}}\boldsymbol{X}_1\boldsymbol{X}_1^{\mathsf{T}}\boldsymbol{\Delta})\right] \\
&= n^{-1/2}E\left[b(Z_1,0)f(0)(\boldsymbol{\lambda}^{\mathsf{T}}\boldsymbol{X}_1\boldsymbol{X}_1^{\mathsf{T}}\boldsymbol{\Delta})\right] + o\left(n^{-1/2}\right),
\end{aligned}$$

where $\xi_n \in \left[-|\boldsymbol{X}_1^{\mathsf{T}}\boldsymbol{\Delta}|n^{-1/2}, |\boldsymbol{X}_1^{\mathsf{T}}\boldsymbol{\Delta}|n^{-1/2}\right]$.

Proof. Upon conditioning on Z_1 and exploiting the independence of Z_1 and ε_2 we obtain the following

$$E\left[w(Z_1,\varepsilon_2)\left(I\left(\varepsilon_2 \le \boldsymbol{X}_1^{\mathsf{T}}\boldsymbol{\Delta} n^{-1/2}\right) - I\left(\varepsilon_2 \le 0\right)\right)\right] = E\left[\int_0^{\boldsymbol{X}_1^{\mathsf{T}}\boldsymbol{\Delta} n^{-1/2}} w(Z_1,x)dF(x)\right].$$

Next, it follows from F1, B2 and the First Mean Value Theorem given on page 230 of Bartle (1976) that

$$\begin{aligned}
E\left[\int_0^{\boldsymbol{X}_1^{\mathsf{T}}\boldsymbol{\Delta} n^{-1/2}} w(Z_1,x)dF(x)\right] &= E\left[w(Z_1,\xi_n)f(\xi_n)\boldsymbol{X}_1^{\mathsf{T}}\boldsymbol{\Delta} n^{-1/2}\right] \\
&= n^{-1/2}E\left[b(Z_1,\xi_n)f(\xi_n)(\boldsymbol{\lambda}^{\mathsf{T}}\boldsymbol{X}_1\boldsymbol{X}_1^{\mathsf{T}}\boldsymbol{\Delta})\right],
\end{aligned}$$

where $\xi_n \in \left[-|\boldsymbol{X}_1^{\mathsf{T}}\boldsymbol{\Delta}|n^{-1/2}, |\boldsymbol{X}_1^{\mathsf{T}}\boldsymbol{\Delta}|n^{-1/2}\right]$. Now, adding in and subtracting out the appropriate quantity we get $n^{-1/2}E\left[b(Z_1,\xi_n)f(\xi_n)(\boldsymbol{\lambda}^{\mathsf{T}}\boldsymbol{X}_1\boldsymbol{X}_1^{\mathsf{T}}\boldsymbol{\Delta})\right] = E_{n1} + E_{n2}$ where

$$\begin{aligned}
E_{n1} &= n^{-1/2}E\left[b(Z_1,0)f(0)(\boldsymbol{\lambda}^{\mathsf{T}}\boldsymbol{X}_1\boldsymbol{X}_1^{\mathsf{T}}\boldsymbol{\Delta})\right] \quad \text{and} \\
E_{n2} &= n^{-1/2}E\left[(b(Z_1,\xi_n)f(\xi_n) - b(Z_1,0)f(0))(\boldsymbol{\lambda}^{\mathsf{T}}\boldsymbol{X}_1\boldsymbol{X}_1^{\mathsf{T}}\boldsymbol{\Delta})\right].
\end{aligned}$$

Note that F2, F4 and B1 imply that the expectation in E_{n1} is finite. Thus, consider E_{n2}. It follows from F4 that $\xi_n = o_p(1)$. Furthermore, it follows from F1, F2, B1 and B2 that $b(Z_1, \xi_n)f(\xi_n) - b(Z_1, 0)f(0)$ is a bounded random variable that is $o_p(1)$. Hence, F4 and a variant of the Dominated Convergence Theorem imply that $E_{n2} = o\left(n^{-1/2}\right)$, which completes the proof. □

Lemma 5.2. *Let $w(Z_1, \varepsilon_2) = b(Z_1, \varepsilon_2)\boldsymbol{\lambda}^{\mathsf{T}}\boldsymbol{X}_1$, where $\boldsymbol{X}_1^{\mathsf{T}} = (1, Z_1)$. Then, assuming F1, F2, F4, B1 and B2, we have the following:*

$$E\left[w^2(Z_1, \varepsilon_2)\left(I\left(\varepsilon_2 \le \boldsymbol{X}_1^{\mathsf{T}}\boldsymbol{\Delta}n^{-1/2}\right) - I\left(\varepsilon_2 \le 0\right)\right)^2\right]$$
$$\le 2n^{-1/2}E\left[b^2(Z_1, \xi_n)f(\xi_n)|\boldsymbol{\lambda}^{\mathsf{T}}\boldsymbol{X}_1|^2|\boldsymbol{X}_1^{\mathsf{T}}\boldsymbol{\Delta}|\right] = O(n^{-1/2}),$$

where $\xi_n \in \left[-|\boldsymbol{X}_1^{\mathsf{T}}\boldsymbol{\Delta}|n^{-1/2}, |\boldsymbol{X}_1^{\mathsf{T}}\boldsymbol{\Delta}|n^{-1/2}\right]$.

Proof. To begin, note that

$$E\left[w^2(Z_1, \varepsilon_2)\left(I\left(\varepsilon_2 \le \boldsymbol{X}_1^{\mathsf{T}}\boldsymbol{\Delta}n^{-1/2}\right) - I\left(\varepsilon_2 \le 0\right)\right)^2\right]$$
$$\le E\left[w^2(Z_1, \varepsilon_2)I\left(|\varepsilon_2| \le |\boldsymbol{X}_1^{\mathsf{T}}\boldsymbol{\Delta}n^{-1/2}|\right)\right].$$

Once more, upon conditioning on Z_1 and exploiting the independence of Z_1 and ε_2 we obtain the following

$$E\left[w^2(Z_1, \varepsilon_2)I\left(|\varepsilon_2| \le |\boldsymbol{X}_1^{\mathsf{T}}\boldsymbol{\Delta}n^{-1/2}|\right)\right] = E\left[\int_{-|\boldsymbol{X}_1^{\mathsf{T}}\boldsymbol{\Delta}n^{-1/2}|}^{|\boldsymbol{X}_1^{\mathsf{T}}\boldsymbol{\Delta}n^{-1/2}|} w^2(Z_1, x)dF(x)\right].$$

It now follows from F1, B2 and the First Mean Value Theorem given on page 230 of Bartle (1976) that

$$E\left[\int_{-|\boldsymbol{X}_1^{\mathsf{T}}\boldsymbol{\Delta}n^{-1/2}|}^{|\boldsymbol{X}_1^{\mathsf{T}}\boldsymbol{\Delta}n^{-1/2}|} w^2(Z_1, x)dF(x)\right] = 2E\left[w^2(Z_1, \xi_n)f(\xi_n)|\boldsymbol{X}_1^{\mathsf{T}}\boldsymbol{\Delta}n^{-1/2}|\right]$$
$$= \frac{2}{\sqrt{n}}E\left[b^2(Z_1, \xi_n)f(\xi_n)|\boldsymbol{\lambda}^{\mathsf{T}}\boldsymbol{X}_1|^2|\boldsymbol{X}_1^{\mathsf{T}}\boldsymbol{\Delta}|\right] \quad (5.14)$$

where $\xi_n \in \left[-|\boldsymbol{X}_1^{\mathsf{T}}\boldsymbol{\Delta}|n^{-1/2}, |\boldsymbol{X}_1^{\mathsf{T}}\boldsymbol{\Delta}|n^{-1/2}\right]$. Lastly, the bounds given in F2, F4 and B1 imply that the expectation in (5.14) is bounded by a finite constant that is free of n. Hence, (5.14) is $O(n^{-1/2})$, which completes the proof. □

Lemma 5.3. *Let $w_i = w(Z_{[i/2]}, \varepsilon_i) = b(Z_{[i/2]}, \varepsilon_i)\boldsymbol{\lambda}^{\mathsf{T}}\boldsymbol{X}_{[i/2]}$, where $\boldsymbol{X}_{[i/2]}^{\mathsf{T}} = (1, Z_{[i/2]})$, and $d_{ni} = I\left(\varepsilon_i \le \boldsymbol{X}_{[i/2]}^{\mathsf{T}}\boldsymbol{\Delta}n^{-1/2}\right) - I\left(\varepsilon_i \le 0\right)$. Furthermore, suppose $i < j$ and that $(Z_{[i/2]}, \varepsilon_i)^{\mathsf{T}}$ and $(Z_{[j/2]}, \varepsilon_j)^{\mathsf{T}}$ are such that all random variables are independent except for $Z_{[i/2]}$ and $Z_{[j/2]}$. Then, assuming F1, F2, F4, B1 and B2, we have the* *following:*

$$|COV\left[w_i d_{ni}, w_j d_{nj}\right]| \le n^{-1} K \lambda^{g_i+g_j-2} + o(n^{-1}),$$

where $K \in (0,\infty)$, $\lambda \in (0,1)$ *and the remainder term are all free of i and j and* g_i *and* g_j *denote the number of generations from the nearest common ancestor of* $Z_{[i/2]}$ *and* $Z_{[j/2]}$, *respectively.*

Proof. To begin, recall from Lemma 5.1 that

$$\mu_{ni} \stackrel{\text{def}}{=} E\left[w_i d_{ni}\right] \;=\; n^{-1/2} E\left[b(Z_1, \xi_n) f(\xi_n)(\boldsymbol{\lambda}^{\mathsf{T}} \boldsymbol{X}_1 \boldsymbol{X}_1^{\mathsf{T}} \boldsymbol{\Delta})\right].$$

Now, upon conditioning on $Z_{[i/2]}$ and $Z_{[j/2]}$ and then exploiting the independence relationships we obtain the following

$$\begin{aligned} &COV\left[w_i d_{ni}, w_j d_{nj}\right] \\ &= E\left[E\left[(w_i d_{ni} - \mu_{ni})(w_j d_{nj} - \mu_{nj}) \mid Z_{[i/2]}, Z_{[j/2]}\right]\right] \\ &= E\left[E\left[w_i d_{ni} - \mu_{ni} \mid Z_{[i/2]}, Z_{[j/2]}\right] E\left[w_j d_{nj} - \mu_{nj} \mid Z_{[i/2]}, Z_{[j/2]}\right]\right]. \end{aligned} \tag{5.15}$$

Furthermore, recall from the proof of Lemma 5.1 that

$$E\left[w_i d_{ni} \mid Z_{[i/2]}, Z_{[j/2]}\right] = n^{-1/2} b(Z_{[i/2]}, \xi_{ni}) f(\xi_{ni})(\boldsymbol{\lambda}^{\mathsf{T}} \boldsymbol{X}_{[i/2]} \boldsymbol{X}_{[i/2]}^{\mathsf{T}} \boldsymbol{\Delta}). \tag{5.16}$$

Applying (5.16) to the two conditional expectations given in (5.15) yields the following

$$COV\left[w_i d_{ni}, w_j d_{nj}\right] = n^{-1} COV\left[Q_i(\xi_{ni}) R(Z_{[i/2]}), Q_j(\xi_{nj}) R(Z_{[j/2]})\right],$$

where $Q_i(\xi) = b(Z_{[i/2]}, \xi) f(\xi)$ and $R(Z_{[i/2]}) = \boldsymbol{\lambda}^{\mathsf{T}} \boldsymbol{X}_{[i/2]} \boldsymbol{X}_{[i/2]}^{\mathsf{T}} \boldsymbol{\Delta}$. Now appropriately add in and subtract out $Q_i(0)$ and $Q_j(0)$ so that

$$COV\left[Q_i(\xi_{ni}) R(Z_{[i/2]}), Q_j(\xi_{nj}) R(Z_{[j/2]})\right] = \sum_{k=1}^{4} C_{nij}(k),$$

where the four covariance terms are given by

$$\begin{aligned} C_{nij}(1) &= COV\left[(Q_i(\xi_{ni}) - Q_i(0)) R(Z_{[i/2]}), (Q_j(\xi_{nj}) - Q_j(0)) R(Z_{[j/2]})\right], \\ C_{nij}(2) &= COV\left[(Q_i(\xi_{ni}) - Q_i(0)) R(Z_{[i/2]}), Q_j(0) R(Z_{[j/2]})\right], \\ C_{nij}(3) &= COV\left[Q_i(0) R(Z_{[i/2]}), (Q_j(\xi_{nj}) - Q_j(0)) R(Z_{[j/2]})\right] \quad \text{and} \\ C_{nij}(4) &= COV\left[Q_i(0) R(Z_{[i/2]}), Q_j(0) R(Z_{[j/2]})\right]. \end{aligned}$$

Consider $C_{nij}(1)$ first. Since the distribution of $Z_{[i/2]}$ is the same for all i and $C_{nij}(1)$ is a covariance term it follows that

$$|C_{nij}(1)| \leq V\left[(Q_i(\xi_{ni}) - Q_i(0))R(Z_{[i/2]})\right]$$
$$\leq E\left[(b(Z_{[i/2]}, \xi_{ni})f(\xi_{ni}) - b(Z_{[i/2]}, 0)f(0))^2(\boldsymbol{\lambda}^{\mathsf{T}}\boldsymbol{X}_{[i/2]}\boldsymbol{X}_{[i/2]}^{\mathsf{T}}\boldsymbol{\Delta})^2\right]. \quad (5.17)$$

Now, using an argument similar to that used for E_{n2} in Lemma 5.1 it can be shown that (5.17) is $o(1)$. Hence, $C_{nij}(1) = o(1)$. Next, consider $C_{nij}(2)$ and $C_{nij}(3)$. In doing so, note that the bounds given in F2, F4 and B1 imply that $V\left[Q_i(0)R(Z_{[i/2]})\right] < \infty$. Hence, an argument similar to that in (5.17) implies that $C_{nij}(2) = o(1)$ and $C_{nij}(3) = o(1)$. Finally, consider $C_{nij}(4)$ and note that this term does not actually depend on n. It follows that $|C_{nij}(4)| \leq f^2(0)|COV\left[h(Z_{[i/2]}), h(Z_{[j/2]})\right]|$, where $h(Z_{[i/2]}) = b(Z_{[i/2]}, 0)(\boldsymbol{\lambda}^{\mathsf{T}}\boldsymbol{X}_{[i/2]}\boldsymbol{X}_{[i/2]}^{\mathsf{T}}\boldsymbol{\Delta})$. Moreover, it follows from B1 and B2 that

$$|h(x) - h(y)| \leq K|x - y|\left(\sum_{i=0}^{1}|x|^i + \sum_{i=0}^{2}|y|^i\right), \quad (5.18)$$

where K is a constant that only depends on $\boldsymbol{\lambda}$, $\boldsymbol{\Delta}$ and the bounds given in B1 and B2. Note that (5.18) satisfies (5.4). In fact, it can be shown that (5.18) along with F4 are also sufficient for Lemma 3.1 of Terpstra and Elbayoumi (2012). Hence, upon making these modifications, Lemma 3.1 of Terpstra and Elbayoumi (2012) implies that $|C_{nij}(4)| \leq f^2(0)K\lambda^{g_i+g_j-2}$. Finally, combining the results for $C_{nij}(k)$, $k = 1, 2, 3, 4$, leads to the following: $|COV\left[w_i d_{ni}, w_j d_{nj}\right]| \leq n^{-1}K\lambda^{g_i+g_j-2} + o(n^{-1})$, which completes the proof. □

Lemma 5.4. *Define $b_t(\boldsymbol{\theta}) = b(Z_{[t/2]}, \varepsilon_t; \boldsymbol{\theta})$. Furthermore, let $\boldsymbol{S}_n^*(\boldsymbol{\Delta})$ denote the negative of the gradient vector evaluated with $b_t(\hat{\boldsymbol{\theta}})$ and let $\boldsymbol{S}_n(\boldsymbol{\Delta})$ be defined analogously with $b_t(\boldsymbol{\theta}_0)$, where $\boldsymbol{\theta}_0$ represents the true $(p \times 1)$ nuisance parameter. Then,*

I. $\|(\boldsymbol{S}_n^*(\boldsymbol{\Delta}) - \boldsymbol{S}_n^*(\boldsymbol{0})) - (\boldsymbol{S}_n(\boldsymbol{\Delta}) - \boldsymbol{S}_n(\boldsymbol{0}))\| = o_p(1)$ *and*
II. $\|\boldsymbol{S}_n^*(\boldsymbol{0}) - \boldsymbol{S}_n(\boldsymbol{0})\| = o_p(1)$

provided the following conditions are satisfied:

A1. $\sqrt{n}(\hat{\boldsymbol{\theta}} - \boldsymbol{\theta}_0) = \boldsymbol{O}_p(1)$,
A2. $\boldsymbol{D}b(z, \varepsilon; \boldsymbol{\theta}) = \frac{\partial}{\partial \boldsymbol{\theta}} b(z, \varepsilon; \boldsymbol{\theta})$ *exists for all* $(z, \varepsilon)^{\mathsf{T}}$,
A3. $\|\boldsymbol{D}b(z, \varepsilon; \boldsymbol{\theta})\| \leq B$ *for all* $(z, \varepsilon)^{\mathsf{T}}$ *and* $\boldsymbol{\theta}$,
A4. The family of functions $\mathcal{D} = \{\boldsymbol{D}b(z, \varepsilon; \boldsymbol{\theta}) : (z, \varepsilon)^{\mathsf{T}} \in \Re^2\}$ *is equicontinuous at* $\boldsymbol{\theta}_0$,
A5. $E\left[\boldsymbol{D}b(Z_1, \varepsilon_2; \boldsymbol{\theta}_0)sgn(\varepsilon_2) \mid Z_1\right] = \boldsymbol{0}$,
A6. $\|\boldsymbol{D}b(x, \varepsilon_2; \boldsymbol{\theta}_0) - \boldsymbol{D}b(y, \varepsilon_2; \boldsymbol{\theta}_0)\| \leq C(\varepsilon_2)|x - y|$, *where* $E\left[C(\varepsilon_2)\right] < \infty$.

Proof. Consider Part I of the lemma first. Assumption A2 along with the Multivariate Mean Value Theorem given on page 365 of Bartle (1976) implies that

$$\begin{aligned} \boldsymbol{U}_n &= (\boldsymbol{S}_n^*(\boldsymbol{\Delta}) - \boldsymbol{S}_n^*(\boldsymbol{0})) - (\boldsymbol{S}_n(\boldsymbol{\Delta}) - \boldsymbol{S}_n(\boldsymbol{0})) \\ &= -2n^{-1/2}\sum_{t=2}^{n}\left(b_t(\hat{\boldsymbol{\theta}}) - b_t(\boldsymbol{\theta}_0)\right)\boldsymbol{X}_{[t/2]}\left(I\left(\varepsilon_t \le \boldsymbol{X}_{[t/2]}^{\mathsf{T}}\boldsymbol{\Delta}n^{-1/2}\right) - I(\varepsilon_t \le 0)\right) \\ &= -2n^{-1/2}\sum_{t=2}^{n}\boldsymbol{Db}_t^{\mathsf{T}}(\boldsymbol{\theta}_t)\left(\hat{\boldsymbol{\theta}} - \boldsymbol{\theta}_0\right)\boldsymbol{X}_{[t/2]}\left(I\left(\varepsilon_t \le \boldsymbol{X}_{[t/2]}^{\mathsf{T}}\boldsymbol{\Delta}n^{-1/2}\right) - I(\varepsilon_t \le 0)\right), \end{aligned}$$

where $\boldsymbol{Db}_t(\boldsymbol{\theta}) = \boldsymbol{Db}(Z_{[t/2]}, \varepsilon_t; \boldsymbol{\theta})$ and $\boldsymbol{\theta}_t$ belongs to the line segment between $\hat{\boldsymbol{\theta}}$ and $\boldsymbol{\theta}_0$. It now follows from A3 of the lemma that

$$\begin{aligned} \|\boldsymbol{U}_n\| &\le 2B\|\sqrt{n}(\hat{\boldsymbol{\theta}} - \boldsymbol{\theta}_0)\|\left((1/n)\sum_{t=2}^{n}\|\boldsymbol{X}_{[t/2]}\|\left|I\left(\varepsilon_t \le \boldsymbol{X}_{[t/2]}^{\mathsf{T}}\boldsymbol{\Delta}n^{-1/2}\right) - I(\varepsilon_t \le 0)\right|\right) \\ &\le 2B\|\sqrt{n}(\hat{\boldsymbol{\theta}} - \boldsymbol{\theta}_0)\|\left((1/n)\sum_{t=2}^{n}\|\boldsymbol{X}_{[t/2]}\|I\left(|\varepsilon_t| \le |\boldsymbol{X}_{[t/2]}^{\mathsf{T}}\boldsymbol{\Delta}|n^{-1/2}\right)\right) \\ &\stackrel{\text{def}}{=} 2B\|\sqrt{n}(\hat{\boldsymbol{\theta}} - \boldsymbol{\theta}_0)\|\overline{U}_n. \end{aligned}$$

Since A1 of the lemma implies that $2B\|\sqrt{n}(\hat{\boldsymbol{\theta}} - \boldsymbol{\theta}_0)\| = O_p(1)$ we need only show that $\overline{U}_n = o_p(1)$ to complete the proof. That said, we note that $E\left[\|\boldsymbol{X}_1\|^2\right] < \infty$ (implied by F4), the continuity of F at 0 (see F1), and the Dominated Convergence Theorem imply that $E\left[(\|\boldsymbol{X}_1\|I\left(|\varepsilon_2| \le |\boldsymbol{X}_1^{\mathsf{T}}\boldsymbol{\Delta}n^{-1/2}|\right))^2\right] = o(1)$. Finally, since the joint distribution of $(Z_{[t/2]}, \varepsilon_t)^{\mathsf{T}}$ is the same for all t it readily follows that $E\left[\overline{U}_n^2\right] = o(1)$, which completes the proof of Part I of the lemma. Let us now consider Part II of the lemma. Once again, it follows from assumption A2 along with the Multivariate Mean Value Theorem that

$$\begin{aligned} (\boldsymbol{S}_n^*(\boldsymbol{0}) - \boldsymbol{S}_n(\boldsymbol{0})) &= n^{-1/2}\sum_{t=2}^{n}\left(b_t(\hat{\boldsymbol{\theta}}) - b_t(\boldsymbol{\theta}_0)\right)\boldsymbol{X}_{[t/2]}\mathrm{sgn}(\varepsilon_t) \\ &= n^{-1/2}\sum_{t=2}^{n}\boldsymbol{Db}_t^{\mathsf{T}}(\boldsymbol{\theta}_t)\left(\hat{\boldsymbol{\theta}} - \boldsymbol{\theta}_0\right)\boldsymbol{X}_{[t/2]}\mathrm{sgn}(\varepsilon_t) \\ &= (1/n)\sum_{t=2}^{n}(\boldsymbol{Db}_t(\boldsymbol{\theta}_t) - \boldsymbol{Db}_t(\boldsymbol{\theta}_0))^{\mathsf{T}}\sqrt{n}\left(\hat{\boldsymbol{\theta}} - \boldsymbol{\theta}_0\right)\boldsymbol{X}_{[t/2]}\mathrm{sgn}(\varepsilon_t) \\ &+ \left((1/n)\sum_{t=2}^{n}\boldsymbol{X}_{[t/2]}\boldsymbol{Db}_t(\boldsymbol{\theta}_0)^{\mathsf{T}}\mathrm{sgn}(\varepsilon_t)\right)\sqrt{n}\left(\hat{\boldsymbol{\theta}} - \boldsymbol{\theta}_0\right) \\ &\stackrel{\text{def}}{=} \boldsymbol{U}_{n1} + \boldsymbol{U}_{n2}. \end{aligned}$$

Consider $\boldsymbol{U}_{n1}$ first. Since $\text{sgn}(\varepsilon_t)$ is bounded by 1, it follows that

$$\|\boldsymbol{U}_{n1}\| \leq \max_{2\leq t\leq n} \|\boldsymbol{Db}_t(\boldsymbol{\theta}_t) - \boldsymbol{Db}_t(\boldsymbol{\theta}_0)\| \left((1/n) \sum_{t=2}^{n} \|\boldsymbol{X}_{[t/2]}\| \right) \|\sqrt{n}\left(\hat{\boldsymbol{\theta}} - \boldsymbol{\theta}_0\right)\|.$$

Now, A1 along with A4 imply that $\max_{2\leq t\leq n} \|\boldsymbol{Db}_t(\boldsymbol{\theta}_t) - \boldsymbol{Db}_t(\boldsymbol{\theta}_0)\| = o_p(1)$. Next, note that $\|\boldsymbol{X}_{[t/2]}\| = h(Z_{[t/2]})$, where $h(z) = \sqrt{1+z^2}$ is a Lipschitz function. Hence, by Theorem 5.1, it follows that $(1/n)\sum_{t=2}^{n} \|\boldsymbol{X}_{[t/2]}\| = E\left[\|\boldsymbol{X}_1\|\right] + o_p(1) = O_p(1)$ (e.g. $a_1 = a_2 = 0$, $w_{00} = 1$, and $\gamma = 2$). Lastly, A1 implies that $\|\sqrt{n}\left(\hat{\boldsymbol{\theta}} - \boldsymbol{\theta}_0\right)\| = O_p(1)$. It follows from these results that $\boldsymbol{U}_{n1} = o_p(1)$. Thus, to complete the proof we need only show that $\boldsymbol{U}_{n2} = o_p(1)$. However, given A1, this will follow if we can show the following result

$$\boldsymbol{U}_{n2}^{ij} = \frac{1}{n}\sum_{t=2}^{n} Z_{[t/2]}^{i} \boldsymbol{Db}_{tj}(\boldsymbol{\theta}_0)\text{sgn}(\varepsilon_t) = o_p(1),$$

where $\boldsymbol{Db}_{tj}(\boldsymbol{\theta}_0)$ denotes the jth component of $\boldsymbol{Db}_t(\boldsymbol{\theta}_0)$, $j = 1, 2, \ldots, p$, and $i = 0, 1$. To begin, fix both i and j and let $h_1(Z_1, \varepsilon_2) = Z_1^i \boldsymbol{Db}_{2j}(\boldsymbol{\theta}_0)\text{sgn}(\varepsilon_2)$. The notation for i and j has been suppressed for the sake of convenience. Clearly, assumption A5 implies that $E\left[h_1(Z_1, \varepsilon_2)\right] = 0$. Moreover, the bound for $\text{sgn}(\varepsilon_t)$, assumption A3, and the finite second moment for the process (see F4) imply that $E\left[|h_1(Z_1, \varepsilon_2)|^{1+\delta}\right] < \infty$ for any $\delta \in (0, 1]$. Lastly, it follows from assumptions A3 and A6 that

$$|h_1(x, \varepsilon_2) - h_1(y, \varepsilon_2)| \leq \max\{B, C(\varepsilon_2)\}|x - y|(1 + |y|)$$

for all i and j. These three results imply that the special case of Theorem 5.2 given in (5.7) (with $a_1 = 0$, $a_2 = 1$, $w_{00} = 1$, $w_{01} = 1$, and $\gamma = 2$) is applicable. In particular, $(1/n)\sum_{t=2}^{n} h_1(Z_{[t/2]}, \varepsilon_t) = \boldsymbol{U}_{n2}^{ij} = o_p(1)$ for all i and j. This completes the proof of Part II and the lemma. □

References

Bartle, R. G. (1976). *The elements of real analysis* (2nd ed.). New York/London/Sydney: Wiley.

Basawa, I.V., & Zhou, J. (2004). Non-Gaussian bifurcating models and quasi-likelihood estimation. *Journal of Applied Probability, 41A*, 55–64.

Bui, Q. M., & Huggins, R. M. (1998). Robust inference for the bivariate bifurcating autoregressive model. *Australian & New Zealand Journal of Statistics, 40*(2), 151–163.

Hall, P., & Heyde, C. C. (1980). *Martingale limit theory and its application*. New York: Academic.

Hamilton, J. D. (1994). *Time series analysis*. Princeton, NJ: Princeton University Press.

Heiler, S., & Willers, R. (1988). Asymptotic normality of R-estimates in the linear model. *Statistics, 19*(2), 173–184.

Hettmansperger, T. P., & McKean, J. W. (2011). Robust nonparametric statistical methods. In *Monographs on statistics and applied probability* (Vol. 119, 2nd ed.). Boca Raton, FL: CRC Press.
Huggins, R. M. (1996). Robust inference for variance components models for single trees of cell lineage data. *The Annals of Statistics, 24*(3), 1145–1160.
Huggins, R. M., & Marschner, I. C. (1991). Robust analysis of the bifurcating autoregressive model in cell lineage studies. *The Australian Journal of Statistics, 33*(2), 209–220.
Jaeckel, L. A. (1972). Estimating regression coefficients by minimizing the dispersion of the residuals. *The Annals of Mathematical Statistics, 43*(5), 1449–1458.
Jurečková, J., & Sen, P. K. (1996). *Robust statistical procedures*. Wiley series in probability and statistics: Applied probability and statistics. New York: Wiley. *Asymptotics and interrelations*. A Wiley-Interscience Publication.
Marschner, I. C. (1992). Robust methods for dependent cell pedigree data. *The Australian Journal of Statistics, 34*(2), 181–198.
Terpstra, J. T., & Elbayoumi, T. (2012). A law of large numbers result for a bifurcating process with an infinite moving average representation. *Statistics & Probability Letters, 82*(1), 123–129.
Terpstra, J. T., McKean, J. W., & Naranjo, J. D. (2000). Highly efficient weighted Wilcoxon estimates for autoregression. *Statistics, 35*(1), 45–80.
Terpstra, J. T., McKean, J. W., & Naranjo, J. D. (2001). Weighted Wilcoxon estimates for autoregression. *Australian & New Zealand Journal of Statistics, 43*(4), 399–419.
Zhou, J., & Basawa, I. V. (2005). Least-squares estimation for bifurcating autoregressive processes. *Statistics & Probability Letters, 74*(1), 77–88.

Chapter 6
Applications of Robust Regression to "Big" Data Problems

Simon J. Sheather

Abstract Robust regression methods have many potential applications in big data problems. In this paper, we consider two such applications using publicly available data. The first application looks at modeling taxi fares based on the trip distance of $n = 49,800$ taxi rides in New York City on Tuesday January 15, 2013. The second application focuses on modeling the airfare from the miles flown of $n = 78,905$ round trip itineraries for single passengers which consisted of 2 direct one-way flights within the contiguous domestic US market on Southwest Airlines in the fourth quarter of 2014. The robust estimates were obtained for both applications using PROC ROBUSTREG in SAS 9.4. In both cases, we find that the confidence intervals around the robust estimates of the parameters in the regression models are very narrow, typically \$0.01 or lower. With these confidence intervals being so narrow, one is left with the impression that these robust estimates differ in some meaningful way across at least some of the robust methods. Finally, utilizing findings in Cox (Biometrika, 102:712–716, 2015) we argue that in such applications it is not surprising that the confidence intervals around the robust estimates are very narrow, thus producing the illusion of apparently very high precision.

Keywords Big data • Robust regression • SAS • Illusion of very high precision

6.1 Introduction

In recent times there has been an explosion of interest in big data. An advanced search of books available on Amazon.com published just in 2014 using the search terms "big data" returned a list of 1,657 books.

Mayer-Schönberger and Cukier (2013) provide many examples of the use of big data in many fields including banking, energy, finance, government, healthcare, retail, sports and travel. According to Mayer-Schönberger and Cukier (2013, p. 26):

S.J. Sheather (✉)
Department of Statistics, Texas A&M University, College Station, TX, USA
e-mail: sheather@stat.tamu.edu

© Springer International Publishing Switzerland 2016

R.Y. Liu, J.W. McKean (eds.), *Robust Rank-Based and Nonparametric Methods*, Springer Proceedings in Mathematics & Statistics 168,
DOI 10.1007/978-3-319-39065-9_6

In many areas, ... a shift is taking place from collecting some data to gathering as much as possible, and if feasible getting everything: N = all.

In this paper we consider two applications of robust regression methods in big data settings. In both applications we report the values of the following robust regression estimates and associated 95 % confidence intervals obtained from PROC ROBUSTREG in SAS 9.4 (with each method based on the default settings) for an M-estimate, the LTS estimate, the so-called LTS FWLS estimate, an S-estimate and an MM-estimate. Details of these five robust estimation methods are provided in Sect. 6.2.

In Sect. 6.3, we consider the first application which looks at modeling taxi fares from the trip distance of $n = 49,800$ taxi rides in New York City on Tuesday January 15, 2013. The second application, detailed in Sect. 6.4, focuses on modeling the airfare from the miles flown for $n = 78,905$ round trip itineraries for single passengers which consisted of two direct one-way flights within the contiguous domestic US market on Southwest Airlines in the fourth quarter of 2014. In both cases, we find that the confidence intervals around the robust estimates of the parameters in the regression models are very narrow, typically \$0.01 or lower. With these confidence intervals being so narrow, one is left with the impression that these robust estimates differ in some meaningful way across at least some of the robust methods.

Finally, in Sect. 6.5, utilizing findings in Cox (2015) we argue that it is not surprising that the confidence intervals around the robust estimates are very narrow, thus producing the illusion of what Cox refers to as "apparently very high precision".

6.2 Robust Regression Methods

Let $X = (x_{ij})$ denote an $n \times p$ matrix of predictors, with the first column typically consisting of 1's, $y = (y_1, y_2, \ldots y_n)^T$denote an $n \times 1$ vector of responses, $e = (e_1, e_2, \ldots e_n)^T$denote an $n \times 1$ vector of errors and $\theta = (\theta_1, \theta_2, \ldots \theta_p)^T$denote an $p \times 1$ vector of unknown regression parameters. In this section, we briefly review robust estimates of θ for the linear model

$$y = X\,\theta + e.$$

Let $r = (r_1, r_2 \ldots r_n)^T$denote an $n \times 1$ vector of residuals. Let x_i^T denote the ith row of X. In what follows, we mostly follow the approach taken by Chen (2002).

6.2.1 M-Estimates

An M-estimate $\hat{\theta}_M$ of θ (Huber 1973) minimizes the following sum

$$Q_M(\theta) = \sum_{i=1}^{n} \rho\left(\frac{r_i}{\sigma}\right).$$

For least squares estimation, ρ is the square function, $\rho(z) = z^2$. If σ is known, then taking derivatives with respect to θ, $\hat{\theta}_M$ is also a solution of the following system of p equations

$$\sum_{i=1}^{n} \psi\left(\frac{r_i}{\sigma}\right) x_{ij} = 0 \;\; j = 1, \ldots, p,$$

where $\psi = \rho'$, the derivative of ρ. If ρ is convex, $\hat{\theta}_M$ is the unique solution.

PROC ROBUSTREG solves this system by using iteratively reweighted least squares (IRLS). The weight function $w(x)$ is defined by

$$w(x) = \frac{\psi(x)}{x}$$

PROC ROBUSTREG provides ten kinds of weight functions (corresponding to ten ψ-functions) through the WEIGHTFUNCTION = option in the MODEL statement. The default weight function is the bisquare function, which is given by

$$w(x, c) = \begin{cases} \frac{\sin\left(\frac{x}{c}\right)}{\frac{x}{c}}, & |x| \le \pi c \\ 0, & |x| > \pi c \end{cases}$$

The default value of the tuning constant, $c = 4.685$, was chosen to yield 95 % asymptotic efficiency of the resulting bisquare M-estimate when the errors follow a the Gaussian distribution.

If σ is unknown, then the function

$$Q_M(\theta) = \sum_{i=1}^{n} \left[\rho\left(\frac{r_i}{\sigma}\right) + a\right]\sigma$$

is minimized with $a > 0$ over θ and σ by alternatively improving $\hat{\theta}_M$ in a location step and $\hat{\sigma}_M$in a scale step. The scale parameter σ can be specified using the SCALE = option in the PROC statement. PROC ROBUSTREG provides three options to estimate scale. The default scale is the median absolute deviation (MAD) of the residuals divided by 0.6745.

6.2.2 LTS Estimate

The least trimmed squares (LTS) estimate $\hat{\theta}_{\text{LTS}}$ of θ (Rousseeuw 1984) minimizes the following sum

$$Q_{\text{LTS}}(\theta) = \sum_{i=1}^{h} r_{(i)}^2$$

where $r^2_{(1)} \leq r^2_{(2)} \leq \ldots \leq r^2_{(n)}$ are the ordered squared residuals and h is defined in the range $\frac{n}{2}+1 \leq h \leq \frac{3n+p+1}{2}$. The breakdown of the LTS estimate is equal to $\frac{n-h}{n}$. In PROC ROBUSTREG the default value of h is $h = [\frac{3n+p+1}{4}]$. The ROBUSTREG procedure uses the FAST-LTS algorithm that was proposed by Rousseeuw and Van Driessen (2000).

The FWLS-LTS estimate is a weighted least squares estimate, with weight 0 assigned to cases that the LTS-estimate identifies as outliers. The FWLS-LTS estimate is equivalent to the least squares estimate after the detected outliers are deleted.

6.2.3 S Estimate

The S estimate $\hat{\theta}_S$ of θ (Rousseeuw and Yohai 1984) minimizes the dispersion $S(\theta)$ where $S(\theta)$ is the solution of

$$\frac{1}{n-p}\sum_{i=1}^{n}\chi\left(\frac{y_i - x_i^T\theta}{S}\right) = \beta$$

where $\beta = \int \chi(s)\, d\Phi(s)$ so that $\hat{\theta}_S$ and $S(\hat{\theta}_S)$ are asymptotically consistent estimates of θ and σ for the Gaussian regression model. The breakdown value of the S estimate is equal to $\beta / \sup_s \chi(s)$.

PROC ROBUSTREG provides two kinds of weight functions χ through the CHIF = option in the MODEL statement. The default weight function is the Tukey function which is given by

$$\chi_{k_0}(s) = \begin{cases} 3\left(\frac{s}{k_0}\right)^2 - 3\left(\frac{s}{k_0}\right)^4 + \left(\frac{s}{k_0}\right)^6, & |s| \leq k_0 \\ 1, & |s| > k_0 \end{cases}$$

The default $k_0 = 2.9366$ and is such that the breakdown value of the S estimate is 0.25, with a corresponding asymptotic efficiency for the Gaussian model of 75.9 %.

6.2.4 MM Estimate

The MM estimate $\hat{\theta}_{MM}$ of θ (Yohai 1987) is based on a combination of the use of high breakdown estimation and efficient estimation procedures. It is based on the following three steps:

1. Compute an initial (consistent) high breakdown value estimate $\hat{\theta}'$. PROC ROBUSTREG uses the LTS estimate because of its speed, efficiency, and high breakdown value.
2. Find a scale estimate $\hat{\sigma}'$ such that

$$\frac{1}{n-p}\sum_{i=1}^{n}\chi\left(\frac{y_i - x_i^T\theta}{S}\right) = \beta$$

   where $\beta = \int \chi(s)\, d\Phi(s)$. PROC ROBUSTREG uses the Tukey χ function (given above) with $k_0 = 2.9366$ as the default, which produces a scale estimate with a breakdown value of 0.25.
3. Find a local minimum $\hat{\theta}_{\text{MM}}$ of

$$Q_{\text{MM}} = \sum_{i=1}^{n}\rho\left(\frac{y_i - x_i^T\theta}{\hat{\sigma}'}\right)$$

   such that $Q_{\text{MM}}\left(\hat{\theta}_{\text{MM}}\right) \le Q_{\text{MM}}\left(\hat{\theta}'\right)$. PROC ROBUSTREG uses the Tukey χ function (given above) with $k_0 = 3.440$ as the default choice for ρ, which produces an MM estimate with 85 % asymptotic efficiency for the Gaussian model.

In this paper for two "big" data regression problems we report the values of the following robust regression estimates and associated 95 % confidence intervals obtained from PROC ROBUSTREG in SAS 9.4 (with each method based on the default settings):

1. M-estimate with bisquare weight function and scale estimate based on the median absolute deviation of the residuals
2. LTS estimate
3. LTS FWLS estimate
4. S-estimate with the Tukey χ function
5. MM estimate with an LTS initial estimate and Tukey χ function

6.3 New York City Taxi Fare Data

On March 11 2014, Chris Whong lodged via email a FOIL (The Freedom of Information Law) request with the New York City Taxi & Limousine Commission (TLC) seeking "NYC taxi trip data in machine readable format from 1/1/2013 through the most current available date". Seven days later a representative from Legal Affairs at TLC advised that the "request for GPS data has been granted" and that they had "2013 data available from January to December". Mr. Whong was asked to "send or bring an external hard drive with a minimum capacity of 200 GB to the TLC offices". Mr. Whong was further advised that "the hard drive must be brand new, still in the box unopened". After following these instructions, Mr. Whong retrieved the drive from the TLC and found that "the data they had loaded only took up about 50 GB". The data were provided in two folders, "Faredata_2013 and Tripdata_2013" (Whong 2014). Each folder contained monthly files (in csv format) giving details of each NYC taxi ride during 2013. The fare data files include the following fields:

medallion—anonymized taxi id
hack_license—anonymized taxi driver id
vender_id—system used to collect the data, either CMT (Mobile Knowledge Systems Inc) or VTS (Verifone Transportation Systems)
pickup_datetime—date and time taxi trip started
payment_type—CRD (Credit card), CSH (cash), DIS, NOC or UNK (the last 3 codes are unexplained and they make up less than 1 % of the data)
fare_amount—the metered fare
surcharge—fare surcharges of $0.5 for trips between 8 pm and 6 am and $1 for trips between 4 pm and 8 pm on weekdays (excluding holidays)
mta_tax—MTA tax of $0.5 for all trips that end in New York City or Nassau, Suffolk, Westchester, Rockland, Dutchess, Orange or Putnam Counties
tip_amount—tip paid by passenger(s)
tolls_amount—total charge for all tolls incurred during the trip
total_amount—total amount paid by passenger(s).

The trip data files include the following fields:

medallion—anonymized taxi id
hack_license—anonymized taxi driver id
vender_id—system used to collect the data, either CMT (Mobile Knowledge Systems Inc.) or VTS (Verifone Transportation Systems)
rate_code—rate_code 1 corresponds to the standard city rate
store_and_fwd_flag—unexplained attribute
pickup_datetime—start time of the trip
dropoff_datetime—end time of the trip
passenger_count—number of passengers
trip_time_in_secs—trip time in seconds measured by the taximeter
trip_distance—trip distance in miles measured by the taximeter
pickup_longitude and pickup_latitude—GPS coordinates at the start of the trip
dropoff_longitude and dropoff_latitude—GPS coordinates at the end of the trip

The url www.andresmh.com/nyctaxitrips/ contains the two sets of monthly data files for 2013. The first step in any analysis of this data set is to combine the fare and trip data files. This can be done by matching the following fields in the two data files:

medallion, hack_license and pickup_datetime.

The fare data file and trip data file for January 2013 contain data on 14,776,615 taxi trips, of which 14,456,067 correspond to rate code 1 taxi trips. According to the New York City Taxi and Limousine Commission[1] for rate code 1, the initial charge is \$2.50 plus 50 cents per 1/5 mile or 50 cents per 60 s in slow traffic or when the vehicle is stopped. According to Wikipedia[2] "slow traffic" is defined to be travelling under 12 miles an hour.

In this study we shall focus on data for taxi trips taken on a randomly selected day in January, 2013, namely Tuesday January 15, 2013. In particular, we shall consider 49,800 taxi trips with the following characteristics:

- vender_id = CMT (Mobile Knowledge Systems Inc.)
- payment_type = CRD (Credit card) or CSH (cash)
- rate_code = 1, which corresponds to the standard city rate
- rounded_trip_distance < 3 miles, where the rounding was down to the nearest 1/5 mile
- average_trip_speed ≥ 25 miles per hour

Average trip speed was calculated from the trip distance and the trip time fields in the data. A decision was made to consider trips with average speed greater than or equal to 25 miles per hour so that very little, if any, of the fare_amount of any trip would be due travelling under 12 miles per hour. Secondly, a decision was made to consider trips of less than 3 miles to limit the number of x values the data takes in order to simplify the presentation, especially in Tables 6.1, 6.2 and 6.3.

Donovan and Work (2015) have studied the New York City taxi trip data from 2010 through 2013. This corresponds to nearly 700 million taxi trips. They found that "roughly 7.5 % of all trips" contain "data errors. …For example, there are …trips where the reported meter distances are significantly shorter than the straight line distance, violating Euclidean geometry" as well as "trips … that … contain impossible distances, times, or velocities." As we shall see below, the 49,800 taxi trips that we consider contain some data errors or outliers, which have not been removed or altered in any way.

Figure 6.1 shows a box plot of fare_amount for each value of RoundedTripDistance. Examining Fig. 6.1 we see examples of the "data errors" reported by Donovan and Work (2015). For example, there is a trip of 0.6 miles with fare_amount equal to \$52 and another trip of 0.8 miles with fare_amount equal to \$40.

In Fig. 6.1, there are typically a large number of multiple values of the same fare_amount for each value of RoundedTripDistance making Fig. 6.1 difficult to interpret. Jittering the data does not work well given that the sample size of 49,800 is so large. Instead we shall consider Table 6.1, which provides a cross tabulation of the values of rounded_trip_distance (denoted by x) and corresponding values of fare_amount. The column headed N gives the number of times that x, rounded_trip_distance takes each value from 0 to 2.8. For example,

[1] Web site: www.nyc.gov/html/tlc/html/passenger/taxicab_rate.shtml.

[2] Web site: https://en.wikipedia.org/wiki/Taxicabs_of_New_York_City.

Table 6.1 Cross tabulation of the values of rounded_trip_distance (denoted by x) and corresponding values of fare_amount

x	N	\$2.5	\$3.0	\$3.5	\$4.0	\$4.5	\$5.0	\$5.5	\$6.0	\$6.5	\$7.0	\$7.5	\$8.0	\$8.5	\$9.0	\$9.5	>\$9.5
0	242	240	1	0	0	0	0	0	0	0	0	0	0	0	0	0	1
0.2	830	301	511	11	7	0	0	0	0	0	0	0	0	0	0	0	0
0.4	2847	313	301	2167	45	14	3	3	0	0	0	0	0	0	1	0	0
0.6	4587	379	118	491	3462	96	23	8	5	1	2	1	0	0	0	0	1
0.8	5245	290	160	3	475	4109	132	30	20	14	4	3	3	0	1	0	1
1	4966	212	75	17	0	399	3979	182	42	23	15	8	0	8	5	0	1
1.2	4357	142	36	8	3	0	265	3609	201	32	32	7	8	7	2	2	3
1.4	4000	134	26	8	4	3	0	251	3173	300	36	24	12	9	8	6	6
1.6	3779	131	19	10	7	6	2	3	189	2836	468	30	20	23	13	7	15
1.8	3584	111	22	7	5	8	7	4	6	141	2573	582	40	18	18	11	31
2	3229	92	26	4	2	5	5	2	0	7	122	2214	649	46	14	10	31
2.2	2987	72	16	2	4	12	12	4	3	1	1	72	1934	734	60	12	48
2.4	2986	53	16	5	2	3	10	11	7	1	0	2	72	1700	939	105	60
2.6	3130	62	10	6	1	5	9	7	8	4	1	2	1	59	1686	1089	180
2.8	3031	54	11	6	6	10	3	10	9	1	2	1	1	4	43	1410	1460

Table 6.2 Median(fare_amount), [MFA], for each value of rounded_trip_distance, [rtd], x

rtd, x	0	0.2	0.4	0.6	0.8	1.0	1.2	1.4	1.6	1.8	2.0	2.2	2.4	2.6	2.8
MFA	$2.5	$3.0	$3.5	$4.0	$4.5	$5.0	$5.5	$6.0	$6.5	$7.0	$7.5	$8.0	$8.5	$9.0	$9.5

when $x = 0.2$, $N = 830$ which implies that there are 830 trips in the data with rounded_trip_distance equal to 0.2 miles. The other entries in Table 6.1 give the number of times that a given pair of values for rounded_trip_distance and fare_amount occur. For example, there are 2167 trips in the data with rounded_trip_distance equal to 0.4 miles with a fare_amount equal to $3.50.

Examining Table 6.1 we see further examples of the "data errors" reported by Donovan and Work (2015). For example, there are 54 trips in the data of 2.8 miles for which the fare_amount equals just the initial charge of $2.50.

Table 6.2 gives the median(fare_amount) for each value of rounded_trip_distance. Notice that the median(fare_amount) is a linear function of rounded_trip_distance. In particular, notice that

$$\text{median}\,(\textit{fare\_amount}) = \$2.50 + \$2.50 * \textit{rounded\_trip\_distance} \tag{6.1}$$

This is to be expected since the fare structure is such that the initial charge is $2.50 plus 50 cents per 1/5 mile.

Table 6.3 gives for each value of rounded_trip_distance, the percentage of trips with fares less than, equal to and greater than the median(fare_amount) at each value of x. Over all values of rounded_trip_distance, 71.5 % of the fare are equal to the relevant median(fare_amount), 12.2 % are less than and 16.3 % are greater than.

Figures 6.2 and 6.3 show plots of robust estimates of the intercept and the slope from a straight line regression fit to the taxi fare data. Also included in these plots are 95 % confidence intervals based on each robust estimate. Table 6.4 gives the numerical values of the estimates and confidence limits displayed in Figs. 6.2 and 6.3. Also included in Table 6.4 are these quantities for a robust rank-based (R) estimate, based on Wilcoxon scores (Hettmansperger and McKean 2011). The values for the R-estimates were kindly provided by a referee. According to the referee, the R-estimates were obtained using the software described in Kloke and McKean (2012).

Examining Figs. 6.2 and 6.3 and/or Table 6.4 we see that the M-estimates (based on the bisquare weight function and MAD scale estimate) and the R-estimates (based on Wilcoxon scores) are equal to the values of the intercept and the slope in (1), namely, $2.50. When calculating these M-estimates, SAS produces the following message "WARNING: The scale is close to 0. A possible exact fit is detected." In fact, SAS reports the MAD of the residuals based estimate of scale to be zero to four decimal places. After examining Table 6.3, we see that this is not surprising.

It is also clear from Figs. 6.2 and 6.3 and Table 6.4 that the confidence intervals around the robust estimates of the slope and the intercept are very narrow, typically

Table 6.3 Percentage of fares less than, equal to or greater than the median(fare_amount) at each x

rounded_trip_distance, x	0	0.2	0.4	0.6	0.8	1.0	1.2	1.4	1.6	1.8	2.0	2.2	2.4	2.6	2.8	OA
% < Median(fare_amount)	0	36	22	22	18	14	10	11	10	9	8	7	6	6	5	12
% = Median(fare_amount)	99	62	76	75	78	80	83	79	75	72	69	65	57	54	47	72
% > Median(fare_amount)	1	2	2	3	4	6	7	10	15	20	23	29	37	41	48	16

Over all values of rounded_trip_distance is abbreviated by OA

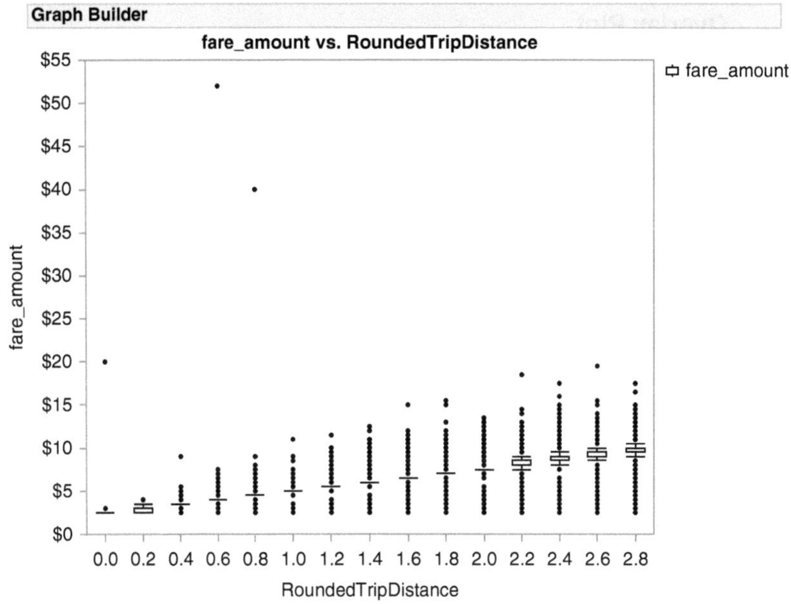

Fig. 6.1 Box plots of fare_amount against RoundedTripDistance

\$0.01 or lower. The narrowness of these confidence intervals seems to be due at least in part to the discrete nature of the Y-variable, fare_amount in addition to phenomena we will discuss in Sect. 6.5. With these confidence intervals being so narrow, one is left with the impression that the robust estimates of the slope and intercept differ in some meaningful way across methods.

6.4 Airline Origin and Destination Survey Ticket Price Data: Southwest Airlines

We next consider data from the Airline Origin and Destination Survey (DB1B). This data base has been studied by many authors. For example, a search on google scholar (www.scholar.google.com) using the following terms:

> "Airline Origin and destination survey (DB1B)" AND regression

returned 160 articles and books. The DB1B is "a 10 % sample of airline tickets from reporting carriers collected by the Office of Airline Information of the Bureau of Transportation Statistics. Data includes origin, destination and other itinerary details of passengers transported. This database is used to determine air traffic patterns, air carrier market shares and passenger flows." The data have been collected quarterly since 1993. Each quarter, the data are reported in three separate tables referred to

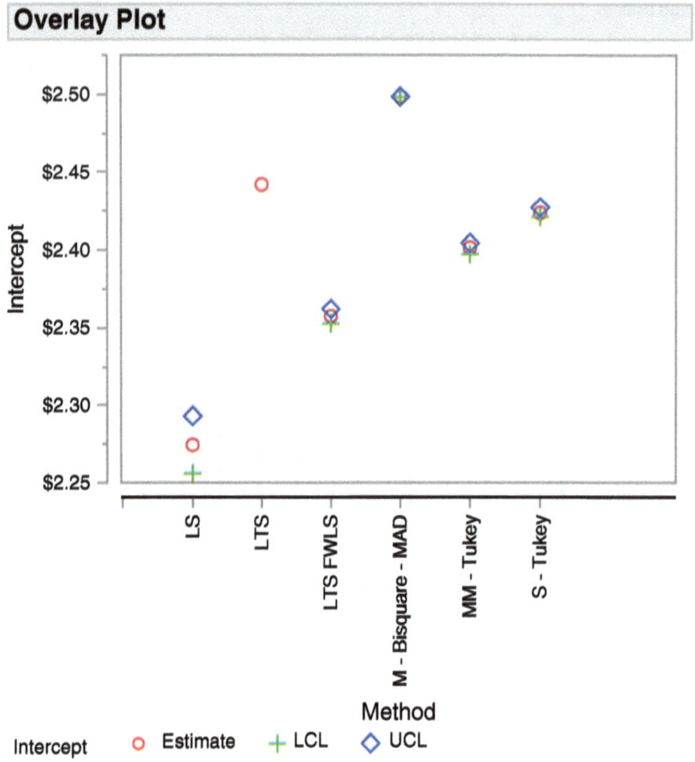

Fig. 6.2 A plot of robust estimates of the intercept along with the associated 95 % confidence intervals

as DB1BCoupon, DB1BMarket and DB1BTicket. In this section we shall consider the DB1BTicket file which contains "summary characteristics of each domestic itinerary on the Origin and Destination Survey, including the reporting carrier, itinerary fare, number of passengers, originating airport, roundtrip indicator, and miles flown".[3]

We shall consider the most recent data, namely, that from Quarter 4, 2014.[4]

The DB1BTicket file contains the following fields:

ItinID—Itinerary ID
Coupons—Number of Coupons in the Itinerary
Year—Year
Quarter—Quarter (1–4)

[3]Source: http://www.transtats.bts.gov/DatabaseInfo.asp?DB_ID=125&DB_Name=Airline%20Origin%20and%20Destination%20Survey%20(DB1B)).

[4]Source: http://www.transtats.bts.gov/DL_SelectFields.asp?Table_ID=272&DB_Short_Name=Origin%20and%20Destination%20Survey.

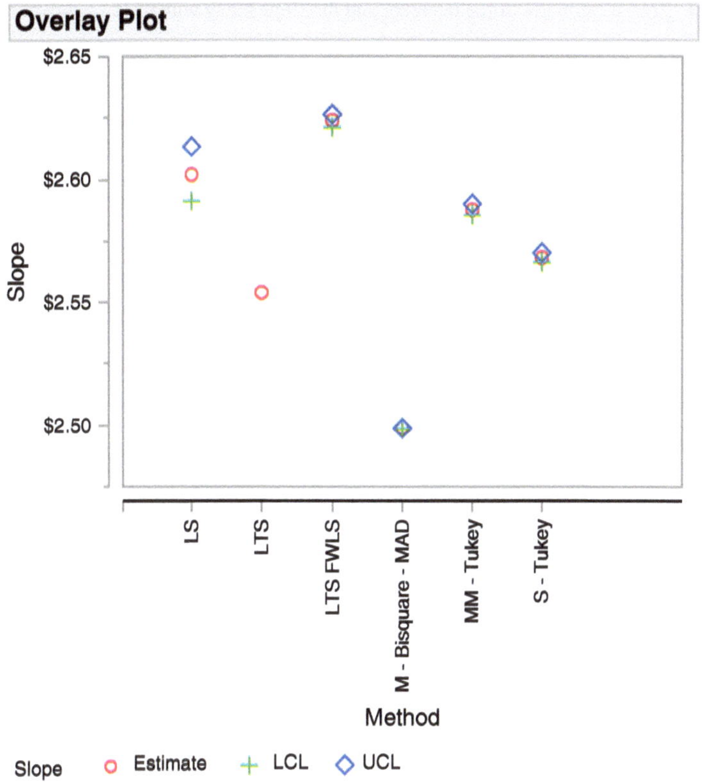

Fig. 6.3 A plot of robust estimates of the slope along with the associated 95 % confidence intervals

Origin—Origin Airport Code
OriginAirportID—An identification number assigned by US DOT to identify a unique airport
OriginAirportSeqID—An identification number assigned by US DOT to identify a unique airport at a given point of time. Airport attributes, such as airport name or coordinates, may change over time.
OriginCityMarketID—An identification number assigned by US DOT to identify a city market.
OriginCountry—Country that the airport is located in
OriginStateFips—Numerical code for the state or territory that the airport is located in
OriginState—Two letter abbreviation for the state or territory that the airport is located in
OriginStateName—Name of the state or territory that the airport is located in
OriginWac—World Area Code for the origin airport
RoundTrip—Round Trip Ticket Indicator (1 = Yes, 0 = No)
OnLine—Single Carrier Indicator (1 = Yes, 0 = No)

Table 6.4 Robust estimates of the intercept and the slope along with 95 % confidence intervals

Method	Intercept Estimate	LCL Intercept	UCL Intercept	Slope Estimate	LCL Slope	UCL Slope	Scale Estimate
LS	2.276	2.258	2.294	2.603	2.592	2.614	0.924
LTS	2.444			2.555			0.182
LTS FWLS	2.359	2.355	2.363	2.625	2.622	2.628	0.205
M-Bisquare-MAD	2.500	2.500	2.500	2.500	2.500	2.500	0.000
MM-Tukey	2.403	2.400	2.406	2.589	2.587	2.591	0.246
S-Tukey	2.425	2.423	2.428	2.569	2.568	2.571	0.245
R-Wilcoxon	2.500	2.500	2.500	2.500	2.500	2.500	0.000

DollarCred—Dollar Credibility Indicator (1 = Fare value is credible, 0 = Fare value is questionable)
FarePerMile—Itinerary Fare per Miles Flown in Dollars (ItinFare/MilesFlown is the calculated value).
RPCarrier—Reporting Carrier
Passengers—Number of Passengers
ItinFare—Itinerary Fare per Person
BulkFare—Bulk Fare Indicator (1 = Yes, 0 = No)
Distance—Itinerary Distance (Including Ground Transport)
DistanceGroup—Distance Group, in 500 Mile Intervals (1 = less than 500 miles, 2 = 500–999 miles, ...)
MilesFlown—Miles flown according to the flight itinerary
ItinGeoType—Itinerary Geography Type (0 = international, 1 = non-contiguous domestic (eg Hawaii and Alaska), 2 = contiguous domestic)

The DB1BTicket file contains data on 3,588,928 flight itineraries involving 7,021,913 passengers. In this paper we shall focus on 78,905 itineraries with the following characteristics:

- RPCarrier = WN (Southwest Airlines)
- RoundTrip = 1
- Coupons = 2
- Passengers = 1
- ItinGeoType = 2
- Online = 1
- DollarCred = 1
- BulkFare = 0

In other words, we shall focus on 78,905 single passenger nonstop round trip flight itineraries on Southwest Airlines in the contiguous domestic market. Our aim was to make the resulting statistical model as simple as possible while still being realistic. For example, Southwest Airlines was chosen since it does not have business or first class fares.

We seek to build a model for ItinFare, the itinerary fare per person from MilesFlown, the miles flown according to the flight itinerary. In this case, the

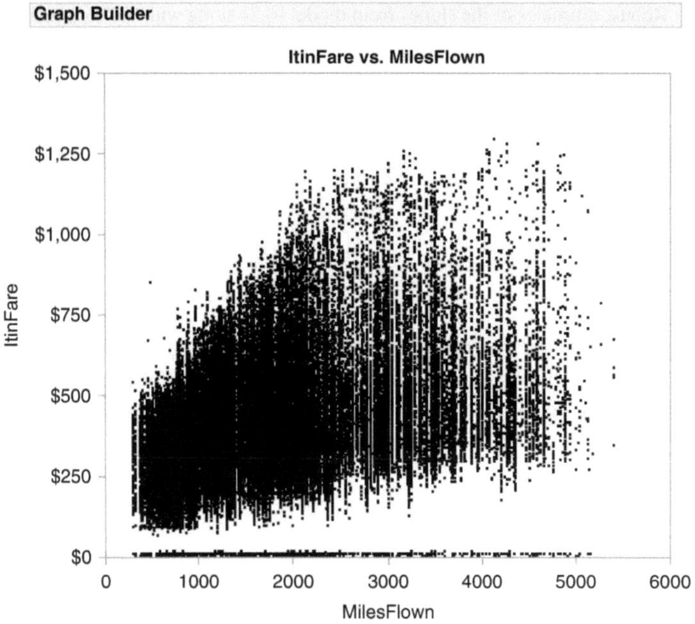

Fig. 6.4 A plot of ItinFare against MilesFlown

MilesFlown equals twice the distance between the cities flown, since we are considering itineraries based on nonstop round trip flights. Figure 6.4 shows a scatter plot of ItinFare against MilesFlown.

Inspecting Fig. 6.4, we see that a number of flight itineraries exist across the range of values taken by MilesFlown with very low values of ItinFare. Examining the data from Fig. 6.4 we find that there are 1150 flight itineraries with ItinFare values less or equal to $20, with 274 flight itineraries with ItinFare equal to $5 and 842 flight itineraries with ItinFare equal to $11. After $20, the next cheapest round trip fare in the data is $66. It is likely that these 1150 flight itineraries correspond to "free" tickets obtained from points in the Southwest Airlines Rapid Rewards frequent flyer program. Presumably these low fare amounts correspond to charges that rewards points do not cover (e.g., pets, extra bag or security fees). Since our focus is on robust regression methods, we will leave these low fares in the data set.

Initial analyses of the data in Fig. 6.4 quickly revealed that a single straight line provides a poor fit to the data. Instead what seems to be necessary are regression splines, in other words, a series of connected straight lines. Use of ADAPTIVEREG procedure in SAS 9.4 (which fits adaptive regression splines using the method developed by Friedman 1991) reveals a knot at around 1500 miles (i.e., 750 miles each way on a round trip). This method combines both regression splines and model selection. It constructs spline basis functions in an adaptive way by automatically selecting appropriate knot values for different variables, and it obtains reduced

Table 6.5 Robust estimates of the slopes from model (6.2) along with 95 % confidence intervals

Method	Slope up to 1500	LCL	UCL	Slope after 1500	LCL	UCL
LS	\$0.1457	\$0.1416	\$0.1497	\$0.0434	\$0.0415	\$0.0452
LTS	\$0.0669			\$0.0216		
LTS FWLS	\$0.1368	\$0.1332	\$0.1404	\$0.0258	\$0.0241	\$0.0275
M—Bisquare—Huber	\$0.1355	\$0.1316	\$0.1394	\$0.0313	\$0.0295	\$0.0330
MM—Tukey	\$0.1254	\$0.1216	\$0.1292	\$0.0242	\$0.0224	\$0.0260
S—Tukey	\$0.1169	\$0.1130	\$0.1208	\$0.0208	\$0.0189	\$0.0226

models by applying model selection techniques. The method does not assume parametric model forms and does not require specification of knot values.

Denote ItinFare by Y and MilesFlown by x. We considered regression spline models of the form

$$Y = \beta_0 + \beta_1 (1500 - x)_- + \beta_2 (x - 1500)_+ \tag{6.2}$$

where

$$(1500 - x)_- = \begin{cases} x - 1500, & x < 1500 \\ 0, & x \geq 1500 \end{cases}$$

and

$$(x - 1500)_+ = \begin{cases} 0, & x < 1500 \\ x - 1500, & x \geq 1500 \end{cases}$$

In what follows, we shall focus our discussion on the estimates of the two slope parameters in model (6.2), since most studies involving the DB1B consider the rate of change of fare with distance.

Table 6.5 gives the values of robust estimates of the slopes from model (6.2) along with 95 % confidence intervals. The first thing that is apparent from Table 6.5 is that the values of the LTS estimate of β_1 differ dramatically from the other robust estimates. In particular, apart from the LTS estimate, the robust estimates of β_1, the slope up to 1500 miles, range from 11.7 cents per mile to 13.7 cents per mile, while the LTS estimate of β_1 is equal to 6.7 cents, just over half the value of the other robust estimates. Such large differences in the estimated rate of change in fare per mile have dramatic practical consequences. On the other hand, the differences in the value of the robust estimates of β_2 compared to the other robust estimates are not as great, since they range from 2.1 cents per mile to 3.1 cents per mile.

Figures 6.5 and 6.6 show plots of robust estimates of the slopes from the regression spline line fit to the airfare data. Also included in these plots are 95 % confidence intervals based on each robust estimate.

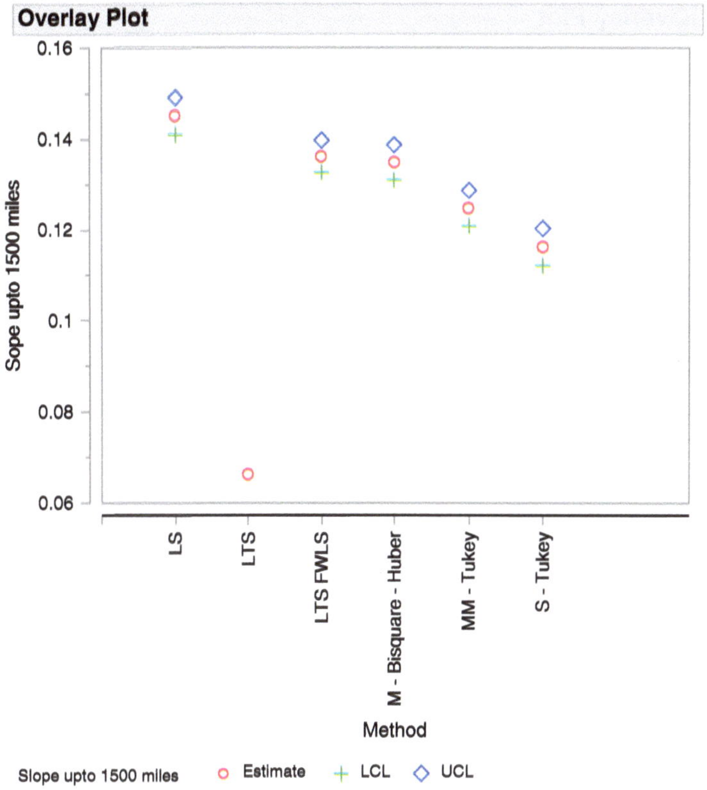

Fig. 6.5 A plot of robust estimates of the slope up to 1500 miles along with the associated 95 % confidence intervals

Examining Figs. 6.5 and 6.6 we see that the confidence intervals around the robust estimates of the slopes in model (6.2) are very narrow, typically \$0.01 or lower. With these confidence intervals being so narrow, one is left with the impression that the robust estimates of the slopes differ in some meaningful way across at least some of the robust methods. We shall examine this issue further in Sect. 6.5.

6.5 Conclusions and Discussion

In the examples considered in Sects. 6.3 and 6.4 the confidence intervals around the robust estimates of regression parameters were very narrow, typically \$0.01 or lower. On face value when these confidence intervals do not overlap, one is left with the impression that the robust estimates differ across methods.

In a recent paper, Cox (2015) finds that

> So-called big data are likely to have complex structure, in particular implying that estimates of precision obtained by applying standard statistical procedures are likely to be misleading.

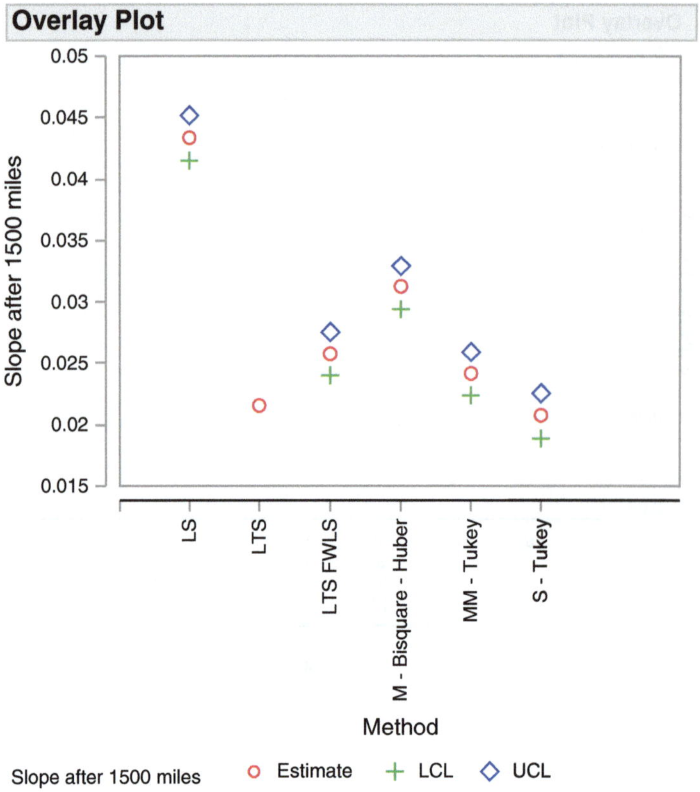

Fig. 6.6 A plot of robust estimates of the slope after 1500 miles along with the associated 95 % confidence intervals

> ... With very large amounts of data, direct use of standard statistical methods ... will tend to produce estimates of apparently very high precision, essentially because of strong explicit or implicit assumptions of at most weak dependence underlying such methods. ... The most serious possibility of misinterpretation arises when the regression coefficient takes very different values in the different base processes.

In addition, Cox (2015) recommends that

We ... "consider big data as evolving in a possibly notional time-frame. At various time-points new sources of variability enter" ... and that we ... "represent the main sources of variation in an explicit model and thereby produce both improved estimates and more relevant assessments of precision".

In the case of the ticket data from the Airline Origin and Destination Survey (DB1B) recall that the data is based on a 10 % sample of airline tickets from reporting carriers. In our example in Sect. 6.4 we considered the airfare and the miles flown of $n = 78,905$ round trip itineraries for single passengers which consisted of 2 direct one-way flights within the contiguous domestic US market on Southwest Airlines in the fourth quarter of 2014. In the analyses presented, no account was taken of the fact that airfares vary across many factors including:

- Time of the day
- Day of the week
- The two airports that the flights are between
- The number of days before the flight during which the ticket was purchased
- How many vacant seats exist on the flight at the time of booking

Thus, it is reasonable to conclude that the regression coefficients in model (6.2) can be expected to take very different values in different combinations of these factors. For example, compare and contrast the airfare for a ticket that is purchased the day of the flight with very few vacant seats at the busiest time of the day between two airports between which there is little competition between carriers the airfare for a ticket that is purchased long before the day of the flight with very many vacant seats at the least busy time of the day between two airports between which there is a great deal of competition between carriers. There is likely to be a very substantial difference between these two airfares. In addition, there is likely to be strong dependence between the airfare of tickets purchased with similar combinations of these factors.

In view of the discussion of the previous paragraph and the findings in Cox (2015) it is not surprising that the confidence intervals around the robust estimates of the slopes in model (6.2) are very narrow, thus producing the illusion of apparently very high precision.

References

Chen, C. (2002). Robust regression and outlier detection with the ROBUSTREG procedure. In *Paper 265, SUGI 27*. Available online http://www2.sas.com/proceedings/sugi27/p265-27.pdf. Cited 24 July 2015.

Cox D. R. (2015). Big data and precision. *Biometrika, 102*, 712–716.

Donovan, B., & Work, D. B. (2015). Using coarse GPS data to quantify city-scale transportation system resilience to extreme events. In *Presented at the transportation research board 94th annual meeting*, Washington, DC, 11–15 January 2015.

Friedman, J. (1991). Multivariate adaptive regression splines. *Annuals of Statistics, 19*, 1–67.

Hettmansperger, T. P., & McKean, J. W. (2011) *Robust nonparametric statistical methods* (2nd ed.). Boca Rotan: Chapman and Hall.

Huber, P. J. (1973). Robust regression: Asymptotics, conjectures and Monte Carlo. *Annuals of Statistics, 1*, 799–821.

Kloke, J. D., & McKean, J. W. (2012). Rfit: Rank-based estimation for linear models. *The R Journal, 4*, 57–64.

Mayer-Schönberger V., & Cukier C. (2013) *Big data: A revolution that will transform how we live, work, and think.* New York: Houghton Mifflin.

Rousseeuw, P. J. (1984). Least median of squares regression. *Journal of the American Statistical Association, 79*, 871–880.

Rousseeuw, P. J., & Van Driessen, K. (2000). An algorithm for positive-breakdown regression based on concentration steps. In W. Gaul, O. Opitz, & M. Schader (Eds.), *Data analysis: Scientific modeling and practical application* (pp. 335–346). New York: Springer.

Rousseeuw, P. J., & Yohai, V. (1984). Robust regression by means of S estimators. In J. Franke, W. Härdle, & R. D. Martin (Eds.), *Lecture notes in statistics 26: Robust and nonlinear time series analysia* (pp. 256–274). New York: Springer.

Whong, C. (2014). *FOILing NYC's taxi trip data*. Available online http://chriswhong.com/open-data/foil_nyc_taxi/. Cited 24 July 2015.

Yohai, V. J. (1987). High breakdown point and high efficiency robust estimates for regression. *Annuals of Statistics, 15*, 642–656.

Chapter 7
Rank-Based Inference for Multivariate Data in Factorial Designs

Arne C. Bathke and Solomon W. Harrar

Abstract We introduce fully nonparametric, rank-based test statistics for inference on multivariate data in factorial designs, and derive their asymptotic sampling distribution. The focus here is on the asymptotic setting where the number of levels of one factor tends to infinity, while the number of levels of the other factor, as well as the replication size per factor level combination, are fixed. The resulting test statistics can be calculated directly, they don't involve any iterative computational procedures. To our knowledge, they provide the first viable approach to a fully nonparametric analysis of, for example, multivariate ordinal responses, or a mix of ordinal with other response variables, in a factorial design setting.

Keywords Asymptotics • Multivariate statistics • Nonparametric method • Ordinal data • Rank test

7.1 Introduction

Multivariate data in factorial designs with few replications arise in agricultural, behavioral and biomedical studies, just to mention a few. However, due to the lack of appropriate inference procedures, such data are often analyzed using simplistic univariate methods or questionable model assumptions (e.g., multivariate normality). In this article, we develop fully nonparametric methods for the analysis of such data. These nonparametric methods allow for the analysis of data with ordinal responses, and they contain desirable invariance properties, as each variable (endpoint) can be monotonically transformed without changing the outcome of the analysis. Furthermore, the proposed methods represent the first nonparametric

A.C. Bathke (✉)
Fachbereich Mathematik, Universität Salzburg, 5020 Salzburg, Austria

Department of Statistics, University of Kentucky, Lexington, KY 40536, USA
e-mail: Arne.Bathke@sbg.ac.at

S.W. Harrar
Department of Statistics, University of Kentucky, Lexington, KY 40536, USA
e-mail: Solomon.Harrar@uky.edu

© Springer International Publishing Switzerland 2016

R.Y. Liu, J.W. McKean (eds.), *Robust Rank-Based and Nonparametric Methods*, Springer Proceedings in Mathematics & Statistics 168,
DOI 10.1007/978-3-319-39065-9_7

approach that is asymptotically valid when the number (a) of different samples or treatment groups (more generally, the number of levels of one of the factors) is large. In order to illustrate application of the procedure, we use the following data example.

7.1.1 Agricultural Field Trial

In an agricultural experiment that stands here exemplary for many similarly conducted field trials, several varieties of crabapples are examined with regard to their disease resistance (Chatfield et al. 2000). The response variable is a rating of tree health, on an ordinal scale from 0 to 5. Trees are evaluated at different times during the growing season, generating a multivariate observation vector per tree. When the experiment is repeated in a different year or at a different location, a second treatment factor is introduced whose main effect and interaction with the plant variety need to be considered, in addition to the variety effect. In Chatfield et al. (2000), the number of crabapple varieties was $a = 63$, justifying the use of methods derived for the asymptotic situation of $a \to \infty$. The number n_{ij} of replicates per variety were between 3 and 5. If we assume that the same study is performed at two different agricultural experiment stations or in two different years, we would be in the situations with $b = 2$.

7.1.2 Model

We describe the model using a two-factor layout corresponding to the data example. Generalization to higher-way layouts can be done using the techniques described here. On each experimental unit, a p-dimensional response vector is observed. These vectors are described by

$$\mathbf{X}_{ijr} = (X_{ijr}^{(1)}, X_{ijr}^{(2)}, \ldots, X_{ijr}^{(p)})'.$$

Here, the first two indices, $i = 1, \ldots, a$, and $j = 1, \ldots, b$, denote the levels of the two explanatory factors considered (in the example, year or location and variety, respectively). The index $r = 1, \ldots, n_{ij}$ stands for the replication or experimental unit within a factor level combination, and the super-index $d = 1, \ldots, p$ denotes the respective variable, among the total of p response variables considered (p-dimensional response). A possible multivariate additive linear model for $\mathbf{X}_{ijr}$ could be:

$$\mathbf{X}_{ijr} = \boldsymbol{\mu} + \boldsymbol{\xi}_i + \boldsymbol{\lambda}_j + \boldsymbol{\gamma}_{ij} + \boldsymbol{\varepsilon}_{ijr},$$

where $\boldsymbol{\xi}$, $\boldsymbol{\lambda}$, $\boldsymbol{\gamma}$ are the effects due to experimental condition, variety, and interaction between experimental condition and variety, and $\boldsymbol{\varepsilon}$ is the random variation assumed to be independently distributed with mean vector $\mathbf{0}$ and covariance matrix $\boldsymbol{\Sigma}_{ij}$.

Some drawbacks of the linear model approach are that the results depend on the type of transformation used and can be heavily influenced by outliers. In this manuscript, we are proposing a completely nonparametric alternative to the linear model approach. This nonparametric model can be written as

$$\mathbf{X}_{ijr} \sim F_{ij}, \tag{7.1}$$

where F_{ij} is the multivariate p-dimensional distribution of the response vector for factor level combination (i,j). This model imposes no restriction on distributions or correlations of error terms or random effects. The dependence in the data, induced by observing several outcome variables on the same subject, is entirely absorbed by modeling them as multivariate observation vectors, allowing for arbitrary, unspecified dependence structures among the response variables. The vectors $\mathbf{X}_{ijr}$ are independent for different indices i, j, or r, but the components of the vectors are possibly dependent.

In this manuscript, we are proposing a completely nonparametric model for the analysis of multivariate data from factorial experiments, applicable in a variety of situations. Inferential methods for the two-factor heteroscedastic model have relatively been well developed in the univariate case in the parametric as well as nonparametric contexts (for the latter, see for example, the monograph Brunner et al. (2002), and the references therein). There is some recent work for the semiparametric multivariate counterpart (Harrar and Bathke 2012; Konietschke et al. 2015; Van Aelst and Willems 2011), and several procedures have been proposed under the assumption of multivariate normality (Belloni and Didier 2008; Girón and del Castillo 2010; Kawasaki and Seo 2012; Krishnamoorthy and Lu 2010; Krishnamoorthy and Yu 2004, 2012; Nel and Van der Merwe 1986; Zhang 2011, 2012; Zhang and Liu 2013). However, not much has been done under the nonparametric paradigm, in particular under the asymptotic framework of a large number of factor levels. This asymptotic setup is becoming increasingly popular due to high throughput diagnostics and other bioinformatics tools which generate massive amounts of data. More motivations for this type of asymptotics in agriculture, health sciences, and other disciplines are found in Boos and Brownie (1995), Akritas and Arnold (2000), Bathke (2002), Bathke (2004) and Harrar and Gupta (2007) in univariate settings, and Gupta et al. (2006), Gupta et al. (2008), Bathke and Harrar (2008) and Harrar and Bathke (2008) in the multivariate setting. Whereas the work of Gupta et al. (2006, 2008) is restricted to the equal covariance matrix case, Bathke and Harrar (2008) and Harrar and Bathke (2008) consider the single factor nonparametric situation.

In the following sections, hypotheses and corresponding test statistics are introduced, and their asymptotic properties are derived. A section is devoted to the cumbersome task of consistent estimation of the variance-covariance matrix, and one section shows empirical evidence regarding the performance of the proposed tests, based on a simulation study.

Regarding the notation, a block diagonal matrix with blocks $\mathbf{A}$ and $\mathbf{B}$ will be written as $\mathbf{A} \oplus \mathbf{B}$, and the Kronecker product of matrices will be denoted as $\mathbf{A} \otimes \mathbf{B}$. See, for example, Schott (2005, Sect. 8.2) or Harville (2008, Sect. 16.1) for the definition and basic properties of the Kronecker product.

7.2 Hypotheses and Test Statistics

In general notation, the two experimental factors are denoted as factor A and factor B, respectively. Based on the nonparametric model (7.1), we will test hypotheses pertaining to these factors. The hypotheses are formulated in terms of the distribution functions F_{ij}. To this end, define the vector $\mathbf{F} = (F_{11}, \ldots, F_{1b}, F_{21}, \ldots, F_{ab})$ of cumulative distribution functions. Here, we assume the normalized versions of the distribution functions, allowing naturally for ties (Kruskal 1952; Lévy 1925; Ruymgaart 1980) and thus not restricting the methodology to absolutely continuous distributions.

Particular hypotheses of interest will be of the form $\mathscr{H}^{\psi} : D_{\psi}\mathbf{F} = \mathbf{0}$, postulating absence of the effect ψ. More specifically,

$$D_{\psi} = \begin{cases} \mathbf{P}_a \otimes \frac{1}{b}\mathbf{J}_b, & \text{for } \psi = A, \\ \mathbf{P}_a \otimes \mathbf{I}_b & \text{for } \psi = A|B, \\ \frac{1}{a}\mathbf{J}_a \otimes \mathbf{P}_b & \text{for } \psi = B, \\ \mathbf{I}_a \otimes \mathbf{P}_b & \text{for } \psi = B|A, \\ \mathbf{P}_a \otimes \mathbf{P}_b & \text{for } \psi = AB, \end{cases}$$

where $\mathbf{I}_d$ is the $d \times d$ identity matrix, $\mathbf{J}_d$ is the $d \times d$ unity matrix (matrix of ones), and $\mathbf{P}_d = \mathbf{I}_d - d^{-1}\mathbf{J}_d$.

These nonparametric hypotheses imply their corresponding parametric counterparts (see, e.g., Brunner et al. 2002; Harrar and Bathke 2008). As an illustration in the univariate context, the interaction effect in a parametric linear model can be expressed as $(\mathbf{P}_a \otimes \mathbf{P}_b)\boldsymbol{\mu}$, where $\boldsymbol{\mu}$ is the lexicographically arranged vector of cell means, $\boldsymbol{\mu} = (\mu_{11}, \ldots, \mu_{1b}, \mu_{21}, \ldots, \mu_{ab})'$. The implication between nonparametric and parametric hypotheses is immediately clear when expressing $(\mathbf{P}_a \otimes \mathbf{P}_b)\boldsymbol{\mu}$ in terms of the distribution functions as $(\mathbf{P}_a \otimes \mathbf{P}_b) \int x d\mathbf{F}(x)$. The same relation holds between the multivariate nonparametric and parametric analogs. The converse of this relation is not true: the parametric hypotheses do not imply their nonparametric counterparts.

In order to define nonparametric (rank-based) test statistics, let $\mathbf{R}_{ij} = (\mathbf{R}_{ij1}, \mathbf{R}_{ij2}, \ldots, \mathbf{R}_{ijn_{ij}})$ where $\mathbf{R}_{ijk} = (R_{ijk}^{(1)}, \ldots, R_{ijk}^{(p)})'$ and $R_{ijk}^{(l)}$ is the (mid-)rank of $X_{ijk}^{(l)}$ among all $N = \sum_{i=1}^{a}\sum_{j=1}^{b} n_{ij}$ random variables $X_{111}^{(l)}, \ldots, X_{abn_{ab}}^{(l)}$. Use of mid-ranks follows naturally from the normalized version of the cumulative distribution function (see above). Arranging these mid-ranks $R_{ijk}^{(l)}$ into a $p \times N$ matrix, put $\mathbf{R} = (\mathbf{R}_1, \mathbf{R}_2, \ldots, \mathbf{R}_a)$ where $\mathbf{R}_i = (\mathbf{R}_{i1}, \mathbf{R}_{i2}, \ldots, \mathbf{R}_{ib})$. Then, denote the $p \times p$ hypothesis and error sum of squares and cross product matrices based on the ranks as $\mathbf{H}^{(A)}(\mathbf{R})$, $\mathbf{H}^{(A|B)}(\mathbf{R})$, $\mathbf{H}^{(AB)}(\mathbf{R})$ and $\mathbf{G}(\mathbf{R})$. The corresponding matrices for testing main and simple effects of factor B can be written analogously to those of factor A. However, due to the large a asymptotics considered in this manuscript, we will not consider tests for the main effect of factor B in detail here.

$$\begin{aligned}
\mathbf{H}^{(A)} &= \frac{1}{a-1}\sum_{i=1}^{a}\sum_{j=1}^{b}(\tilde{\mathbf{R}}_{i..}-\tilde{\mathbf{R}}_{...})(\tilde{\mathbf{R}}_{i..}-\tilde{\mathbf{R}}_{...})' \\
&= \frac{1}{a-1}\mathbf{R}\left[\left(\bigoplus_{i=1}^{a}\bigoplus_{j=1}^{b}\frac{1}{n_{ij}}\mathbf{1}_{n_{ij}}\right)(\mathbf{P}_a\otimes\frac{1}{b}\mathbf{J}_b)\left(\bigoplus_{i=1}^{a}\bigoplus_{j=1}^{b}\frac{1}{n_{ij}}\mathbf{1}_{n_{ij}}'\right)\right]\mathbf{R}', \\
\mathbf{H}^{(A|B)} &= \frac{1}{(a-1)b}\sum_{i=1}^{a}\sum_{j=1}^{b}(\bar{\mathbf{R}}_{ij.}-\tilde{\mathbf{R}}_{.j.})(\bar{\mathbf{R}}_{ij.}-\tilde{\mathbf{R}}_{.j.})' \\
&= \frac{1}{(a-1)b}\mathbf{R}\left[\left(\bigoplus_{i=1}^{a}\bigoplus_{j=1}^{b}\frac{1}{n_{ij}}\mathbf{1}_{n_{ij}}\right)(\mathbf{P}_a\otimes\mathbf{I}_b)\left(\bigoplus_{i=1}^{a}\bigoplus_{j=1}^{b}\frac{1}{n_{ij}}\mathbf{1}_{n_{ij}}'\right)\right]\mathbf{R}', \\
\mathbf{H}^{(B)} &= \frac{1}{b-1}\sum_{i=1}^{a}\sum_{j=1}^{b}(\tilde{\mathbf{R}}_{.j.}-\tilde{\mathbf{R}}_{...})(\tilde{\mathbf{R}}_{.j.}-\tilde{\mathbf{R}}_{...})' \\
&= \frac{1}{b-1}\mathbf{R}\left[\left(\bigoplus_{i=1}^{a}\bigoplus_{j=1}^{b}\frac{1}{n_{ij}}\mathbf{1}_{n_{ij}}\right)(\frac{1}{a}\mathbf{J}_a\otimes\mathbf{P}_b)\left(\bigoplus_{i=1}^{a}\bigoplus_{j=1}^{b}\frac{1}{n_{ij}}\mathbf{1}_{n_{ij}}'\right)\right]\mathbf{R}', \\
\mathbf{H}^{(B|A)} &= \frac{1}{a(b-1)}\sum_{i=1}^{a}\sum_{j=1}^{b}(\bar{\mathbf{R}}_{ij.}-\tilde{\mathbf{R}}_{i..})(\bar{\mathbf{R}}_{ij.}-\tilde{\mathbf{R}}_{i..})' \\
&= \frac{1}{a(b-1)}\mathbf{R}\left[\left(\bigoplus_{i=1}^{a}\bigoplus_{j=1}^{b}\frac{1}{n_{ij}}\mathbf{1}_{n_{ij}}\right)(\mathbf{I}_a\otimes\mathbf{P}_b)\left(\bigoplus_{i=1}^{a}\bigoplus_{j=1}^{b}\frac{1}{n_{ij}}\mathbf{1}_{n_{ij}}'\right)\right]\mathbf{R}', \\
\mathbf{H}^{(AB)} &= \frac{1}{(a-1)(b=1)}\sum_{i=1}^{a}\sum_{j=1}^{b}(\bar{\mathbf{R}}_{ij.}-\tilde{\mathbf{R}}_{i..}-\tilde{\mathbf{R}}_{.j.}+\tilde{\mathbf{R}}_{...})(\bar{\mathbf{R}}_{ij.}-\tilde{\mathbf{R}}_{i..}-\tilde{\mathbf{R}}_{.j.}+\tilde{\mathbf{R}}_{...}) \\
&= \frac{1}{(a-1)(b=1)}\mathbf{R}\left[\left(\bigoplus_{i=1}^{a}\bigoplus_{j=1}^{b}\frac{1}{n_{ij}}\mathbf{1}_{n_{ij}}\right)(\mathbf{P}_a\otimes\mathbf{P}_b)\left(\bigoplus_{i=1}^{a}\bigoplus_{j=1}^{b}\frac{1}{n_{ij}}\mathbf{1}_{n_{ij}}'\right)\right]\mathbf{R}', \quad \text{and} \\
\mathbf{G} &= \frac{1}{ab}\sum_{i=1}^{a}\sum_{j=1}^{b}\frac{1}{n_{ij}(n_{ij}-1)}\sum_{k=1}^{n_{ij}}(\mathbf{R}_{ijk}-\bar{\mathbf{R}}_{ij.})(\mathbf{R}_{ijk}-\bar{\mathbf{R}}_{ij.})' = \frac{1}{ab}\sum_{i=1}^{a}\sum_{j=1}^{b}\frac{1}{n_{ij}}\mathbf{S}_{ij} \\
&= \frac{1}{ab}\mathbf{R}\left(\bigoplus_{i=1}^{a}\bigoplus_{j=1}^{b}\frac{1}{n_{ij}(1-n_{ij})}\mathbf{P}_{n_{ij}}\right)\mathbf{R}',
\end{aligned}$$

where $\bar{\mathbf{R}}_{ij.} = \frac{1}{n_{ij}} \sum_{k=1}^{n_{ij}} \mathbf{R}_{ijk}$, $\tilde{\mathbf{R}}_{i..} = \frac{1}{b} \sum_{j=1}^{b} \bar{\mathbf{R}}_{ij.}$, $\tilde{\mathbf{R}}_{.j.} = \frac{1}{a} \sum_{i=1}^{a} \bar{\mathbf{R}}_{ij.}$, $\tilde{\mathbf{R}}_{...} = \frac{1}{ab} \sum_{i=1}^{a} \sum_{j=1}^{b} \bar{\mathbf{R}}_{ij.}$, and $\mathbf{S}_{ij} = \frac{1}{(n_{ij}-1)} \sum_{k=1}^{n_{ij}} (\mathbf{R}_{ijk} - \bar{\mathbf{R}}_{ij.})(\mathbf{R}_{ijk} - \bar{\mathbf{R}}_{ij.})'$.

These sum of squares and cross product matrices constitute essentially a nonparametric multivariate unweighted means analysis. The matrix notation above shows the pattern after which they can also be defined in higher-way layouts. To keep this manuscript concise, this extension to higher-way layouts is not carried out in detail here. Under the hypothesis $\mathscr{H}^{\psi}$, the expectation of $\mathbf{H}^{(\psi)}$ is equal to the expectation of $\mathbf{G}$, thus allowing for the following way to construct multivariate test statistics.

Let ψ be one of the effects under consideration: AB, $A|B$, A, B, or $B|A$. We propose the following multivariate test statistics for testing $\mathscr{H}^{\psi}$.

(a) Dempster's ANOVA Type criterion: $T_{\mathrm{D}}^{(\psi)} = \mathrm{tr}(\mathbf{H}^{(\psi)})/\mathrm{tr}(\mathbf{G})$.
(b) Wilks' Λ criterion: $T_{\mathrm{LR}}^{(\psi)} = \log|\mathbf{I} + \mathbf{H}^{(\psi)}\mathbf{G}^{-}|$.
(c) The Lawley-Hotelling criterion: $T_{\mathrm{LH}}^{(\psi)} = \mathrm{tr}(\mathbf{H}^{(\psi)}\mathbf{G}^{-})$.
(d) The Bartlett-Nanda-Pillai criterion: $T_{\mathrm{BNP}}^{(\psi)} = \mathrm{tr}\left(\mathbf{H}^{(\psi)}\mathbf{G}^{-}(\mathbf{I} + \mathbf{H}^{(\psi)}\mathbf{G}^{-})\right)$.

These test statistics are similar to the four test statistics considered in Harrar and Bathke (2012) in the context of a two-factor semiparametric MANOVA under heteroscedasticity. Their use in this manuscript is distinct in two important ways. In the present article, the sums of squares and cross products $\mathbf{H}^{(\psi)}$ and $\mathbf{G}$ are computed from the ranks which can not be assumed to be independent across subjects. Due to the discreteness of the rankings, it may not be reasonable to assume non-singularity of the matrices $\mathbf{G}$ and $\mathbf{H}^{(\psi)} + \mathbf{G}$. Thus we use here Moore-Penrose generalized inverses in defining the test statistics. The Moore-Penrose generalized inverse has the useful continuity property (Schott 2005, Sect. 5.7; for a proof see, e.g., Penrose 1955).

7.3 Asymptotic Results

For the asymptotic derivations in this section, we will assume that $a \to \infty$, b bounded, and $\forall i,j : n_{ij}$ bounded. The asymptotics are somewhat involved as the quadratic forms $\mathbf{H}^{(\psi)}$ and $\mathbf{G}$ are based on a matrix of ranks $\mathbf{R}$ which has both row-wise and column-wise dependence.

For the mathematical derivations in the technical proofs of this manuscript, it is convenient to use the so-called "asymptotic rank transforms" (ART) and "rank transforms" (RT). They are formally introduced in the following definition. For the concept of ART, see also Brunner et al. (2002, p. 77).

Definition 7.1. Let $\mathbf{X}_{ijk} = (X_{ijk}^{(1)}, \ldots, X_{ijk}^{(p)})'$, $i = 1, \ldots, a$, $j = 1, \ldots, b$, and $k = 1, \ldots, n_{ij}$, be independent random vectors with possibly dependent components $X_{ijk}^{(l)}$ whose marginal distribution is $F_{ij}^{(l)}$, $l = 1, \ldots, p$. Let $N = \sum_{i=1}^{a} \sum_{j=1}^{b} n_{ij}$. Further let

$$H^{(l)}(x) = \frac{1}{N} \sum_{i=1}^{a} \sum_{j=1}^{b} n_{ij} F_{ij}^{(l)}(x)$$

denote the *average cdf* for variable (l),

$$\hat{H}^{(l)}(x) = \frac{1}{N} \sum_{i=1}^{a} \sum_{j=1}^{b} \sum_{k=1}^{n_{ij}} c(x - X_{ijk}^{(l)}),$$

where $c(t) = 0, 1/2, 1$ if $t < 0, t = 0, t > 0$, respectively, denotes the *average empirical cdf*, and $\mathbf{Y} = (\mathbf{Y}_1, \ldots, \mathbf{Y}_a)$ where $\mathbf{Y}_i = (\mathbf{Y}_{i1}, \ldots, \mathbf{Y}_{ib})$, $\mathbf{Y}_{ij} = (\mathbf{Y}_{ij1}, \ldots, \mathbf{Y}_{ijn_{ij}})$ and $\mathbf{Y}_{ijk} = (Y_{ijk}^{(1)}, \ldots, Y_{ijk}^{(p)})'$ where $Y_{ijk}^{(l)} = H^{(l)}(X_{ijk}^{(l)})$ is known as the *asymptotic rank transform (ART)* of $X_{ijk}^{(l)}$. The matrix of *rank transforms (RT)*, $\hat{\mathbf{Y}}$, is defined analogously, with elements $\hat{Y}_{ijk}^{(l)} = \hat{H}^{(l)}(X_{ijk}^{(l)})$.

The expression "rank transform" pays tribute to the fact that $\hat{Y}_{ijk}^{(l)}$ is related to the (mid-)rank $R_{ijk}^{(l)}$ by $\hat{Y}_{ijk}^{(l)} = N^{-1}(R_{ijk}^{(l)} - \frac{1}{2})$. However, the "asymptotic rank transforms" are technically more tractable than the "rank transforms", due to the simpler covariance structure of $\mathbf{Y}$ as compared to $\hat{\mathbf{Y}}$. Note that the ART of independent random variables are independent, but the RT are not.

Denote $\mathrm{Var}(\mathbf{Y}_{ij1}) = \boldsymbol{\Sigma}_{ij}$ and assume that the following limit exists:

$$\lim_{a \to \infty} \frac{1}{ab} \sum_{i=1}^{a} \sum_{j=1}^{b} \boldsymbol{\Sigma}_{ij} = \boldsymbol{\Sigma}.$$

For later use, we also introduce the notation $\mathbf{M} = (\boldsymbol{\mu}_1, \boldsymbol{\mu}_2, \ldots, \boldsymbol{\mu}_a)$, $\boldsymbol{\mu}_i = (\boldsymbol{\mu}_{i1}, \ldots, \boldsymbol{\mu}_{ib})$, $\boldsymbol{\mu}_{ij} = (\boldsymbol{\mu}_{ij1}, \ldots, \boldsymbol{\mu}_{ijn_{ij}})$, where $\boldsymbol{\mu}_{ijk} = (\mu_{ijk}^{(1)}, \ldots, \mu_{ijk}^{(p)})'$ is the vector of expectations of the ART vector $\mathbf{Y}_{ijk}$, that is $\mu_{ijk}^{(l)} = E(Y_{ijk}^{(l)})$, and $\mathbf{Y}_{\mu} = \mathbf{Y} - \mathbf{M}$, $\hat{\mathbf{Y}}_{\mu} = \hat{\mathbf{Y}} - \mathbf{M}$.

For $\psi \in \{A, B, A|B, B|A, AB\}$, we denote the ART analogs of the matrices $\mathbf{H}^{(\psi)}$ and $\mathbf{G}$ defined in Sect. 7.2 by $\tilde{\mathbf{H}}^{(\psi)}$ and $\tilde{\mathbf{G}}$, respectively. In order to prove asymptotic normality results for the rank-based test statistics considered in this paper, we need to first establish the asymptotic equivalence of certain quadratic forms defined in terms of $(\mathbf{H}^{(\psi)}, \mathbf{G})$ (based on "rank transforms") and the corresponding quadratic forms defined in terms of $(\tilde{\mathbf{H}}^{(\psi)}, \tilde{\mathbf{G}})$ (based on "asymptotic rank transforms").

We begin this task by showing the asymptotic equivalence between certain matrix differences in "rank transforms' and the corresponding ones in "asymptotic rank transforms". Recall that the ranks $R_{ijk}^{(l)}$ take values in $[1, N]$, while the "rank transforms" $\hat{Y}_{ijk}^{(l)} = N^{-1}(R_{ijk}^{(l)} - \frac{1}{2})$ and the "asymptotic rank transforms" $Y_{ijk}^{(l)}$ take values within the unit interval, making it necessary to divide the rank matrices by N^2 in order to be able to establish asymptotic equivalence.

Proposition 7.1. *Assume b, p and n are bounded. Then*

(i) Under the hypothesis $\mathscr{H}^{\psi}$ *for* $\psi \in \{A, A|B, AB\}$

$$\sqrt{a}\left\{\frac{1}{N^2}(\mathbf{H}^{(\psi)} - \mathbf{G}) - (\tilde{\mathbf{H}}^{(\psi)} - \tilde{\mathbf{G}})\right\} = o_p(1) \quad \text{as } a \to \infty\,.$$

(ii) Under the hypothesis $\mathscr{H}^{\psi}$ *for* $\psi \in \{B, B|A\}$

$$\frac{1}{N^2}\mathbf{H}^{(\psi)} - \tilde{\mathbf{H}}^{(\psi)} = o_p(1) \quad \text{as } a \to \infty\,.$$

and

(iii) $N^{-2}\mathbf{G} - \tilde{\mathbf{G}} = o_p(1)$ *as* $a \to \infty$.

Proof. The proof can be established using the same techniques as in the proof of Theorem 4 in Harrar and Bathke (2008).

The following proposition asserts that the difference $N^{-2}\mathbf{G} - \boldsymbol{\Sigma}$ is asymptotically ($a \to \infty$) stochastically negligible.

Proposition 7.2. *Assume that the* n_{ij} *are bounded. Then* $N^{-2}\mathbf{G} - \boldsymbol{\Sigma} \xrightarrow{p} 0$ *as* $a \to \infty$.

Proof. Since $N^{-2}\mathbf{G} - \tilde{\mathbf{G}} = o_p(1)$ by (iii) of Proposition 7.1, it suffices to show that $\tilde{\mathbf{G}} - \boldsymbol{\Sigma} \xrightarrow{p} 0$. This follows from Theorem 1 of Harrar and Bathke (2012) if we show that $\sum_{i=1}^{a}\sum_{j=1}^{b} n_{ij}^{-2}(n_{ij} - 1)^{-1}\boldsymbol{\Sigma}_{ij} \otimes \boldsymbol{\Sigma}_{ij} = o(a^2)$ and $\sum_{i=1}^{a}\sum_{j=1}^{b} n_{ij}^{-3}K_4(\mathbf{Y}_{ij1}) = o(a^2)$ as $a \to \infty$. These two follow from the fact that the components of $\mathbf{Y}_{ij1}$ are uniformly bounded random variables.

Next, we obtain the asymptotic null distributions of the four test statistics for testing the main, simple, and interaction effects. Since the results for testing $\mathscr{H}^{(AB)}$, $\mathscr{H}^{(A)}$, $\mathscr{H}^{(A|B)}$ and $\mathscr{H}^{(B|A)}$ are similar in form and their derivations proceed along the same lines, we consider them together.

We know from Proposition 7.2 that $N^{-2}\mathbf{G} - \boldsymbol{\Sigma} = o_p(1)$ as $a \to \infty$, and it is established in Theorem 7.1 below that $\sqrt{a}(\mathbf{H}^{(\psi)} - \mathbf{G})\boldsymbol{\Omega} = O_p(1)$ as $a \to \infty$, for any matrix of constants $\boldsymbol{\Omega}$. All four test statistics, scaled and centered suitably, can be expressed as

$$\sqrt{a}\,(\ell T_{\mathscr{G}}^{(\psi)} - h) = \sqrt{a}\,\mathrm{tr}(\mathbf{H}^{(\psi)} - \mathbf{G})\boldsymbol{\Omega} + o_p(1), \tag{7.2}$$

where $\ell = 1, 2, 1, 4$, $h = 1$, $2p\log 2$, p, $2p$ and $\boldsymbol{\Omega} = (1/\mathrm{tr}\boldsymbol{\Sigma})\mathbf{I}_p, \boldsymbol{\Sigma}^{-}, \boldsymbol{\Sigma}^{-}, \boldsymbol{\Sigma}^{-}$ for $\mathscr{G} = \mathrm{D}, \mathrm{LR}, \mathrm{LH}, \mathrm{BNP}$, respectively (see Harrar and Bathke 2012, for more details). Therefore, the null distributions of the four test statistics can be derived in a unified manner by obtaining the null distribution of $\sqrt{a}\,\mathrm{tr}(\mathbf{H}^{(\psi)} - \mathbf{G})\boldsymbol{\Omega}$ for any fixed matrix $\boldsymbol{\Omega}$. The null distribution of this latter quantity is given in Theorem 7.1.

Theorem 7.1. *Let* $\psi = AB, A, A|B,$ *or* $B|A$. *Under the hypothesis* $\mathscr{H}_0^{(\psi)}$, $\sqrt{a}\,\mathrm{tr}(\mathbf{H}^{(\psi)} - \mathbf{G})\boldsymbol{\Omega} \xrightarrow{\mathscr{L}} N\left(0, \tau_\psi^2(\boldsymbol{\Omega})\right)$ *as* $a \to \infty$ *and* n_{ij} *and* b *bounded, where*

$$\tau_\psi^2(\boldsymbol{\Omega}) = \begin{cases} \frac{2}{b}\left\{v_1(\boldsymbol{\Omega}) + \frac{v_2(\boldsymbol{\Omega})}{(b-1)^2}\right\} & \text{when} \quad \psi = AB, \\ \frac{2}{b}\left\{v_1(\boldsymbol{\Omega}) + v_2(\boldsymbol{\Omega})\right\} & \text{when} \quad \psi = A, \\ \frac{2}{b} v_1(\boldsymbol{\Omega}) & \text{when} \quad \psi = A|B, \\ \frac{2}{b^2}\left\{v_1(\boldsymbol{\Omega}) + \frac{v_2(\boldsymbol{\Omega})}{(b-1)^2}\right\} & \text{when} \quad \psi = B|A. \end{cases}$$

Here,

$$v_1(\boldsymbol{\Omega}) = \lim_{a\to\infty} \frac{1}{ab} \sum_{i=1}^{a} \sum_{j=1}^{b} \frac{\mathrm{tr}(\boldsymbol{\Omega}\boldsymbol{\Sigma}_{ij})^2}{n_{ij}(n_{ij}-1)},$$

and

$$v_2 = \lim_{a\to\infty} \frac{1}{ab} \sum_{i=1}^{a} \sum_{j\neq j'}^{b} \frac{\mathrm{tr}(\boldsymbol{\Omega}\boldsymbol{\Sigma}_{ij}\boldsymbol{\Omega}\boldsymbol{\Sigma}_{ij'})}{n_{ij}n_{ij'}},$$

assuming the limits exist.

Proof. Considering Proposition 7.1, it is enough to show that $N^{-2}\sqrt{a}\,\mathrm{tr}(\mathbf{H}^{(\psi)} - \mathbf{G})\boldsymbol{\Omega} \xrightarrow{\mathscr{L}} N\left(0, \tau_\psi^2(\boldsymbol{\Omega})\right)$ as $a \to \infty$ and n_{ij} and b bounded. This follows from Theorem 2 of Harrar and Bathke (2012) if for some $\delta > 0$, $E|(\mathbf{Y}_{ij1} - \frac{1}{2}\mathbf{1})'\boldsymbol{\Sigma}_{ij}^{-1}(\mathbf{Y}_{ij1} - \frac{1}{2}\mathbf{1})|^{2+\delta} < \infty$ and

$$\lim_{a\to\infty} \frac{1}{a} \sum_{i=1}^{a} \sum_{j=1}^{b} \frac{1}{n_{ij}^{1+\delta/2}(n_{ij}-1)^{1+\delta/2}} \mathrm{tr}(\boldsymbol{\Omega}\boldsymbol{\Sigma}_{ij})^{2+\delta} < \infty \text{ and}$$

$$\lim_{a\to\infty} \frac{1}{a} \sum_{i=1}^{a} \sum_{j\neq j'}^{b} \frac{1}{n_{ij}^{1+\delta/2} n_{ij'}^{1+\delta/2}} \mathrm{tr}(\boldsymbol{\Omega}\boldsymbol{\Sigma}_{ij}\boldsymbol{\Omega}\boldsymbol{\Sigma}_{ij'})^{1+\delta/2} < \infty.$$

Recalling again that the components of $\mathbf{Y}_{ij1}$ are bounded random variables completes the proof.

Under the assumptions and notations of Theorem 7.1, the asymptotic distribution of Dempster's ANOVA type criterion can be obtained by setting $\boldsymbol{\Omega} = (1/\mathrm{tr}\boldsymbol{\Sigma})\mathbf{I}_p$. For the other three criteria, we set $\boldsymbol{\Omega} = \boldsymbol{\Sigma}^{-1}$ to get the asymptotic null distributions.

Needless to say, the asymptotic null distributions of T_{LR}, T_{LH} and T_{BNP}, scaled and centered as in (7.2), are the same up to the order $O(a^{-1/2})$. A comparison of the asymptotic variances in Theorem 7.1 reveals that the test statistic for the interaction effect has smaller variance than that of the main effect. Also we see from the asymptotic variances in Theorem 7.1 that the test statistic for the simple effect of A has smaller variance compared to that of either the interaction or main effects.

7.4 Consistent Variance and Covariance Matrix Estimation

Multivariate data in factorial designs present a major technical difficulty considering the derivation of valid nonparametric test statistics: Unlike in the multivariate one-way design discussed in Harrar and Bathke (2008), the covariance matrices do not simplify under the null hypotheses that are considered here. Therefore, it is more complicated to devise consistent variance estimators.

The following theorem provides an asymptotic result formulated in terms of the unobservable "asymptotic rank transforms". The expression is analogous to the variance estimator defined in Theorem 2.3 of Harrar and Bathke (2012) in the semiparametric context. However, due to the fact that the "asymptotic rank transforms" are per definition bounded between 0 and 1, it is not necessary to require a moment condition as in Harrar and Bathke (2012).

Theorem 7.2. *Let the model and assumptions be as in Theorem 7.1. Define*

$$\widetilde{\boldsymbol{\Psi}}_{ij}(\boldsymbol{\Omega}) = \frac{1}{4c_{ij}} \sum_{(k_1,k_2,k_3,k_4)\in\mathscr{K}}^{n_{ij}} \boldsymbol{\Omega}(\mathbf{Y}_{ijk_1} - \mathbf{Y}_{ijk_2})(\mathbf{Y}_{ijk_1} - \mathbf{Y}_{ijk_2})'$$
$$\times \boldsymbol{\Omega}(\mathbf{Y}_{ijk_3} - \mathbf{Y}_{ijk_4})(\mathbf{Y}_{ijk_3} - \mathbf{Y}_{ijk_4})',$$

where $\mathscr{K}$ is the set of all quadruples $\kappa = (k_1, k_2, k_3, k_4)$ where no element in κ is equal to any other element in κ, and $c_{ij} = n_{ij}(n_{ij}-1)(n_{ij}-2)(n_{ij}-3)$. Also, define

$$\tilde{\mathbf{S}}_{ij} = \frac{1}{(n_{ij}-1)} \sum_{k=1}^{n_{ij}} (\mathbf{Y}_{ijk} - \bar{\mathbf{Y}}_{ij.})(\mathbf{Y}_{ijk} - \bar{\mathbf{Y}}_{ij.})'.$$

Then,

$$\frac{1}{ab}\sum_{i=1}^{a}\sum_{j=1}^{b}\frac{1}{n_{ij}(n_{ij}-1)}\mathrm{tr}(\widetilde{\boldsymbol{\Psi}}_{ij}(\boldsymbol{\Omega})) - \frac{1}{ab}\sum_{i=1}^{a}\sum_{j=1}^{b}\frac{1}{n_{ij}(n_{ij}-1)}\mathrm{tr}(\boldsymbol{\Omega}\boldsymbol{\Sigma}_{ij})^2 = o_p(1),$$

and

$$\frac{1}{ab}\sum_{i=1}^{a}\sum_{j\neq j'}^{b}\frac{1}{n_{ij}n_{ij'}}\text{tr}(\boldsymbol{\Omega}\tilde{\mathbf{S}}_{ij}\boldsymbol{\Omega}\tilde{\mathbf{S}}_{ij'}) - \frac{1}{ab}\sum_{i=1}^{a}\sum_{j\neq j'}^{b}\frac{1}{n_{ij}n_{ij'}}\text{tr}(\boldsymbol{\Omega}\boldsymbol{\Sigma}_{ij}\boldsymbol{\Omega}\boldsymbol{\Sigma}_{ij'}) = o_p(1),$$

as $a\to\infty$.

The proof follows similar to that of Theorem 2.3 in Harrar and Bathke (2012), or rather from the theory of U-statistics (see, e.g., Serfling 1980).

Since the "variance estimator" presented in the previous theorem is not observable and therefore can not be used in practice, in the next two theorems we are introducing observable rank-based estimators and establish their asymptotic equivalence to corresponding expressions formulated in terms of the "asymptotic rank transforms".

Theorem 7.3. *Let* $\widetilde{\boldsymbol{\Psi}}_{ij}(\boldsymbol{\Omega})$ *be defined as in Theorem 7.2. Define* $\widehat{\boldsymbol{\Psi}}_{ij}(\boldsymbol{\Omega})$ *analogously, but using rank transforms instead of asymptotic rank transforms (see Definition 7.1). Then,*

$$D = \frac{1}{ab}\sum_{i=1}^{a}\sum_{j=1}^{b}\frac{1}{n_{ij}(n_{ij}-1)}\text{tr}(\widehat{\boldsymbol{\Psi}}_{ij}(\boldsymbol{\Omega})) - \frac{1}{ab}\sum_{i=1}^{a}\sum_{j=1}^{b}\frac{1}{n_{ij}(n_{ij}-1)}\text{tr}(\widetilde{\boldsymbol{\Psi}}_{ij}(\boldsymbol{\Omega})) = o_p(1),$$

as $a\to\infty$.

Proof. Without loss of generality, assume that $\boldsymbol{\Omega} = \mathbf{I}$. Define

$$\tilde{D} = \frac{1}{ab}\sum_{i=1}^{a}\sum_{j=1}^{b}\frac{1}{n_{ij}(n_{ij}-1)}\frac{1}{4c_{ij}}$$

$$\sum_{(k_1,k_2,k_3,k_4)\in\mathscr{K}}^{n_{ij}}\Big[(\hat{\mathbf{Y}}_{ijk_1}-\hat{\mathbf{Y}}_{ijk_2})(\hat{\mathbf{Y}}_{ijk_1}-\hat{\mathbf{Y}}_{ijk_2})'\otimes(\hat{\mathbf{Y}}_{ijk_3}-\hat{\mathbf{Y}}_{ijk_4})(\hat{\mathbf{Y}}_{ijk_3}-\hat{\mathbf{Y}}_{ijk_4})'$$

$$-(\mathbf{Y}_{ijk_1}-\mathbf{Y}_{ijk_2})(\mathbf{Y}_{ijk_1}-\mathbf{Y}_{ijk_2})'\otimes(\mathbf{Y}_{ijk_3}-\mathbf{Y}_{ijk_4})(\mathbf{Y}_{ijk_3}-\mathbf{Y}_{ijk_4})'\Big],$$

where the c_{ij} are as defined in Theorem 7.2, and consider an arbitrary element of this $p^2\times p^2$ matrix. Each element $\tilde{D}_{q_1,q_2,q_3,q_4}$ is uniquely determined by a combination of four indices q_1,q_2,q_3,q_4, where $q_r = 1,\ldots,p$, $r = 1,\ldots,4$. Then, we have

$$\tilde{D}_{q_1,q_2,q_3,q_4} = \frac{1}{ab}\sum_{i=1}^{a}\sum_{j=1}^{b}\frac{1}{n_{ij}(n_{ij}-1)}\frac{1}{4c_{ij}}\sum_{(k_1,k_2,k_3,k_4)\in\mathscr{K}}^{n_{ij}}$$

$$\Big[(\hat{Y}_{ijk_1}^{(q_1)}-\hat{Y}_{ijk_2}^{(q_1)})(\hat{Y}_{ijk_1}^{(q_2)}-\hat{Y}_{ijk_2}^{(q_2)})(\hat{Y}_{ijk_3}^{(q_3)}-\hat{Y}_{ijk_4}^{(q_3)})(\hat{Y}_{ijk_3}^{(q_4)}-\hat{Y}_{ijk_4}^{(q_4)})$$

$$-(Y_{ijk_1}^{(q_1)}-Y_{ijk_2}^{(q_1)})(Y_{ijk_1}^{(q_2)}-Y_{ijk_2}^{(q_2)})(Y_{ijk_3}^{(q_3)}-Y_{ijk_4}^{(q_3)})(Y_{ijk_3}^{(q_4)}-Y_{ijk_4}^{(q_4)})\Big]$$

$$= \frac{1}{ab}\sum_{i=1}^{a}\sum_{j=1}^{b}\frac{1}{n_{ij}(n_{ij}-1)4c_{ij}}\sum_{(k_1,k_2,k_3,k_4)\in\mathscr{K}}^{n_{ij}}$$

$$\Big[(\hat{H}^{(q_1)}(X_{ijk_1}^{(q_1)}) - \hat{H}^{(q_1)}(X_{ijk_2}^{(q_1)}))(\hat{H}^{(q_2)}(X_{ijk_1}^{(q_2)}) - \hat{H}^{(q_2)}(X_{ijk_2}^{(q_2)}))$$

$$\times (\hat{H}^{(q_3)}(X_{ijk_3}^{(q_3)}) - \hat{H}^{(q_3)}(X_{ijk_4}^{(q_3)}))(\hat{H}^{(q_4)}(X_{ijk_3}^{(q_4)}) - \hat{H}^{(q_4)}(X_{ijk_4}^{(q_4)}))$$

$$- (H^{(q_1)}(X_{ijk_1}^{(q_1)}) - H^{(q_1)}(X_{ijk_2}^{(q_1)}))(H^{(q_2)}(X_{ijk_1}^{(q_2)}) - H^{(q_2)}(X_{ijk_2}^{(q_2)}))$$

$$\times (H^{(q_3)}(X_{ijk_3}^{(q_3)}) - H^{(q_3)}(X_{ijk_4}^{(q_3)}))(H^{(q_4)}(X_{ijk_3}^{(q_4)}) - H^{(q_4)}(X_{ijk_4}^{(q_4)}))\Big]$$

$$= \frac{1}{ab}\sum_{i=1}^{a}\sum_{j=1}^{b}\frac{1}{n_{ij}(n_{ij}-1)4c_{ij}}\sum_{(k_1,k_2,k_3,k_4)\in\mathscr{K}}^{n_{ij}}\frac{1}{N^4}\sum_{s_1=1}^{N}\sum_{s_2=1}^{N}\sum_{s_3=1}^{N}\sum_{s_4=1}^{N}$$

$$\zeta(X_{ijk_1}^{(q_1)}, X_{ijk_2}^{(q_1)}, X_{ijk_1}^{(q_2)}, X_{ijk_2}^{(q_2)}, X_{ijk_3}^{(q_3)}, X_{ijk_4}^{(q_3)}, X_{ijk_3}^{(q_4)}, X_{ijk_4}^{(q_4)}, X_{s_1}, X_{s_2}, X_{s_3}, X_{s_4}),$$

where $\zeta(X_{ijk_1}^{(q_1)}, X_{ijk_2}^{(q_1)}, X_{ijk_1}^{(q_2)}, X_{ijk_2}^{(q_2)}, X_{ijk_3}^{(q_3)}, X_{ijk_4}^{(q_3)}, X_{ijk_3}^{(q_4)}, X_{ijk_4}^{(q_4)}, X_{s_1}, X_{s_2}, X_{s_3}, X_{s_4})$

$$= \Big([c(X_{ijk_1}^{(q_1)} - X_{s_1}) - c(X_{ijk_2}^{(q_1)} - X_{s_1})][c(X_{ijk_1}^{(q_2)} - X_{s_2}) - c(X_{ijk_2}^{(q_2)} - X_{s_2})]$$

$$\times [c(X_{ijk_3}^{(q_3)} - X_{s_3}) - c(X_{ijk_4}^{(q_3)} - X_{s_3})][c(X_{ijk_3}^{(q_4)} - X_{s_4}) - c(X_{ijk_4}^{(q_4)} - X_{s_4})]$$

$$- [F_{s_1}(X_{ijk_1}^{(q_1)}) - F_{s_1}(X_{ijk_2}^{(q_1)})][F_{s_2}(X_{ijk_1}^{(q_2)}) - F_{s_2}(X_{ijk_2}^{(q_2)})]$$

$$\times [F_{s_3}(X_{ijk_3}^{(q_3)}) - F_{s_3}(X_{ijk_4}^{(q_3)})][F_{s_4}(X_{ijk_3}^{(q_4)}) - F_{s_4}(X_{ijk_4}^{(q_4)})]\Big),$$

$c(\cdot)$ denotes the normalized counting function $c(x) = (I\{x > 0\} + I\{x \geq 0\})$, and F_t denotes the cdf of X_t.

Note that $E(\zeta) = 0$ if all indices s_1, s_2, s_3, s_4 are different from each other, and the corresponding random variables independent of the other eight random variables. This holds because the first part of ζ, integrated over $(X_{s_1}, X_{s_2}, X_{s_3}, X_{s_4})$, equals the second part. Therefore, $E(\tilde{D}) \to 0$ since the number of (s_1, s_2, s_3, s_4) index combinations resulting in nonzero expectation is of order N^3, but the sum is divided by N^4. Consider now

$$\tilde{D}^2_{q_1,q_2,q_3,q_4} = \frac{1}{a^2b^2}\sum_{i_1=1}^{a}\sum_{i_2=1}^{a}\sum_{j_1=1}^{b}\sum_{j_2=1}^{b}\frac{1}{n_{i_1j_1}n_{i_2j_2}(n_{i_1j_1}-1)(n_{i_2j_2}-1)16c_{i_1j_1}c_{i_2j_2}}$$

$$\sum_{(k_1,k_2,k_3,k_4)\in\mathscr{K}}^{n_{i_1j_1}}\sum_{(l_1,l_2,l_3,l_4)\in\mathscr{K}}^{n_{i_2j_2}}\frac{1}{N^8}\sum_{s_1=1}^{N}\sum_{s_2=1}^{N}\sum_{s_3=1}^{N}\sum_{s_4=1}^{N}\sum_{t_1=1}^{N}\sum_{t_2=1}^{N}\sum_{t_3=1}^{N}\sum_{t_4=1}^{N}$$

$$\zeta(X_{i_1j_1k_1}^{(q_1)}, X_{i_1j_1k_2}^{(q_1)}, X_{i_1j_1k_1}^{(q_2)}, X_{i_1j_1k_2}^{(q_2)}, X_{i_1j_1k_3}^{(q_3)}, X_{i_1j_1k_4}^{(q_3)}, X_{i_1j_1k_3}^{(q_4)}, X_{i_1j_1k_4}^{(q_4)}, X_{s_1}, X_{s_2}, X_{s_3}, X_{s_4})$$

$$\times\, \zeta(X_{i_2j_2l_1}^{(q_1)}, X_{i_2j_2l_2}^{(q_1)}, X_{i_2j_2l_1}^{(q_2)}, X_{i_2j_2l_2}^{(q_2)}, X_{i_2j_2l_3}^{(q_3)}, X_{i_2j_2l_4}^{(q_3)}, X_{i_2j_2l_3}^{(q_4)}, X_{i_2j_2l_4}^{(q_4)}, X_{t_1}, X_{t_2}, X_{t_3}, X_{t_4}).$$

Again, when all involved random variables with indices $s_1, s_2, s_3, s_4, t_1, t_2, t_3, t_4$ are independent of each other, and of the remaining random variables, the expectation of each ζ-function is zero, and therefore also the expectation of the product. Similar to above, this can be seen by first integrating over the random variables with indices (s_1, s_2, s_3, s_4), conditional on those with indices (k_1, k_2, k_3, k_4). The number of cases with nonzero expectation is again of smaller order, in this case N^7, while division is by N^8. It follows that $E(\tilde{D}^2_{q_1,q_2,q_3,q_4}) \to 0$ and therefore $\tilde{D}_{q_1,q_2,q_3,q_4} = o_p(1)$ for each element of $\tilde{D}$, which proves $\tilde{D} = o_p(1)$.

Theorem 7.4. *Let $\widetilde{\mathbf{S}}_{ij}$ be defined as in Theorem 7.2, and define $\widehat{\mathbf{S}}_{ij}$ analogously, but using rank transforms instead of asymptotic rank transforms. Then,*

$$K = \frac{1}{ab}\sum_{i=1}^{a}\sum_{j\neq j'}^{b}\frac{1}{n_{ij}n_{ij'}}\mathrm{tr}(\boldsymbol{\Omega}\widehat{\mathbf{S}}_{ij}\boldsymbol{\Omega}\widehat{\mathbf{S}}_{ij'}) - \frac{1}{ab}\sum_{i=1}^{a}\sum_{j\neq j'}^{b}\frac{1}{n_{ij}n_{ij'}}\mathrm{tr}(\boldsymbol{\Omega}\widetilde{\mathbf{S}}_{ij}\boldsymbol{\Omega}\widetilde{\mathbf{S}}_{ij'}) = o_p(1),$$

as $a \to \infty$.

Proof. As in the proof of Theorem 7.3, assume without loss of generality that $\boldsymbol{\Omega} = \mathbf{I}$, and define

$$\tilde{\mathbf{K}} = \frac{1}{ab}\sum_{i=1}^{a}\sum_{j\neq j'}^{b}\frac{1}{n_{ij}(1-n_{ij})n_{ij'}(1-n_{ij'})}\sum_{k=1}^{n_{ij}}\sum_{k'=1}^{n_{ij'}}$$

$$\big[(\hat{\mathbf{Y}}_{ijk} - \hat{\bar{\mathbf{Y}}}_{ij\cdot})(\hat{\mathbf{Y}}_{ijk} - \hat{\bar{\mathbf{Y}}}_{ij\cdot})' \otimes (\hat{\mathbf{Y}}_{ij'k'} - \hat{\bar{\mathbf{Y}}}_{ij'\cdot})(\hat{\mathbf{Y}}_{ij'k'} - \hat{\bar{\mathbf{Y}}}_{ij'\cdot})'$$
$$- (\mathbf{Y}_{ijk} - \bar{\mathbf{Y}}_{ij\cdot})(\mathbf{Y}_{ijk} - \bar{\mathbf{Y}}_{ij\cdot})' \otimes (\mathbf{Y}_{ij'k'} - \bar{\mathbf{Y}}_{ij'\cdot})(\mathbf{Y}_{ij'k'} - \bar{\mathbf{Y}}_{ij'\cdot})'\big],$$

and consider again an arbitrary element $\tilde{K}_{q_1,q_2,q_3,q_4}$ of this $p^2 \times p^2$ matrix that is determined by a combination of four indices q_1, q_2, q_3, q_4, where $q_r = 1, \ldots, p$, $r = 1, \ldots, 4$.

$$\tilde{K}_{q_1,q_2,q_3,q_4} = \frac{1}{ab}\sum_{i=1}^{a}\sum_{j\neq j'}^{b}\frac{1}{n_{ij}(1-n_{ij})n_{ij'}(1-n_{ij'})}\sum_{k=1}^{n_{ij}}\sum_{k'=1}^{n_{ij'}}$$

$$\big[(\hat{Y}^{(q_1)}_{ijk} - \hat{\bar{Y}}^{(q_1)}_{ij\cdot})\hat{Y}^{(q_2)}_{ijk}(\hat{Y}^{(q_3)}_{ij'k'} - \hat{\bar{Y}}^{(q_3)}_{ij'\cdot})\hat{Y}^{(q_4)}_{ij'k'} - (Y^{(q_1)}_{ijk} - \bar{Y}^{(q_1)}_{ij\cdot})Y^{(q_2)}_{ijk}(Y^{(q_3)}_{ij'k'} - \bar{Y}^{(q_3)}_{ij'\cdot})Y^{(q_4)}_{ij'k'}\big]$$

$$= \frac{1}{ab}\sum_{i=1}^{a}\sum_{j\neq j'}^{b}\frac{1}{n_{ij}(1-n_{ij})n_{ij'}(1-n_{ij'})}\sum_{k=1}^{n_{ij}}\sum_{k'=1}^{n_{ij'}}$$

$$\Big\{\big[\hat{Y}^{(q_1)}_{ijk}\hat{Y}^{(q_2)}_{ijk}\hat{Y}^{(q_3)}_{ij'k'}\hat{Y}^{(q_4)}_{ij'k'} - Y^{(q_1)}_{ijk}Y^{(q_2)}_{ijk}Y^{(q_3)}_{ij'k'}Y^{(q_4)}_{ij'k'}\big]$$
$$- \big[\hat{\bar{Y}}^{(q_1)}_{ij\cdot}\hat{Y}^{(q_2)}_{ijk}(\hat{Y}^{(q_3)}_{ij'k'} - \hat{\bar{Y}}^{(q_3)}_{ij'\cdot})\hat{Y}^{(q_4)}_{ij'k'} - \bar{Y}^{(q_1)}_{ij\cdot}Y^{(q_2)}_{ijk}(Y^{(q_3)}_{ij'k'} - \bar{Y}^{(q_3)}_{ij'\cdot})Y^{(q_4)}_{ij'k'}\big]$$
$$- \big[\hat{Y}^{(q_1)}_{ijk}\hat{Y}^{(q_2)}_{ijk}\hat{\bar{Y}}^{(q_3)}_{ij'\cdot}\hat{Y}^{(q_4)}_{ij'k'} - Y^{(q_1)}_{ijk}Y^{(q_2)}_{ijk}\bar{Y}^{(q_3)}_{ij'\cdot}Y^{(q_4)}_{ij'k'}\big]\Big\}.$$

The terms in each of the three square brackets can be considered separately, using basically the same techniques for each. We show details of the proof for the first term.

$$\begin{aligned}
&\hat{Y}_{ijk}^{(q_1)}\hat{Y}_{ijk}^{(q_2)}\hat{Y}_{ij'k'}^{(q_3)}\hat{Y}_{ij'k'}^{(q_4)} - Y_{ijk}^{(q_1)}Y_{ijk}^{(q_2)}Y_{ij'k'}^{(q_3)}Y_{ij'k'}^{(q_4)} \\
&= \hat{H}^{(q_1)}(X_{ijk}^{(q_1)})\hat{H}^{(q_2)}(X_{ijk}^{(q_2)})\hat{H}^{(q_3)}(X_{ij'k'}^{(q_3)})\hat{H}^{(q_4)}(X_{ij'k'}^{(q_4)}) \\
&- H^{(q_1)}(X_{ijk}^{(q_1)})H^{(q_2)}(X_{ijk}^{(q_2)})H^{(q_3)}(X_{ij'k'}^{(q_3)})H^{(q_4)}(X_{ij'k'}^{(q_4)}) \\
&= \frac{1}{N^4}\sum_{s_1=1}^{N}\sum_{s_2=1}^{N}\sum_{s_3=1}^{N}\sum_{s_4=1}^{N}\big[c(X_{ijk}^{(q_1)} - X_{s_1})c(X_{ijk}^{(q_2)} - X_{s_2})c(X_{ij'k'}^{(q_3)} - X_{s_3})c(X_{ij'k'}^{(q_4)} - X_{s_4}) \\
&\qquad - F_{s_1}(X_{ijk}^{(q_1)})F_{s_2}(X_{ijk}^{(q_2)})F_{s_3}(X_{ij'k'}^{(q_3)})F_{s_4}(X_{ij'k'}^{(q_4)})\big].
\end{aligned}$$

Clearly, the expected value of this expression is 0 when all indices s_1, s_2, s_3, s_4 are different from each other, and the corresponding random variables independent of the other four random variables. This can be seen by integrating over $(X_{s_1}, X_{s_2}, X_{s_3}, X_{s_4})$ first. The number of (s_1, s_2, s_3, s_4) index combinations resulting in nonzero expectation is of order N^3, while the sum is divided by N^4. Using similar techniques for the remaining components of $\tilde{K}_{q_1,q_2,q_3,q_4}$, it follows that $E(\tilde{K}_{q_1,q_2,q_3,q_4}) \to 0$. Consider next

$$\begin{aligned}
\tilde{K}^2_{q_1,q_2,q_3,q_4} &= \frac{1}{a^2b^2}\sum_{i_1=1}^{a}\sum_{i_2=1}^{a}\sum_{j_1\neq j_1'}^{b}\sum_{j_2\neq j_2'}^{b} \\
&\frac{1}{n_{i_1j_1}n_{i_2j_2}(1-n_{i_1j_1})(1-n_{i_2j_2})n_{i_1j_1'}n_{i_2j_2'}(1-n_{i_1j_1'})(1-n_{i_2j_2'})} \\
&\sum_{k_1=1}^{n_{i_1j_1}}\sum_{k_2=1}^{n_{i_2j_2}}\sum_{k'=1}^{n_{i_1j_1'}}\sum_{k_2'=1}^{n_{i_2j_2'}}\big[(\hat{Y}_{i_1j_1k_1}^{(q_1)} - \hat{\bar{Y}}_{i_1j_1\cdot}^{(q_1)})\hat{Y}_{i_1j_1k_1}^{(q_2)}(\hat{Y}_{i_1j_1'k_1'}^{(q_3)} - \hat{\bar{Y}}_{i_1j_1'\cdot}^{(q_3)})\hat{Y}_{i_1j_1'k_1'}^{(q_4)} \\
&\qquad - (Y_{i_1j_1k_1}^{(q_1)} - \bar{Y}_{i_1j_1\cdot}^{(q_1)})Y_{i_1j_1k_1}^{(q_2)}(Y_{i_1j_1'k_1'}^{(q_3)} - \bar{Y}_{i_1j_1'\cdot}^{(q_3)})Y_{i_1j_1'k_1'}^{(q_4)}\big] \\
&\big[(\hat{Y}_{i_2j_2k_2}^{(q_1)} - \hat{\bar{Y}}_{i_2j_2\cdot}^{(q_1)})\hat{Y}_{i_2j_2k_2}^{(q_2)}(\hat{Y}_{i_2j_2'k_2'}^{(q_3)} - \hat{\bar{Y}}_{i_2j_2'\cdot}^{(q_3)})\hat{Y}_{i_2j_2'k_2'}^{(q_4)} \\
&\qquad - (Y_{i_2j_2k_2}^{(q_1)} - \bar{Y}_{i_2j\cdot}^{(q_1)})Y_{i_2j_2k_2}^{(q_2)}(Y_{i_2j_2'k_2'}^{(q_3)} - \bar{Y}_{i_2j_2'\cdot}^{(q_3)})Y_{i_2j_2'k_2'}^{(q_4)}\big].
\end{aligned}$$

The product of the square brackets can be decomposed into the following and similar terms, using the same decomposition as in the first part of this proof.

$$\frac{1}{N^8}\sum_{s_1=1}^{N}\sum_{s_2=1}^{N}\sum_{s_3=1}^{N}\sum_{s_4=1}^{N}\sum_{t_1=1}^{N}\sum_{t_2=1}^{N}\sum_{t_3=1}^{N}\sum_{t_4=1}^{N}$$
$$\Big[c(X^{(q_1)}_{i_1j_1k_1}-X_{s_1})c(X^{(q_2)}_{i_1j_1k_1}-X_{s_2})c(X^{(q_3)}_{i_1j_1'k_1'}-X_{s_3})c(X^{(q_4)}_{i_1j_1'k_1'}-X_{s_4})$$
$$-F_{s_1}(X^{(q_1)}_{i_1j_1k_1})F_{s_2}(X^{(q_2)}_{i_1j_1k_1})F_{s_3}(X^{(q_3)}_{i_1j_1'k_1'})F_{s_4}(X^{(q_4)}_{i_1j_1'k_1'})\Big]$$
$$\Big[c(X^{(q_1)}_{i_2j_2k_2}-X_{t_1})c(X^{(q_2)}_{i_2j_2k_2}-X_{t_2})c(X^{(q_3)}_{i_2j_2'k_2'}-X_{t_3})c(X^{(q_4)}_{i_2j_2'k_2'}-X_{t_4})$$
$$-F_{t_1}(X^{(q_1)}_{i_2j_2k_2})F_{t_2}(X^{(q_2)}_{i_2j_2k_2})F_{t_3}(X^{(q_3)}_{i_2j_2'k_2'})F_{t_4}(X^{(q_4)}_{i_2j_2'k_2'})\Big].$$

As above, it can be seen that when all involved random variables with indices $s_1, s_2, s_3, s_4, t_1, t_2, t_3, t_4$ are independent of each other, and of the remaining random variables, the expectation of this expression is zero. The number of cases with nonzero expectation is of order N^7, while division is by N^8. A tedious calculation verifies that this is also the case for the remaining components of $\tilde{K}^2_{q_1,q_2,q_3,q_4}$. Thus, $E(\tilde{K}^2_{q_1,q_2,q_3,q_4}) \to 0$ and $\tilde{K}_{q_1,q_2,q_3,q_4} = o_p(1)$ for each element of $\tilde{K}$, proving $\tilde{K} = o_p(1)$.

The three previous theorems together establish the consistency of a rank-based estimator of the asymptotic variances. Aggregating the results so far, we can take advantage of the results from Harrar and Bathke (2012) and formulate Theorem 7.5.

Theorem 7.5. *Let* $\psi = AB, A, A|B,$ *or* $B|A$. *Under the hypothesis* $\mathscr{H}_0^{(\psi)}$, $\sqrt{a}\,\mathrm{tr}(\mathbf{H}^{(\psi)} - \mathbf{G})\hat{\boldsymbol{\Omega}}\,\hat{\tau}_\psi^{-1}(\hat{\boldsymbol{\Omega}}) \xrightarrow{\mathscr{L}} N(0,1)$ *as* $a \to \infty$ *and* n_{ij} *and* b *bounded, where* $\hat{\boldsymbol{\Omega}}$ *is the consistent estimator of* $\boldsymbol{\Omega}$ *obtained by replacing* $-\boldsymbol{\Sigma}$ *with* $N^{-2}\mathbf{G}$ *(see Proposition 7.2), and where*

$$\hat{\tau}^2_\psi(\hat{\boldsymbol{\Omega}}) = \begin{cases} \frac{2}{b}\left\{\hat{\upsilon}_1(\hat{\boldsymbol{\Omega}}) + \frac{\hat{\upsilon}_2(\hat{\boldsymbol{\Omega}})}{(b-1)^2}\right\} & \text{when } \psi = AB, \\ \frac{2}{b}\left\{\hat{\upsilon}_1(\hat{\boldsymbol{\Omega}}) + \hat{\upsilon}_2(\hat{\boldsymbol{\Omega}})\right\} & \text{when } \psi = A, \\ \frac{2}{b}\hat{\upsilon}_1(\hat{\boldsymbol{\Omega}}) & \text{when } \psi = A|B, \\ \frac{2}{b^2}\left\{\hat{\upsilon}_1(\hat{\boldsymbol{\Omega}}) + \frac{\hat{\upsilon}_2(\hat{\boldsymbol{\Omega}})}{(b-1)^2}\right\} & \text{when } \psi = B|A. \end{cases}$$

Here, $\hat{\upsilon}_1(\hat{\boldsymbol{\Omega}}) = \frac{1}{ab}\sum_{i=1}^{a}\sum_{j=1}^{b}\frac{\mathrm{tr}(\widehat{\boldsymbol{\Psi}}_{ij}(\hat{\boldsymbol{\Omega}}))}{n_{ij}(n_{ij}-1)}$ *and* $\hat{\upsilon}_2(\hat{\boldsymbol{\Omega}}) = \frac{1}{ab}\sum_{i=1}^{a}\sum_{j\neq j'}^{b}\frac{\mathrm{tr}(\hat{\boldsymbol{\Omega}}\hat{\mathbf{S}}_{ij}\hat{\boldsymbol{\Omega}}\hat{\mathbf{S}}_{ij'})}{n_{ij}n_{ij'}}$.

7.5 Simulation Study

In order to investigate the finite sample performance of the proposed inference methods under the exemplary setting of dimension $p = 3$, number of levels of factor A between $a = 6$ and $a = 50$, number of levels of factor B set to $b = 3$, sample sizes

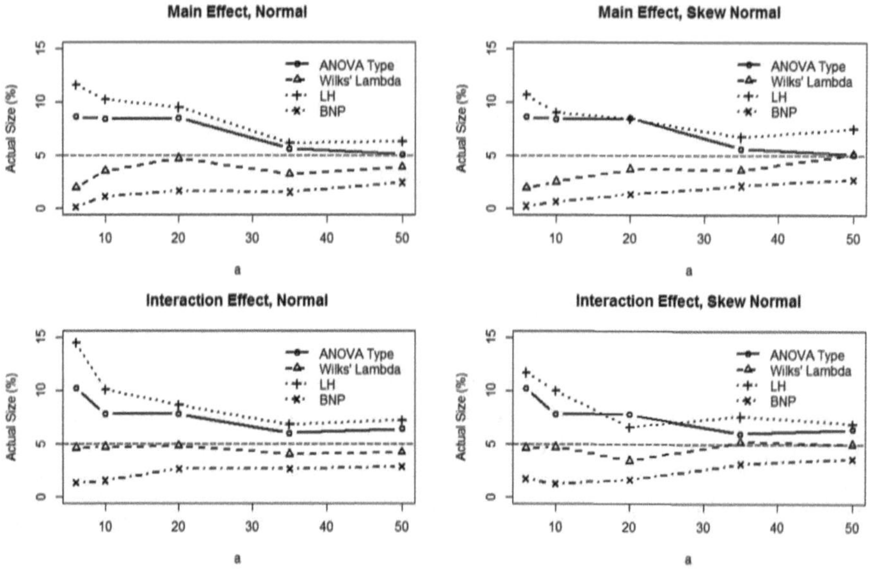

Fig. 7.1 Simulated α under null hypothesis for $p = 3$, $b = 3$, $a = 6$ to 50, $n_{ij} = 4, 5, 6$. Normal and skew normal underlying data distributions, nominal α 5 %. Main effect of A and interaction between A and B

per cell between $n_{ij} = 4$, 5, and 6. Underlying distributions chosen were normal and skew normal. The multivariate skew normal data were generated according to Proposition 6 in Azzalini and Dalla Valle (1996) where we used (in their notation) $\boldsymbol{\delta} = \frac{\sqrt{2}}{\sqrt{p(p+1)+2}}(1, \ldots, p)'$ and $\boldsymbol{\Omega} = \mathrm{I}_p + \frac{1}{2\pi}\boldsymbol{\delta}\boldsymbol{\delta}'$.

The results under null hypothesis are shown in Fig. 7.1. As expected from a fully nonparametric rank-based approach, the underlying distribution does not have a major effect on the performance. In all cases considered, Wilks' Λ type statistic performed best, in the sense of the simulated level being closest to the nominal level, while not exceeding it.

Due to its best performance under null hypothesis, Wilks' Λ type test statistic was selected for a power simulation. Here, the statistic based on variablewise ranks, as proposed in the present article, was compared to the power of the analogous procedure using the original observations instead of the ranks (justified by Harrar and Bathke 2012). While there were no visible differences for underlying normal distributions, the power gain of the nonparametric rank-based method became quite pronounced when the underlying distribution was chosen as contaminated normal. Figure 7.2 shows simulation results for an exemplary situation with heteroscedastic contaminated multivariate normal distributions $0.9N_3(\mathbf{0}, \boldsymbol{\Sigma}_{ij}) + 0.1N_3(10 \cdot \mathbf{1}, \boldsymbol{\Sigma}_{ij})$. Here, $\boldsymbol{\Sigma}_{ij}$ were different compound symmetric variance-covariance matrices with off-diagonal elements $\rho_{ij} = \sqrt{ij}/(1 + ij)$ and diagonal elements $1 - \rho_{ij}$.

Alternatives were modeled by location shifts. Specifically, in the main effects power simulation, expected values were shifted up by δ units for levels 10–20 of

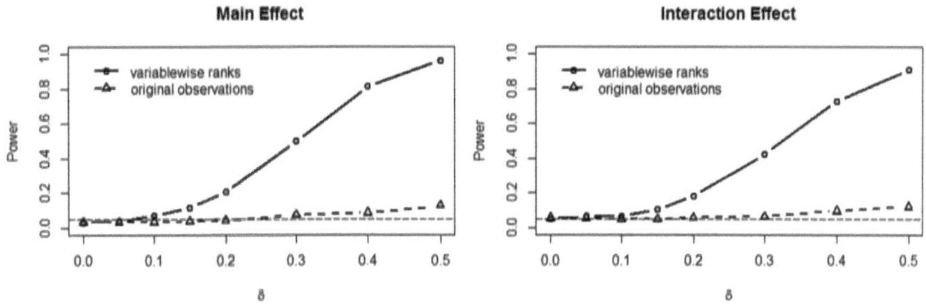

Fig. 7.2 Simulated power of Wilks' Λ type test statistic using ranks, and using raw data, $p = 3$, $a = 20$, $b = 3$, $n_{ij} = 4, 5, 6$. Contaminated normal underlying data distributions, nominal α 5 %. Main effect of A and interaction between A and B. Location shifts for main and interaction effects as described in the text

factor A, for all variables, while they were shifted down by δ units for the other levels 1–9. In the interaction effects power simulation, the upwards shift was for the factor level combinations (i, j) with $i \geq 10, j \geq 2$, whereas the downwards shift was for $i < 10, j < 2$. In both cases, $a = 20$, $b = 3$, $p = 3$.

The results show the rather striking advantages of a nonparametric rank-based approach over its semiparametric competitor using the original observations instead of ranks.

7.6 Discussions and Conclusions

In this somewhat theoretical manuscript, we have introduced fully nonparametric, rank-based test statistics for inference on multivariate data in factorial designs. To our knowledge, no comparable results in such general applicability (for example for fully ordinal data) have been established yet. Due to the rather cumbersome technicalities, the work has only been carried out here for a design with two factors, but it can be extended in a straightforward way to higher-way layouts. Also, we have focused here on large (a) asymptotics (number of factor levels of factor A tends to infinity) and only considered those test statistics in detail that yield asymptotic normality under this type of asymptotic setting. The asymptotic distribution of the test for main effect of factor B will be that of a weighted sum of χ^2 random variables.

It should be pointed out that the test statistics can be calculated directly, they don't involve any iterative computational procedures. The test statistics presented here can be taken as a basis for small sample approximations based on moment estimators or expansions. In future work, it would be interesting to compare their performance with resampling based methods such as those from Konietschke et al. (2015), or with other robust procedures based on semiparametric models.

Acknowledgements Dedicated to Joe McKean on the occasion of his 70th birthday.

References

Akritas, M., & Arnold S. (2000). Asymptotics for analysis of variance when the number of levels is large. *Journal of the American Statistical association, 95*(449), 212–226.

Azzalini, A., & Dalla Valle, A. (1996). The multivariate skew-normal distribution. *Biometrika, 83*(4), 715–726.

Bathke, A. C. (2002). Anova for a large number of treatments. *Mathematical Methods of Statistics, 11*(1), 118–132.

Bathke, A. C. (2004). The anova F test can still be used in some balanced designs with unequal variances and nonnormal data. *Journal of Statistical Planning and Inference, 126*(2), 413–422.

Bathke, A. C., & Harrar, S.W. (2008). Nonparametric methods in multivariate factorial designs for large number of factor levels. *Journal of Statistical Planning and Inference, 138*(3), 588–610.

Belloni, A., & Didier, G. (2008). On the Behrens–Fisher problem: A globally convergent algorithm and a finite-sample study of the Wald, LR and LM tests. *The Annals of Statistics, 36*, 2377–2408.

Boos, D.D., & Brownie, C. (1995). Anova and rank tests when the number of treatments is large. *Statistics & Probability Letters, 23*(2), 183–191.

Brunner, E., Domhof, S., & Langer, F. (2002). *Nonparametric analysis of longitudinal data in factorial experiments*. New York: Wiley.

Chatfield, J. A., Draper, E. A., Cochran, K. D., & Herms, D. A. (2000). Evaluation of crabapples for apple scab at the Secrest Arboretum in Wooster, Ohio. *Special circular 177*, The Ohio State University, Ohio Agricultural Research and Development Center.

Girón, F. J., & del Castillo, C. (2010). The multivariate Behrens–Fisher distribution. *Journal of Multivariate Analysis, 101*(9), 2091–2102.

Gupta, A. K., Harrar, S. W., & Fujikoshi, Y. (2006). Asymptotics for testing hypothesis in some multivariate variance components model under non-normality. *Journal of Multivariate Analysis, 97*, 148–178.

Gupta, A. K., Harrar, S. W., & Fujikoshi, Y. (2008). Manova for large hypothesis degrees of freedom under non-normality. *Test, 17*(1), 120–137.

Harrar, S. W., & Bathke, A. C. (2008). Nonparametric methods for unbalanced multivariate data and many factor levels. *Journal of Multivariate Analysis, 99*(8), 1635–1664.

Harrar, S. W., & Bathke, A. C. (2012). A modified two-factor multivariate analysis of variance: asymptotics and small sample approximations (and erratum). *Annals of the Institute of Statistical Mathematics, 64*(1 & 5), 135–165 & 1087.

Harrar, S. W., & Gupta, A. K. (2007) Asymptotic expansion for the null distribution of the f-statistic in one-way anova under non-normality. *Annals of the Institute of Statistical Mathematics, 59*(3), 531–556.

Harville, D. A. (2008). *Matrix algebra from a statistician's perspective*. Berlin: Springer.

Kawasaki, T., & Seo, T. (2015). A two sample test for mean vectors with unequal covariance matrices. *Communications in Statistics-Simulation and Computation, 44*, 1850–1866.

Konietschke, F., Bathke, A. C., Harrar, S. W., & Pauly, M. (2015). Parametric and nonparametric bootstrap methods for general MANOVA. *Journal of Multivariate Analysis, 140*, 291–301.

Krishnamoorthy, K., & Lu, F. (2010). A parametric bootstrap solution to the MANOVA under heteroscedasticity. *Journal of Statistical Computation and Simulation, 80*(8), 873–887.

Krishnamoorthy, K., & Yu, J. (2004). Modified Nel and Van der Merwe test for the multivariate Behrens–Fisher problem. *Statistics & Probability Letters, 66*(2), 161–169.

Krishnamoorthy, K., & Yu, J. (2012). Multivariate Behrens–Fisher problem with missing data. *Journal of Multivariate Analysis, 105*(1), 141–150.

Kruskal, W. H. (1952). A nonparametric test for the several sample problem. *Annals of Mathematical Statistics, 23*, 525–540.

Lévy, P. (1925). *Calcul des Probabilités*. Paris: Gauthiers-Villars.

Nel, D.G., & Van der Merwe, C. A. (1986). A solution to the multivariate Behrens-Fisher problem. *Communications in Statistics-Theory and Methods, 15*(12), 3719–3735.

Penrose, R. (1955). A generalized inverse for matrices. *Proceedings of the Cambridge Philosophical Society, 51*, 17–19.

Ruymgaart, F. H. (1980). A unified approach to the asymptotic distribution theory of certain midrank statistics. In *Statistique non parametrique asymptotique*. *Lecture notes on mathematics* (Vol. 821). Berlin: Springer.

Schott, J. R. (2005). *Matrix analysis for statistics*. Hoboken: Wiley.

Serfling, R. (1980). *Approximation theorems of mathematical statistics*. New York: Wiley.

Van Aelst, S., & Willems, G. (2011). Robust and efficient one-way MANOVA tests. *Journal of the American Statistical Association, 106*(494), 706–718.

Zhang, J. -T. (2011). Two-way MANOVA with unequal cell sizes and unequal cell covariance matrices. *Technometrics, 53*(4), 426–439.

Zhang, J. -T. (2012). An approximate Hotelling T^2-test for heteroscedastic one-way MANOVA. *Open Journal of Statistics, 2*(1), 1–11.

Zhang, J. -T., & Liu, X. (2013). A modified Bartlett test for heteroscedastic one-way MANOVA. *Metrika, 76*(1), 135–152.

Chapter 8
Two-Sample Rank-Sum Test for Order Restricted Randomized Designs

Yiping Sun and Omer Ozturk

Abstract This paper develops a new nonparametric test for the location shift between two populations based on order restricted randomized design (ORRD). The ORRD exploits the use of subjective, imprecise or rough information among experimental units to create a blocking factor. The blocking factor, in a given set of H experimental units, is constructed by ranking the units from smallest to largest and then assigning them into H ranking classes (judgment blocks). The design then uses a restricted randomization to assign the treatment regimes to experimental units across these judgment blocks. This randomization scheme induces a positive correlation structure among within-set response measurements. The positive correlation structure then acts as a variance reduction technique in the inference of a contrast parameter in an ORRD. The paper develops a rank-sum test to test the difference between two treatment medians. It is shown that the test performs better than its competitors regardless of the accuracy of the ranking information of within-set units. The paper also constructs point and interval estimators for the contrast parameter. For set sizes $H > 2$, there are more than one ORRDs. The paper constructs an optimal design that maximizes the asymptotic Pitman efficacy of the proposed test among all possible ORRDs. The proposed test is applied to ACTG 320 clinical trial data.

Keywords Order restricted randomized design • Ranking error • Wilcoxon test • Ranked set sampling • Pitman efficiency

Y. Sun
PTC Therapeutics, INC., 674 Roosevelt Avenue, Piscataway, NJ 08854, USA
e-mail: suny1963@yahoo.com

O. Ozturk (✉)
The Ohio State University, 404 Cockins Hall, 1958 Neil Avenue, Columbus, OH 43210, USA
e-mail: ozturk.4@stat.osu.edu

© Springer International Publishing Switzerland 2016

R.Y. Liu, J.W. McKean (eds.), *Robust Rank-Based and Nonparametric Methods*, Springer Proceedings in Mathematics & Statistics 168,
DOI 10.1007/978-3-319-39065-9_8

8.1 Introduction

In a design of experiment, natural variation among experimental units can be accounted in several ways. One of the common approach is to identify an auxiliary variable either to fit a covariance model or to create a blocking variable to construct homogeneous groups. In certain settings, there may not exits a clearly defined auxiliary variable and the natural variation can be assessed with informal subjective observations. Although subjective observations may be questionable in quality, it is extremely useful in reducing the total variation in an experiment.

As an example, consider a study of efficacy comparison between a new drug and an existing one. In this setting, formal measurements on experimental units may include age, gender, weight, etc., which can be used as covariates or as a blocking factor in the design of the experiments. Informal measurements may include general health status, pre-medical history of patients, etc. This information in general may be incomplete, subjective, imprecise, even biased, and it may not be converted to numerical numbers or categories easily. Nonetheless, the information is very useful to improve statistical inference if it is used properly at the design and analysis stages of the experiment.

Ozturk and MacEachern (2004, 2007) proposed order restricted randomized design (ORRD) to use this informal and subjective information to construct judgment ranking blocks to estimate contrast parameters in a design of experiment. The following model describes the underlying structure between the responses and model parameters in an additive model

$$Z_{ij} = \theta_i + \gamma_{ij} \text{ for } i = 1, 2 \text{ and } j = 1, \cdots, n, \tag{8.1}$$

where Z_{ij} is the response measurement from the experimental unit j with treatment i; θ_i is the median effect of treatment i; γ_{ij}'s are independent and identically distributed (iid) random errors from a distribution F with finite Fisher information. This model indicates that the error term γ_{ij} is the property of an experimental unit. Thus, the heterogeneity among experimental units is explained by these random error terms. The premise of ORRD is to use this heterogeneity by ranking the residual terms in model (8.1) at the design stage of an experiment.

The order restricted randomized design can be constructed through a two-step procedure, ranking and randomization. In ranking step, we select $2H$ experimental units at random from an infinite population and divide them into two sets, each of size H. Units in each set are judgment ranked based on their anticipated error terms, γ_{ij}, with available subjective auxiliary information. Let $R_1, \ldots, R_H$ denote the subjective ranks of the experimental units in each set. Note that ranking of experimental units, similar to the construction of blocks in randomized block design, is performed pre-experimentally based on inherent variations among the experimental units since experiment has not been conducted yet. Hence, there may be ranking error in the construction of judgment ranks.

Table 8.1 Illustration of an ORRD with replication size n, set size H, sets α and β

Replication	Set[a]	Control	Treatment
1	1	$X_{[\alpha_1]1}, \cdots, X_{[\alpha_u]1}$	$Y_{[\beta_1]1}, \cdots, Y_{[\beta_{H-u}]1}$
	2	$X_{[\beta_1]1}, \cdots, X_{[\beta_{H-u}]1}$	$Y_{[\alpha_1]1}, \cdots, Y_{[\alpha_u]1}$
2	1	$X_{[\alpha_1]2}, \cdots, X_{[\alpha_u]2}$	$Y_{[\beta_1]2}, \cdots, Y_{[\beta_{H-u}]2}$
	2	$X_{[\beta_1]2}, \cdots, X_{[\beta_{H-u}]2}$	$Y_{[\alpha_1]2}, \cdots, Y_{[\alpha_u]2}$
⋮	⋮	⋮	⋮
n	1	$X_{[\alpha_1]n}, \cdots, X_{[\alpha_u]n}$	$Y_{[\beta_1]n}, \cdots, Y_{[\beta_{H-u}]n}$
	2	$X_{[\beta_1]n}, \cdots, X_{[\beta_{H-u}]n}$	$Y_{[\alpha_1]n}, \cdots, Y_{[\alpha_u]n}$

[a]Judgment order statistics from the same set are correlated, but judgment order statistics from different sets are independent

In randomization step, we separate the ranks $(R_1, \cdots, R_H)$ into two disjoint non-empty subsets $\boldsymbol{\alpha}$ and $\boldsymbol{\beta}$, where $\boldsymbol{\alpha} = (\alpha_1, \cdots, \alpha_u)$ and $\boldsymbol{\beta} = (\beta_1, \cdots, \beta_{H-u})$. In the first set, we then perform a randomization to decide whether the control or treatment regimes are assigned to the units that have ranks in set $\boldsymbol{\alpha}$ or $\boldsymbol{\beta}$. In the second set, opposite allocation is performed without randomization so that each treatment regime is applied to all ranks. These two steps are called a *replication*. This basic process is repeated n times to increase the sample size. Table 8.1 illustrates the construction of ORRD for replication size n, set size H, and sets $\boldsymbol{\alpha}$ and $\boldsymbol{\beta}$.

Model (8.1) under ORRD can be revised to reflect the ranking structure of within-set residuals

$$Z_{i[h]j} = \theta_i + \gamma_{i[h]j} \text{ for } i = 1, 2; h = 1, \cdots, H, \text{ and } j = 1, \cdots, n, \tag{8.2}$$

where $Z_{i[h]j}$ and $\gamma_{i[h]j}$ are, respectively, the response measurement and error term for treatment i ranked unit h and replication j. The parameter θ_i is the median effect of treatment i. The square brackets are used to indicate that there may be ranking error in the construction of judgment ranks. If there is no ranking error, we replace the square brackets with the round parentheses. In this case, the random variable $Z_{i[h]j}$ and $\gamma_{i[h]j}$ become the hth order statistics in a set of size H. In this paper unless stated otherwise, we use an arbitrary, but consistent ranking scheme. The ranking scheme is called consistent if the following equality holds

$$F(y) = \sum_{h=1}^{H} F_{[h]}(y),$$

where $F_{[h]}(y)$ is the cdf of the hth judgment order statistic, $\gamma_{i[h]j}$.

In Table 8.1, the response variable $Z_{i[h]j}$ reduces to $Z_{i[h]j} = X_{[h]j}$ if $i = 1$ and $Z_{i[h]j} = Y_{[h]j}$ if $i = 2$. It is important to recognize that all X- and Y-measurements are correlated if they belong to the same set since they are judgment order statistics in a set of size H. On the other hand, any two measurements from different sets are independent since they are ordered independently in different sets. Under a perfect ranking model, judgment order statistics $X_{[h]j}$ and $Y_{[h]j}$ become the hth order statistics in a set of size H from control and treatment group in replication j, respectively. In this case, we use the standard notation $X_{(h)j}$ and $Y_{(h)j}$ to denote the order statistics.

For any fixed value of H, the ORRD is determined by the choices of set $\boldsymbol{\alpha}$ and set $\boldsymbol{\beta}$. Thus, the set $D = (\boldsymbol{\alpha}, \boldsymbol{\beta})$ is called the design parameter of ORRD. When H is 2, the design is unique. When H is greater than 2, there are $2^{H-1} - 1$ possible designs depending on the how the integers $1, \cdots, H$ are allocated into sets $\boldsymbol{\alpha}$ and $\boldsymbol{\beta}$. For $H > 2$, an optimal design can be chosen from all possible designs based on some reasonable criteria.

A data set obtained from ORRD has a specific feature. Within set observations are not independent although between set observations are independent. Under some mild assumptions, within-set observations are positively correlated. This specific correlation structure of the data leads to an important characteristic of the ORRD. Since the ranked units from the same set are separated into two treatment groups, the within set positive correlations translate into negative correlations for the estimation of the contrast parameter and leads to a variance reduction for the contrast estimator.

The subjective information has also been successfully used in a ranked set sampling (RSS) design in McIntyre (1952, 2005). As in ORRD, an RSS design ranks EUs in a small set without measurement, but it uses these ranks to select a single EU for measurements to construct homogeneous groups of EUs. Unlike an ORRD where all units in the set are measured, remaining $H-1$ units in the set in an RSS design are unused. Hence, RSS design could be too restrictive in design of an experiment where potential EUs are expensive or limited. The construction of RSS and most current literature review can be found in Wolfe (2012) and Hollander et al. (2014, Chap. 15).

In recent years, there has been increased research activity in ORRD. Ozturk and MacEachern (2007) looked at the design issues of ORRD in a two-sample problem and developed statistical inference for the contrast parameter of a location shift between treatment and control means. Sun (2007) and Ozturk and Sun (2009) constructed rank based inference for a two sample problem. Due and MacEachern (2007) constructed several statistics to estimate the location shift between treatment and control regimes based on judgment post stratified ORR designs. Markiewicz (2008) fitted a linear model to ORRD to estimate the model parameters based on L-1 norm. Recently Ozturk and MacEachern (2013) used generalized linear models to draw inference based on ORRD. Gao and Ozturk (2015) developed inference for linear models based on rank dispersion function.

In this paper, we use ORRD to develop a rank-sum test for location shift between control and treatment populations. Section 8.2 introduces the test statistic and investigates its asymptotic null distribution. It is shown that the test statistic is asymptotically normal and has higher asymptotic Pitman efficacy than its competitors. Section 8.3 discusses the asymptotic null distribution of the test statistic under imperfect ranking. Section 8.4 provides empirical evidence to investigate the properties of the tests and estimators. Section 8.5 applies the proposed tests to a clinical trial data set. Section 8.6 provides a concluding remark. The details of the proofs of the theorems can be found in PhD dissertation in the Department of Statistics at the Ohio state University (Sun 2007)

8.2 Two-Sample Rank-Sum Test

Let $F(x)$ and $G(y) = F(y - \Delta)$ be the cumulative distribution functions (cdfs) of the control and treatment populations, respectively. The parameter, $\Delta = \theta_Y - \theta_X$, represents a location shift between these two distributions, where θ_X and θ_Y are the medians of G and F. Let $F_{[i]}$ and $G_{[i]}$ be the cdfs for $X_{[i]j}$ and $Y_{[i]j}$, the ith judgment order statistics in replication $j; j = 1, \cdots, n$, respectively.

We develop a nonparametric test to test the null hypothesis $H_0 : \Delta = 0$ against the alternative hypothesis $H_1 : \Delta \neq 0$. Let

$$T = \sum_{i=1}^{H}\sum_{j=1}^{n}\sum_{k=1}^{H}\sum_{t=1}^{n} I(X_{[i]j} \leq Y_{[k]t}).$$

Our test rejects the null hypothesis for extreme values of T. For computational easiness, we center the test statistic T and write

$$\begin{aligned} T^* &= \sum_{i=1}^{H}\sum_{j=1}^{n}\sum_{k=1}^{H}\sum_{t=1}^{n} [I(X_{[i]j} \leq Y_{[k]t}) - \tau_{ik}] \\ &= \sum_{i=1}^{H}\sum_{j=1}^{n}\sum_{k=1}^{H}\sum_{t=1}^{n} I(X_{[i]j} \leq Y_{[k]t}) - \frac{n^2H^2}{2}, \end{aligned}$$

where

$$\tau_{ik} = E[I(X_{[i]j} \leq Y_{[k]t})].$$

It is clear that $E[T^*] = 0$. For notational convenience, we also define $\bar{T}^* = \frac{T^*}{(nH)^2}$ and $\bar{T} = \frac{T}{(nH)^2}$.

Standard asymptotic theory is not applicable to derive the asymptotic null distribution of the test statistic due to the fact that T is not a sum of independent random variables. We first project the test statistic onto a space of linear functions of independent random variables.

We rearrange the data in the ORRD as follows

$$\begin{aligned} \mathbf{Z}_{1s} &= (X_{[\alpha_1]s}, \cdots, X_{[\alpha_u]s}, Y_{[\beta_1]s}, \cdots, Y_{[\beta_{H-u}]s}), \qquad s = 1, \cdots, n \\ \mathbf{Z}_{2s} &= (X_{[\beta_1]s}, \cdots, X_{[\beta_{H-u}]s}, Y_{[\alpha_1]s}, \cdots, Y_{[\alpha_u]s}), \qquad s = 1, \cdots, n \\ \mathbf{Z} &= (\mathbf{Z}_{11}, \cdots, \mathbf{Z}_{1n}, \mathbf{Z}_{21}, \cdots, \mathbf{Z}_{2n}), \end{aligned}$$

where $\mathbf{Z}_{1s}$ and $\mathbf{Z}_{2s}$ indicate the set 1 and set 2 observations in the sth replication in Table 8.1, respectively. All observations in vectors $\mathbf{Z}_{1s}$ and $\mathbf{Z}_{2s}$ are positively correlated, but the vectors $\mathbf{Z}_{11}, \cdots, \mathbf{Z}_{1n}, \mathbf{Z}_{21}, \cdots, \mathbf{Z}_{2n}$ are mutually independent

H-dimensional random vectors since the sets are randomly selected in ORRD. The test statistic T^* can be considered as a random variable based on independent random vectors $\mathbf{Z}_{11}, \cdots, \mathbf{Z}_{1n}, \mathbf{Z}_{21}, \cdots, \mathbf{Z}_{2n}$, such that $E[T^*] = 0$. We first construct a projection of $\bar{T}^*$, $\bar{V}_p$, onto the space of linear functions of $\mathbf{Z}_{11}, \cdots, \mathbf{Z}_{1n}$, $\mathbf{Z}_{21}, \cdots, \mathbf{Z}_{2n}$.

Lemma 8.1. *Let $F(x)$ and $G(y) = F(y-\Delta)$ be the cdfs of the control and treatment populations, respectively. Under the null hypothesis, the projection of $\bar{T}^*$ onto the space of linear functions of $\mathbf{Z}_{11}, \cdots, \mathbf{Z}_{1n}, \mathbf{Z}_{21}, \cdots, \mathbf{Z}_{2n}$ is given by*

$$\begin{aligned} \bar{V}_P &= \sum_{s=1}^{n} [E(\bar{T}^*|\mathbf{Z}_{1s}) + E(\bar{T}^*|\mathbf{Z}_{2s})] \\ &= \frac{1}{nH} \sum_{s=1}^{n} \left\{ \sum_{i=1}^{u} [1 - F(X_{[\alpha_i]s}) - \bar{\tau}_{\alpha_i \cdot}] + \sum_{k=1}^{H-u} [F(Y_{[\beta_k]s}) - \bar{\tau}_{\cdot \beta_k}] \right\} \\ &+ \frac{1}{nH} \sum_{s=1}^{n} \left\{ \sum_{i=1}^{H-u} [1 - F(X_{[\beta_i]s}) - \bar{\tau}_{\beta_i \cdot}] + \sum_{k=1}^{u} [F(Y_{[\alpha_k]s}) - \bar{\tau}_{\cdot \alpha_k}] \right\}, \end{aligned} \tag{8.3}$$

where

$$\bar{\tau}_{\alpha_i \cdot} = \sum_{t=1}^{H} \tau_{\alpha_i t}/H, \bar{\tau}_{\beta_i \cdot} = \sum_{t=1}^{H} \tau_{\beta_i t}/H, \bar{\tau}_{\cdot \alpha_k} = \sum_{t=1}^{H} \tau_{t \alpha_k}/H, \bar{\tau}_{\cdot \beta_k} = \sum_{t=1}^{H} \tau_{t \beta_k}/H.$$

Proof. Under the null hypothesis $F(x) = G(x)$. We partition $\bar{T}^*$ into four different sums

$$\begin{aligned} \bar{T}^* &= \frac{1}{nH^2} \sum_{i=1}^{u} \sum_{j=1}^{n} \sum_{k=1}^{u} \sum_{t=1}^{n} [I(X_{[\alpha_i]j} \le Y_{[\alpha_k]t}) - \tau_{\alpha_i \alpha_k}] \\ &+ \frac{1}{nH^2} \sum_{i=1}^{u} \sum_{j=1}^{n} \sum_{k=1}^{H-u} \sum_{t=1}^{n} [I(X_{[\alpha_i]j} \le Y_{[\beta_k]t}) - \tau_{\alpha_i \beta_k}] \\ &+ \frac{1}{nH^2} \sum_{i=1}^{H-u} \sum_{j=1}^{n} \sum_{k=1}^{u} \sum_{t=1}^{n} [I(X_{[\beta_i]j} \le Y_{[\alpha_k]t}) - \tau_{\beta_i \alpha_k}] \\ &+ \frac{1}{nH^2} \sum_{i=1}^{H-u} \sum_{j=1}^{n} \sum_{k=1}^{H-u} \sum_{t=1}^{n} [I(X_{[\beta_i]j} \le Y_{[\beta_k]t}) - \tau_{\beta_i \beta_k}]. \end{aligned}$$

The projection of $\bar{T}^*$ follows from

$$\bar{V}_P = \sum_{s=1}^{n} [E(\bar{T}^*|\mathbf{Z}_{1s}) + E(\bar{T}^*|\mathbf{Z}_{2s})].$$

It can be shown that

$$
\begin{aligned}
E(\bar{T}^*|\mathbf{Z}_{1s}) &= \frac{n-1}{n^2H^2} I(s=j, t \neq s) \left\{ \sum_{i=1}^{u} \sum_{k=1}^{u} [1 - F_{[\alpha_k]}(X_{[\alpha_i]s}) - \tau_{\alpha_i \alpha_k}] \right. \\
&+ \left. \sum_{i=1}^{u} \sum_{k=1}^{H-u} [1 - F_{[\beta_k]}(X_{[\alpha_i]s}) - \tau_{\alpha_i \beta_k}] \right\} \\
&+ \frac{n-1}{n^2H^2} I(s=t, j \neq s) \left\{ \sum_{i=1}^{u} \sum_{k=1}^{H-u} [F_{[\alpha_i]}(Y_{[\beta_k]s}) - \tau_{\alpha_i \beta_k}] \right. \\
&+ \left. \sum_{i=1}^{H-u} \sum_{k=1}^{H-u} [F_{[\beta_k]}(Y_{[\beta_k]s}) - \tau_{\beta_i \beta_k}] \right\} + o(1/n^2)
\end{aligned}
$$

and

$$
\begin{aligned}
E(\bar{T}^*|\mathbf{Z}_{2s}) &= \frac{n-1}{n^2H^2} I(s=j, t \neq s) \left\{ \sum_{i=1}^{H-u} \sum_{k=1}^{u} [1 - F_{[\alpha_k]}(X_{[\beta_i]s}) - \tau_{\beta_i \alpha_k}] \right. \\
&+ \left. \sum_{i=1}^{H-u} \sum_{k=1}^{H-u} [1 - F_{[\beta_k]}(X_{[\beta_i]s}) - \tau_{\beta_i \beta_k}] \right\} \\
&+ \frac{n-1}{n^2H^2} I(s=j, t \neq s) \left\{ \sum_{i=1}^{u} \sum_{k=1}^{u} [F_{[\alpha_i]}(Y_{[\alpha_k]s}) - \tau_{\alpha_i \alpha_k}] \right. \\
&+ \left. \sum_{i=1}^{H-u} \sum_{k=1}^{u} [F_{[\beta_i]}(Y_{[\alpha_k]s}) - \tau_{\beta_i \alpha_k}] \right\} + o(1/n^2)
\end{aligned}
$$

Combining these two conditional expectation and using the equality in consistent ranking scheme, $\sum_{i=1}^{H} F_{[i]}(t) = HF(t)$, we show that

$$
\begin{aligned}
\bar{V}_P &= \sum_{s=1}^{n} [E(\bar{T}^*|\mathbf{Z}_{1s}) + E(\bar{T}^*|\mathbf{Z}_{2s})] \\
&= \frac{1}{nH} \sum_{s=1}^{n} \left\{ \sum_{i=1}^{u} [1 - F(X_{[\alpha_i]s}) - \bar{\tau}_{\alpha_i \cdot}] + \sum_{k=1}^{H-u} [F(Y_{[\beta_k]s}) - \bar{\tau}_{\cdot \beta_k}] \right\} \\
&+ \frac{1}{nH} \sum_{s=1}^{n} \left\{ \sum_{i=1}^{H-u} [1 - F(X_{[\beta_i]s}) - \bar{\tau}_{\beta_i \cdot}] + \sum_{k=1}^{u} [F(Y_{[\alpha_k]s}) - \bar{\tau}_{\cdot \alpha_k}] \right\} + o_p(1/n),
\end{aligned}
$$

The proof is completed by showing that difference between the variances of $\bar{V}_P$ and $\bar{T}^*$ goes to zero as n gets large.

In Eq. (8.3), for each s the random variables in curly brackets belong to the same set and hence, they are positively correlated. On the other hand, the random variables in the first and second sums belong to set type 1 and set type 2 in Table 8.1, respectively. These random variables are independent. We then observe that the statistic $\bar{V}_P$ is a sum of $2n$ independent random variables and limiting distribution of $\overline{V}_P$ follows from the central limit theorem.

Theorem 8.1. *The null distribution of $\sqrt{2nH}(\bar{T}-\frac{1}{2})$, as n goes to infinity, converges to a normal distribution with mean 0 and variance σ^2, where*

$$\sigma^2 = \frac{2}{H}\left\{ Var\left[\sum_{i=1}^{u}(1-F(X_{[\alpha_i]1})-\bar{\tau}_{\alpha_i\cdot}) + \sum_{k=1}^{H-u}(F(Y_{[\beta_k]1})-\bar{\tau}_{\cdot\beta_k})\right]\right.$$
$$\left. + \; Var\left[\sum_{i=1}^{H-u}(1-F(X_{[\beta_i]1})-\bar{\tau}_{\beta_i\cdot}) + \sum_{k=1}^{u}(F(Y_{[\alpha_k]1})-\bar{\tau}_{\cdot\alpha_k})\right]\right\}.$$

Proof. The proof follows from the central limit theorem.

It is clear from Theorem 8.1 that the asymptotic null distribution of $\sqrt{2nH}(\overline{T}-\frac{1}{2})$ depends on ranking mechanism and is not distribution free in general. Under perfect ranking, Theorem 8.1 can be simplified and the proposed rank-sum test statistic T becomes asymptotically distribution free. In this case, the test statistic T becomes functions of order statistics from control and treatment groups

$$T = \sum_{i=1}^{H}\sum_{j=1}^{n}\sum_{k=1}^{H}\sum_{t=1}^{n} I(X_{(i)j} \leq Y_{(k)t}).$$

The asymptotic variance of $\sqrt{2nH}(\bar{T} - 1/2)$ reduces to an explicit expression depending only the design parameters $\boldsymbol{\alpha}$ and $\boldsymbol{\beta}$.

Theorem 8.2. *Under perfect ranking, the asymptotic null distribution of $\sqrt{2nH}(\overline{T} - \frac{1}{2})$ converges to a normal distribution with mean zero and variance σ_p^2, where*

$$\sigma_p^2 = \frac{4}{H}\left\{ \sum_{i=1}^{u} \frac{\alpha_i(H+1-\alpha_i)}{(H+1)^2(H+2)} + 2\sum_{i=1}^{u}\sum_{j=1}^{u} \frac{I(\alpha_i < \alpha_j)\alpha_i(H+1-\alpha_j)}{(H+1)^2\,(H+2)} + \right.$$
$$\sum_{k=1}^{H-u} \frac{\beta_k(H+1-\beta_k)}{(H+1)^2(H+2)} + 2\sum_{k=1}^{H-u}\sum_{t=1}^{H-u} \frac{I(\beta_k < \beta_t)\beta_k(H+1-\beta_t)}{(H+1)^2\,(H+2)}$$
$$\left. -2\sum_{i=1}^{u}\sum_{k=1}^{H-u} \frac{I(\alpha_i < \beta_k)\alpha_i(H+1-\beta_k)}{(H+1)^2\,(H+2)} - 2\sum_{i=1}^{u}\sum_{k=1}^{H-u} \frac{I(\beta_k < \alpha_i)\beta_k(H+1-\alpha_i)}{(H+1)^2\,(H+2)} \right\}.$$

Proof. Under perfect ranking $F(X_{(k)})$ is distributed as beta distribution with parameters k and $H + 1 - k$. The proof is completed by evaluating the moments of beta distributions in σ^2 in Theorem 8.1.

There are more than one ways that an ORRD can be constructed when $H > 2$. In this case, it is desirable to select a design so that the asymptotic Pitman efficacy of the test is larger than any other design among all possible ORRDs. Let $G_{\Delta_n}(t) = F(t - \Delta_N)$, where $\Delta_n = a/\sqrt{n}$, $a > 0$. Under this local alternative, the Pitman efficacy of the test based on the design parameters α and β is given in the following theorem.

Theorem 8.3. *Let X and Y be two random variables with cdfs F(X) and G(Y) = F(Y − Δ), respectively, and f(X) be the probability density function (pdf) of F(x). Assume that* $\int f^2(x)dx$ *is bounded. The asymptotic Pitman efficacy of* $T^*(\Delta)$ *is given by*

$$c^2_{ORRD} = \frac{\{\frac{d}{d\Delta}E_\Delta \bar{T}^*(0)\big|_{\Delta=0}\}^2}{\sigma_p^2} = \frac{(\int f^2(x)dx)^2}{\sigma_p^2}.$$

Proof. For the shift parameter Δ, we compute $E_\Delta \bar{T}^*(0)$ by using the expected values of each term in the partition of $\bar{T}^*(0)$ in the proof of Lemma 8.1

$$\begin{aligned} E_\Delta \bar{T}^*(0) &= \frac{1}{H}\sum_{i=1}^{u}\left\{1 - \int F(x-\Delta)dF_{(\alpha_k)}(x) - \bar{\tau}_{\alpha_i.}\right\} \\ &\quad + \frac{1}{H}\sum_{i=1}^{H-u}\left\{1 - \int F(x-\Delta)dF_{(\beta_i)}(x) - \bar{\tau}_{\beta_i.}\right\} + o(1) \\ &= 1 - \int F(x-\Delta)dF(x) - \bar{\tau}_{..} + o(1). \end{aligned}$$

The proof is completed by taking the derivative with respect to Δ.

We note that c^2_{ORRD} depends on $\boldsymbol{\alpha}$ and $\boldsymbol{\beta}$ through σ_p^2. Hence, maximizing c^2_{ORRD} is equivalent to minimizing σ_p^2 over all $\boldsymbol{\alpha}$ and $\boldsymbol{\beta}$.

Theorem 8.4. *Let* $H > 2$ *be any fixed integer. The null variance of the test statistic* $\sqrt{2nH}(\bar{T} - \frac{1}{2})$, σ_p^2, *is minimized when set* $\boldsymbol{\alpha}$ *takes odd integers only and set* $\boldsymbol{\beta}$ *takes even integers only, or vise verse.*

Proof. The proof is divided into two steps. Step I shows that any design of the form $D_{ki,jr} = (\cdots \beta_k, \alpha_i, \cdots, \alpha_j, \beta r, \cdots)$, $i < j$ can be improved by designs $D_{ki,rj} = (\cdots, \beta_k, \alpha_i, \cdots, \beta_r, \alpha_j, \cdots)$ and $D_{ik,jr} = (\cdots, \alpha_i, \beta_k, \alpha_{i+1}, \cdots, \alpha_j, \beta r, \cdots)$. In another words, the variance of $\bar{T}^*$ based on design $D_{ki,jr}$ is larger than its variance under designs $D_{ki,rj}$ and $D_{ik,jr}$, where $D_{ki,jr}$ is the design that all integers between the ith and jth entries are from set α. By using this result, we search the optimal design with no consecutive αs and βs in the middle ranks. On the other hand, there can still be consecutive αs or βs at the both ends. Then in the step II, we need to show that the designs having two consecutive αs (βs) at the end, can be improved by switching one of the $\alpha(\beta)$. The details of the proof can be found in Sun (2007).

The Theorem 8.4 indicates that the optimal design is the one that distributes integers, $1, \cdots, H$, to set α and β as evenly as possible. This can be achieved by putting odd integers in one set and even integers in the other. The asymptotic variance of the test statistics $\bar{T}$ reduces to a simple form for the optimal design.

Corollary 8.1. *Assume that set α and β contains odd and even integers. Under perfect ranking assumption, the asymptotic null variance of $\bar{T}$ based on optimal design, σ^2_{Opt} reduces to*

$$\sigma^2_{Opt} = \begin{cases} \frac{1}{(H+1)^2} & \text{if } H \text{ is even} \\ \frac{1}{H(H+2)} & \text{if } H \text{ is odd.} \end{cases}$$

There are two competitors of the proposed test in the literature, rank-sum test under simple random samples (SRS) and ranked set samples. Bohn and Wolfe (1992) introduced RSS analog of the Mann-Whitney-Wilcoxon (MWW) statistic, U_{RSS}, to test the null hypothesis that the two populations are stochastically equivalent against the alternative hypothesis that one population is stochastically larger than the other, where

$$U_{RSS} = \sum_{s=1}^{q} \sum_{t=1}^{n} \sum_{i=1}^{k} \sum_{j=1}^{m} I(X_{[i]j} < Y_{[s]t}).$$

Under perfect ranking, the null distribution of U_{RSS} is asymptotically normal and the test based on U_{RSS} has higher asymptotic Pitman efficiency with respect to SRS rank-sum test.

The asymptotic Pitman efficacy of the MWW test based on SRS is available in standard text books, such as Hettmansperger and McKean (2011), Randles and Wolfe (1991). When the sample sizes in X- and Y-samples are equal, it is given by

$$c^2_{SRS} = 3\left(\int f^2(x)dx\right)^2.$$

The Pitman efficacy of RSS rank-sum test is given in Bohn and Wolfe (1992)

$$c^2_{RSS} = \frac{3(H+1)}{2}\left(\int f^2(x)dx\right)^2.$$

We compare the proposed optimal ORRD test to its competitors by matching their sample sizes. Let T_1 and T_2 be two tests. Then the asymptotic Pitman relative efficiency (ARE) of T_1 with respect to T_2 is defined as $\text{ARE}(T_1, T_2) = \frac{c_1^2}{c_2^2}$, where c_1^2 and c_2^2 are the asymptotic Pitman efficacies of the tests T_1 and T_2, respectively. If the $\text{ARE}(T_1, T_2)$ is larger than 1, the test T_1 is superior to the test T_2.

Table 8.2 The asymptotic Pitman relative efficiencies

H	ARE (RSS, SRS)	ARE (ORRD, SRS)	ARE (ORRD, RSS)
Even	$\frac{H+1}{2}$	$\frac{(H+1)^2}{3}$	$\frac{2(H+1)}{3}$
Odd	$\frac{H+1}{2}$	$\frac{H(H+2)}{3}$	$\frac{2H(H+2)}{3(H+1)}$

In order to match the fully measured observations in all three designs, the equal set and replication sizes are selected in each treatment groups. The ARE results for the three designs (SRS, RSS and ORRD) summarized in Table 8.2. Table 8.2 shows that the proposed test outperforms its competitors. It is also clear that the AREs depend on whether the set size H is even or odd, but they are free from the underlying distribution. The designs with odd set sizes yield higher efficiency results than the designs with even set sizes.

8.3 Calibration for Ranking Error

The proposed rank-sum test is not distribution-free under imperfect ranking. The variance of the test statistic depends on the ranking mechanism and it is usually smaller than σ_P^2. Thus, if we use σ_P^2 to conduct the test when there is ranking error, the test statistic will be inflated and lead to larger type I error rate. To reduce the impact of ranking error, we calibrate the test by replacing σ_P^2 with the estimate of the variance (σ^2 in Theorem 8.1) of the test statistic from the data. We construct a consistent estimator for σ^2 under a consistent ranking scheme.

Under the null hypothesis, X- and Y-samples have the same distributions. To increase the sample size in the estimation of σ^2, for each judgment class h, $h = 1, \cdots, H$, we combine X- and Y-sample data as follows

$$Z_{[h]j} = \begin{cases} X_{[h],j} & \text{if } j = 1, \cdots, n, \\ Y_{[h],j-n} & \text{if } j = n+1, \cdots, 2n. \end{cases}$$

The variance σ^2 in Theorem 8.1 can be written in a slightly different form in terms of $\mu_{[i]}$ and $\nu_{[i,j]}$

$$\begin{aligned} \sigma^2 &= \frac{4}{H}\left\{\frac{H}{3} - \sum_{i=1}^{H}\mu_{[i]}^2 + 2\sum_{i=1}^{H}\sum_{j=1}^{H} I(i<j)(\nu_{[i,j]} - \mu_{[i]}\mu_{[j]}) \right. \\ &\quad - 4\sum_{i=1}^{u}\sum_{k=1}^{H-u} I(\alpha_i < \beta_k)(\nu_{[\alpha_i,\beta_k]} - \mu_{[\alpha_i]}\mu_{[\beta_k]}) \\ &\quad \left. - 4\sum_{i=1}^{u}\sum_{k=1}^{H-u} I(\alpha_i > \beta_k)(\nu_{[\alpha_i,\beta_k]} - \mu_{[\alpha_i]}\mu_{[\beta_k]})\right\}, \end{aligned} \tag{8.4}$$

where

$$\mu_{[i]} = E[F(Z_{[i]1})] \text{ and } \nu_{[i,j]} = E[F(Z_{[i]1})F(Z_{[j]1})].$$

A consistent estimator for σ^2 can be constructed through consistent estimators of $\mu_{[i]}, i = 1, \cdots, H$, and $\nu_{[i,j]}, i = 1, \cdots, H, j = 1, \cdots, H$

Lemma 8.2. *Under an arbitrary but consistent ranking procedure the consistent and unbiased estimator of $\mu_{[i]}$ and $\nu_{[i,j]}$ are given by*

$$\widehat{\mu}_{[i]} = \frac{1}{2n(2n-1)H} \sum_{j=1}^{2n} \sum_{k=1}^{2n} \sum_{s=1}^{H} I(k \neq j) I(Z_{[s]k} \leq Z_{[i]j}),$$

$$\widehat{\nu}_{[i,j]} = C_{n,H} \sum_{l=1}^{2n} \sum_{k=1}^{2n} \sum_{t=1}^{2n} I(k \neq l) I(t \neq k, l) [\sum_{s=1}^{H} I(Z_{[s]k} \leq Z_{[i]l}) \sum_{s=1}^{H} I(Z_{[s]t} \leq Z_{[j]l})]$$

where

$$C_{n,H} = \frac{1}{4n(2n-1)(n-1)H^2}.$$

Proof. For the proof of unbiasedness, we take the expected values with respect to $Z_{[s]t}$, $s = 1, \cdots, H$ and $t = 1, \cdots, 2n$

$$\begin{aligned}
E(\hat{\mu}_{[i]}) &= \frac{1}{2n(2n-1)H} \sum_{j=1}^{2n} \sum_{k=1}^{2n} \sum_{s=1}^{H} I(k \neq j) E \left\{ I(Z_{[s]k} \leq Z_{[i]j}) \right\} \\
&= \frac{1}{2n(2n-1)H} \sum_{j=1}^{2n} \sum_{k=1}^{2n} \sum_{s=1}^{H} I(k \neq j) E \left\{ F_{[s]}(Z_{[i]j}) \right\} \\
&= \frac{H}{2n(2n-1)H} \sum_{j=1}^{2n} \sum_{k=1}^{2n} I(k \neq j) EF(Z_{[i]j}) = \mu_{[i]}
\end{aligned}$$

$$\begin{aligned}
E\left\{ \widehat{\nu}_{[i,j]} \right\} &= C_{n,H} \sum_{l=1}^{2n} \sum_{k=1}^{2n} \sum_{t=1}^{2n} I(k \neq l) I(t \neq k, l) E \left[\sum_{s=1}^{H} I(Z_{[s]k} \leq Z_{[i]l}) \sum_{s=1}^{H} I(Z_{[s]t} \leq Z_{[j]l}) \right] \\
&= \frac{2n(2n-1)(2n-2)}{4n(2n-1)(n-1)H^2} \sum_{s=1}^{H} \sum_{v=1}^{H} EI(Z_{[s]2} \leq Z_{[i]1}) I(Z_{[v]3} \leq Z_{[j]1}) \\
&= \frac{1}{H^2} \sum_{s=1}^{H} \sum_{v=1}^{H} EF_{[s]}(Z_{[i]1}) F_{[h]}(Z_{[j]1}) = \nu_{[i,j]}.
\end{aligned}$$

The proof of consistency can be proved by showing that the variances of these estimators go to zero as n gets large. The details of the proof can be found in Sun (2007).

Note that σ^2 is a finite sum and a continuous function of $\mu_{[i]}$ and $\nu_{[i,j]}$. Then an unbiased and consistent estimator of σ^2 is obtained by inserting $\widehat{\mu}_{[i]}$ and $\widehat{\nu}_{[i,j]}$ in Eq. (8.4), which yields

$$\begin{aligned}\widehat{\sigma}^2 = \frac{4}{H}\Bigg\{ & \frac{H}{3} - \sum_{i=1}^{H}\widehat{\mu}_{[i]}^2 + 2\sum_{i=1}^{H}\sum_{j=1}^{H} I(i<j)(\widehat{\nu}_{[i,j]} - \widehat{\mu}_{[i]}\widehat{\mu}_{[j]}) \\ & - 4\sum_{i=1}^{u}\sum_{k=1}^{H-u} I(\alpha_i < \beta_k)(\widehat{\nu}_{[\alpha_i,\beta_k]} - \widehat{\mu}_{[\alpha_i]}\widehat{\mu}_{[\beta_k]}) \\ & - 4\sum_{i=1}^{u}\sum_{k=1}^{H-u} I(\alpha_i > \beta_k)(\widehat{\nu}_{[\alpha_i,\beta_k]} - \widehat{\mu}_{[\alpha_i]}\widehat{\mu}_{[\beta_k]})\Bigg\}.\end{aligned}$$

Theorem 8.5. *For a fixed set size H, under an arbitrary and consistent ranking scheme, the asymptotic null distribution of $\sqrt{2nH}(\overline{T}-\frac{1}{2})/\widehat{\sigma}$ converges to a standard normal distribution as n goes to infinity.*

Proof. The proof of the theorem follows from Slutsky's theorem.

To implement the proposed test under imperfect ranking, centered data should be used to compute the null variance of T since it needs to be estimated under the null hypothesis. Let M_x and M_y be the medians of X- and Y-sample observations, respectively. We then center X and Y observations with $\widetilde{X}_{[i],j} = X_{[i]j} - M_x$ and $\widetilde{Y}_{[i],j} = Y_{[i]j} - M_y$ for $i = 1,\cdots,H, j = 1,\cdots,n$. The estimator $\widehat{\sigma}$ is constructed based on the centered data

$$\widetilde{Z}_{[h]j} = \begin{cases} \widetilde{X}_{[h],j} & \text{if } j = 1,\cdots,n, \\ \widetilde{Y}_{[h],j-n} & \text{if } j = n+1,\cdots,2n. \end{cases}$$

In the next section, we further study the test under imperfect ranking through simulations and show empirically that the null distribution of T can be better approximated by a Student's t-distribution when the sample size is moderately large.

In order to complete the inferential procedures, we develop a point estimator and a distribution-free confidence interval for Δ. The point estimator of Δ is constructed from Hodges-Lehman estimator. Hodges-Lehman estimator is defined as the median of the pair-wise differences of X- and Y-sample observations,

$$\widehat{\Delta} = median\{Y_{[k]t} - X_{[i]j}; i = 1,\cdots,H; j = 1,\cdots,n; k = 1,\cdots,H; t = 1,\cdots,n\}.$$

Since the asymptotic Pitman efficacy of the test is $c^2_{ORRD} = (\int f^2(x)dx)^2/\sigma^2$, $\sqrt{2nH}\widehat{\Delta}$ converges in distribution to a normal distribution with mean zero and variance c^{-2}_{ORRD}.

It is easy to see that under the null hypothesis, the distribution of the proposed test statistic T is symmetric around $(nH)^2/2$. Therefore, the distribution-free confidence interval of Δ follows directly from the inversion of the null distribution of T. Let $D_{(1)} \leq \cdots \leq D_{((nH)^2)}$ be the ordered differences of $Y_{[k]t} - X_{[i]j}$ for $i = 1, \cdots, H$; $j = 1, \cdots, n$; $k = 1, \cdots, H$; $t = 1, \cdots, n$. If $P_0(T \leq k^*) = \alpha/2$, then we have that

$$[D_{(k^*+1)}, D_{((nH)^2-k^*)}) \tag{8.5}$$

is a $100(1-\alpha)\%$ confidence interval for Δ. For large n, the quantity k^* can be approximated from the asymptotic null distribution of T,

$$k^* = (nH)^2/2 - 0.5 - z_{\alpha/2}\hat{\sigma}_T, \tag{8.6}$$

where $\hat{\sigma}^2_T$ is the variance estimate of the asymptotic null distribution of T, and $\hat{\sigma}^2_T = (nH)^3\hat{\sigma}^2/2$. Furthermore, if we use the optimal ORRD with perfect ranking, $\hat{\sigma}^2_T$ is replaced with $\frac{(nH)^3}{2(H+1)^2}$ if H is even and $\frac{n^3H^2}{2(H+2)}$ if H is odd.

8.4 Empirical Evidence

In this section, a simulation study is performed to show how the type I error rates and empirical power of the proposed test behave under various simulation settings when the sample size is relatively small. The simulation settings consist of different set (H) and replication (n) sizes, various degree of ranking information, and some common underlying distributions (F).

There are three different available models to quantify the quality of within-set ranking information, Dell and Clutter (1972), Bohn and Wolfe (1992) and Fligner and MacEachern (2006). In this paper, we use Dell and Clutter model to rank the experimental units. Dell-Clutter model states that the error term ε_i, that is the property of the experimental units, is modeled with

$$u_i = \epsilon_i + \omega_i, i = 1, \cdots, H,$$

where ω_i's are iid draws form a normal distribution with mean zero and variance τ^2. The model above creates a set of vectors (u_i, ϵ_i) for $i = 1, \cdots, H$. The units in these sets are ranked based on the first components of (u_i, ϵ_i) for $i = 1, \cdots, H$ and the second components are selected as the judgment ranked units. The quality of ranking in this model is controlled by the noise variable through its variance τ^2. This dependence can be expressed in terms of the correlation coefficient between u and ϵ

$$\rho = \frac{\phi}{\sqrt{\phi^2 + \tau^2}},$$

where ϕ^2 is the variance of ϵ_i. It is clear that if ω has a degenerate distribution, its variance is 0 and $\rho = 1$. This leads to perfect ranking $\epsilon_i \equiv u_i$. Otherwise, ranking process will contain ranking error. The magnitude of this error depends on the size of τ^2.

Throughout the simulation study, in each replication, the ORRD residuals are generated from the following algorithm for each simulation parameter combination.

Step I: Generate ϵ_i and ω_i independently from the underlying distribution F with mean 0 and variance ϕ^2 and the normal distribution with mean 0 and variance τ^2, respectively, for $i = 1, 2, \cdots, 2nH$. Compute $u_i = \epsilon_i + \omega_i$ for $i = 1, \cdots, 2nH$.
Step II: Randomly separate these u_i's into $2n$ sets, each of size H.
Step III: Rank u_i's in each set from 1 to H.
Step IV: Randomly separate the $2n$ sets into two groups, n sets of each. In one of the groups, perform a randomization to decide whether the control or treatment regime is assigned to units that have ranks in set $\boldsymbol{\alpha}$ or in set $\boldsymbol{\beta}$. In the other group, perform an opposite allocation without a randomization. The residuals in the control group and the treatment group are $\epsilon_{[h]1i}$ and $\epsilon_{[h]2i}$ for $h = 1, \cdots, H$ and $j = 1, \cdots, n$.
Step V: Construct the control and the treatment group response measurements $\boldsymbol{X}$ and $\boldsymbol{Y}$, respectively, from $X_{[h]i} = \epsilon_{[h]1i}$ and $Y_{[h]i} = \Delta + \varepsilon_{[h]2i}$ for $h = 1, \cdots, H$ and $j = 1, \cdots, n$.
Step VI: Under the null hypothesis, $\Delta = 0$, compute $z = \sqrt{2nH}(\bar{T} - \frac{1}{2})/\sigma_p$ under perfect ranking or $z = \sqrt{2nH}(\bar{T} - \frac{1}{2})/\hat{\sigma}$ under imperfect ranking.

The simulation parameters in perfect ranking include the set size $H = 2, 3, 4, 5$, replication size $n = 3, 5, 7, 8, 10$, correlation coefficient $\rho = 1$, and three different underlying distributions. These distributions are standard normal distribution (N(0,1)), Student's t-distribution with 3-degrees of freedom (t(3)), and the log-normal distribution (LN(0,1)). The simulation size is taken to be 5000 replication. The Type I error rates are estimated as the proportion of $|z|$ values that exceed the critical value 1.96 for a 5 % test.

Table 8.3 illustrates the estimated type I error rates under perfect ranking. The entries in Table 8.3 indicate that the estimated type I error rates are reasonably close to nominal size (0.05) for replication size as small as $n = 3$. It appears that heavy tailed and skewed distributions require slightly larger sample sizes (replication size $n \geq 5$).

To see the impact of ranking error on the test, we conducted another simulation study under imperfect ranking. In this part of the simulation, type I errors are estimated from the proportion of $|z| = |\sqrt{2nH}(\bar{T} - \frac{1}{2})/\sigma_p|$ values that exceed the critical value 1.96 for a 5 % test. In this case test is not calibrated for ranking error since it uses variance of the test statistic under perfect ranking. Entries of Table 8.4 shows that although the performance of the test is excellent under perfect ranking, the true type I error rates are inflated seriously under imperfect ranking. It is obvious that even a small amount of ranking error, such as $\rho = 0.9$, has a big impact on type I error. This suggests that a calibration is necessary in practice to conduct a reasonable testing procedure.

Table 8.3 The estimated type I error rates under perfect ranking

F	H	$n = 3$	$n = 5$	$n = 7$	$n = 8$	$n = 10$
$N(0, 1)$	2	0.049	0.047	0.043	0.047	0.048
	3	0.029	0.044	0.049	0.044	0.053
	4	0.035	0.044	0.045	0.045	0.048
	5	0.032	0.042	0.041	0.045	0.049
$t(3)$	2	0.046	0.047	0.043	0.043	0.048
	3	0.031	0.040	0.050	0.045	0.050
	4	0.036	0.044	0.042	0.045	0.045
	5	0.032	0.044	0.041	0.045	0.048
$LN(0, 1)$	2	0.047	0.048	0.043	0.048	0.046
	3	0.030	0.041	0.045	0.043	0.046
	4	0.036	0.048	0.046	0.043	0.043
	5	0.032	0.044	0.039	0.046	0.048

Table 8.4 The estimated type I error rates under imperfect ranking without calibration, $n = 5$

F	H	$\rho = 1$	$\rho = 0.9$	$\rho = 0.75$	$\rho = 0.5$
$N(0, 1)$	2	0.042	0.112	0.178	0.240
	3	0.038	0.161	0.259	0.339
	4	0.043	0.242	0.371	0.469
	5	0.044	0.283	0.447	0.523

A third simulation study is conducted to investigate the size of the calibrated test under imperfect ranking. In this part, the simulation parameters are chosen to be $H = 2, \cdots, 5$, $n = 4, 5, 7, 10$, $\rho = 0.9, 0.75, 0.5$. The underlying distributions are the same as in perfect ranking case. For the brevity of the presentation, we only report the estimated type I error rates when $\rho = 0.75$ in Table 8.5.

Table 8.5 provides two estimates, one based on normal approximation (N_A) in Theorem 8.1, and the other based on Student's t-distribution (t_A) with degrees of freedom computed from the Satterthwaite approximation. We noticed that estimated type I error rates slowly converge to nominal value 0.05 under normal approximation. This is not surprising since the estimation of σ in the calibrated test introduces additional variation and inflates the type I error rates for small sample sizes. The estimated type I error rates based on t-approximation, however, are much closer to the nominal value 0.05 for the three underlying distributions and all sample sizes as small as 4. The difference between both approximations gets smaller as the sample size increases. It is clear from Table 8.5 the calibrated test works reasonably well regardless of ranking quality.

We next investigate the empirical power of the test. Simulation study considered set size $H = 3$, the number of replication $n = 5$ and varying degree of judgment ranking information. Residual for the ORRD are again generated from Dell and Clutter model for $\rho = 1, 0.9, 0.75$, and 0.50. For the alternative hypothesis we considered location shift of $\Delta = 0(0.1)1$. The empirical powers of the rank-sum test of ORRD along with classical Mann-Whitney-Wilcoxon test is given in Fig. 8.1.

Figure 8.1 illustrates that the new test has substantially higher power than the power of the Mann-Whitney-Wilcoxon test as long as there is some information to

Table 8.5 Empirical type I error rates of calibrated test when ρ =0.75

		$N(0,1)$		t(3)		$LN(0,1)$	
H	n	t_A	N_A	t_A	N_A	t_A	N_A
2	4	0.064	0.143	0.064	0.144	0.068	0.146
	5	0.054	0.114	0.056	0.115	0.058	0.117
	7	0.057	0.093	0.055	0.091	0.056	0.092
	10	0.054	0.074	0.058	0.078	0.059	0.078
3	4	0.060	0.148	0.059	0.140	0.052	0.137
	5	0.058	0.118	0.058	0.121	0.063	0.118
	7	0.061	0.095	0.061	0.096	0.057	0.093
	10	0.054	0.072	0.059	0.077	0.057	0.078
4	4	0.058	0.136	0.056	0.139	0.056	0.138
	5	0.054	0.111	0.056	0.116	0.051	0.110
	7	0.060	0.095	0.052	0.087	0.056	0.088
	10	0.060	0.080	0.058	0.079	0.057	0.079
5	4	0.052	0.136	0.053	0.136	0.051	0.123
	5	0.056	0.115	0.058	0.117	0.056	0.113
	7	0.056	0.093	0.059	0.096	0.057	0.089
	10	0.056	0.078	0.055	0.075	0.055	0.075

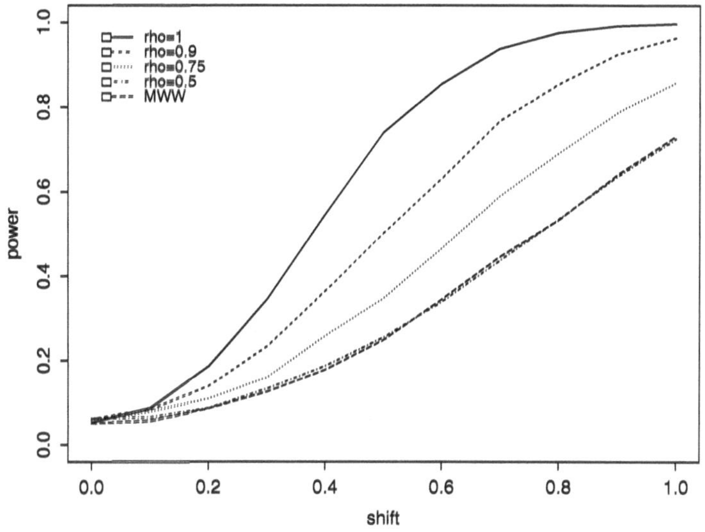

Fig. 8.1 Empirical power curves of the rank-sum tests for selected ρ, set size $H = 3$, replication size $n = 5$ and simulation size 5,000

rank the units prior to experimentation. If the quality of ranking information is poor, the correlation coefficient is less than 0.5, the ORR design is as good as simple random sampling design. This indicates that the proposed test does not loose its power for settings where ranking information leads to a random ranking.

8.5 Application to ACTG 320 Clinical Trial

In this section, we illustrate the use of the proposed test in a data set. Ideally, we should design an experiment based on an ORRD and collect data to apply the procedure. This kind of experiment, however, is not available currently. We therefore apply the proposed test to a well-known AIDS Clinical Trial Group Protocol 320 (ACTG 320) study performed by Hammer et al. (1997).

The researchers in ACTG 320 clinical trial compare two types of treatments for human immunodeficiency virus type I (HIV-1). One treatment is a two-drug regime involving two nucleoside analogues (lamivudine and zidovudine/stavudine), the other is a three-drug regime including the protease inhibitor (indinavir) along with the two nucleoside analogues. The study is designed as a randomized, double-blind, placebo-controlled trial with 1156 HIV-infected patients who have no more than 200 CD4 cells per cubic millimeter at the screening stage and are not previously treated by lamivudine and indinavir, but by zidovudine for at least 3 months. Their conclusions show that the three-drug regime is better than the two-drug regime in term of slowing the development of HIV-1 disease.

In our study, we select an ORRD sample of size 30 from all available patients in this study to compare the CD4 cell counts (CD4-4) at week 4 between two treatment regimes. We denote two-drug regime as treatment A and three-drug regime as treatment B. We predetermine the set size $H = 3$ and the replication size $n = 5$, and use the design $\boldsymbol{\alpha} = \{1, 3\}$ and $\boldsymbol{\beta} = \{2\}$. The correlation coefficient between the base line CD4 cell counts (CD4-B) at the screening stage and CD4 cell counts (CD4-4) at week 4 is 0.798. Therefore, it is reasonable to use the CD4-B cell counts at the screening stage to rank the subjects to create judgment blocs. The histograms of the CD4 cell counts at week 4 for each treatment regime and the pooled data are skewed to right and have similar shape. The difference of medians of the CD4 cell counts from the two treatment regime is $M_B - M_A = 30$ per cubic millimeter. The potential outlier in treatment A and the skewness of distributions suggest that use of the proposed test would be appropriate for this data.

For the purpose of illustration, we treat all 1077 patients as potential population after excluding the missing values in the CD4 cell counts at the screening stage and week 4. Thirty patients are taken from the population through the following steps that resembles the ORRD as closely as possible.

In Step I, each patient is assigned a random number and the data are sorted by assigned random numbers so that the patients in the data set are listed in a random order. In step II, the first three patients are selected from the list and sorted based on their CD4 cell counts at the screening stage (CD4-B). The rankers are only allowed to see the CD4-B cell counts and the treatment variable. If the first and third patients in this sorted sample are from treatment A and the second patient is from treatment B, we then select these three patients as type I set in replication 1. However, if the first and third patients are from treatment B and the second patient is from treatment A, we then select these three patients as type II set in replicate 1. If the first three patients are in neither of the two forms, we discard them and take the next

Table 8.6 The sampled ORRD data for CD4-4 cell counts

Treat.	CD4-B	CD4-4	Repl.	Set	Rank
A	16.0	19	1	1	1
B	29.0	98	1	1	2
A	95.0	100	1	1	3
B	25.0	120	1	2	1
A	157.5	161	1	2	2
B	206.0	430	1	2	3
A	85.5	81	2	1	1
B	115.0	176	2	1	2
A	172.0	189	2	1	3
B	26.0	85	2	2	1
A	55.0	61	2	2	2
B	159.0	207	2	2	3
A	132.5	180	3	1	1
B	156.5	199	3	1	2
A	160.0	170	3	1	3
B	14.0	35	3	2	1
A	59.0	102	3	2	2
B	204.0	227	3	2	3
A	107.0	109	4	1	1
B	139.5	181	4	1	2
A	219.0	238	4	1	3
B	21.5	50	4	2	1
A	50.0	91	4	2	2
B	58.5	135	4	2	3
A	4.5	0	5	1	1
B	39.0	107	5	1	2
A	225.5	261	5	1	3
B	5.0	126	5	2	1
A	10.5	24	5	2	2
B	209.5	220	5	2	3

three patients from the remaining population. We continue to this process until we select one type I and one type II sets. In step III, step II is performed repeatedly on the remaining population units until five replications are selected. Although the experiment is not originally designed as an ORRD, sampled data resembles a valid ORRD structure with $H = 3$ and design parameters $\boldsymbol{\alpha} = \{1, 3\}$ and $\boldsymbol{\beta} = \{2\}$. The sampled data set is listed in Table 8.6

We apply the proposed testing procedure to test if the medians of CD4 cell counts at week 4 for the two treatment regimes are different. The test statistic for the data in Table 8.6 yields that $T = 83$ and p-value $= 0.018$. The value of the Hodges Lehman estimator of the parameter is 31 CD4 cell counts per cubic millimeter with an associated 95 % confidence interval from 24 to 37 CD4 cell counts per cubic millimeter. The confidence interval contains the true value $\Delta = 30$. Based on the statistical evidence, we conclude that the three-drug treatment regime provides a substantial amount of improvement over the two-drug treatment regime.

8.6 Concluding Remarks

This papers develops a distribution-free inference based on order restricted randomized design for the location shift between two distributions. The new design exploits the use of subjective information to rank the experimental units to produce more accurate inference for the contrast parameter. It is shown that the estimators and test statistics have limiting normal distribution regardless of the quality of ranking information. A simulation study shows that the asymptotic results remain valid even for relatively small sample sizes. The proposed testing procedures are applied to a clinical trial.

The approach that we have taken in this paper extends to more complex treatment structure with k-treatments. A test, similar to Kruskal-Wallis test, can be constructed. In this case, interesting design issues may appear.

References

Bohn, L. L., & Wolfe, D. A. (1992). Nonparametric two-sample procedures for ranked-set samples data. *Journal of the American Statistical Association 87*, 552–561.

Dell, T. R., & Clutter, J. L. (1972). Ranked set sampling theory with order statistics background. *Biometrics, 28*, 545–555.

Due, J., & MacEachern, S. N. (2007). Judgment post-stratification for designed experiments. *Biometrics, 64*, 345–354.

Fligner, M. A., & MacEachern, S. N. (2006). Nonparametric two-sample methods for ranked-set sample data. *Journal of the American Statistical Association, 101*(475), 1107–1118.

Gao, J., & Ozturk, O. (2015). Rank regression in order restricted randomized designs. Under review.

Hammer, S. M., Squires, K. E., Hughes, M. D., Grimes, J. M., Demeter, L. M., Currier, J. S., et al. (1997). A controlled trial of two nucleoside analogues plus Indinavir in persons with human immunodeficiency virus infection and CD4 cell counts of 200 per cubic millimeter or less. *The New England Journal of Medicine, 337*, 725–733.

Hettmansperger, T. P., & McKean, J. M., (2011). *Robust nonparametric statistical methods.* New York: Oxford University Press.

Hollander, M., Wolfe, D. A., & Chicken, E. (2014). Nonparametric statistical methods. New Jersey: Wiley.

Markiewicz, S. (2008). *Nonparametric inference using order restricted randomized designs.* Ph.D. dissertation, Department of Statistics, The Ohio State University.

McIntyre, G. A. (1952). A method for unbiased selective sampling using ranked sets. *Australian Journal of Agricultural Research, 3*, 385–390

McIntyre, G. A. (2005). A method for unbiased selective sampling using ranked sets. *The American Statistician, 3*, 230–232.

Ozturk, O., & MacEachern, S. N. (2004). Control versus treatment comparison under order restricted randomization. *Annals of the Institute of Statistical Mathematics, 56*, 701–720.
Ozturk, O., & MacEachern, S. N. (2007). Order restricted randomized designs and two sample inference. *Environmental and Ecological Statistics, 14*, 365–381.
Ozturk, O., & Sun, Y. (2009). Rank-sum test based on order restricted randomized design. In *Proceeding of the World Congress of Engineering* (pp. 1270–1274).
Ozturk, O., & MacEachern, S. N. (2013). Inference based on general linear models for order restricted randomization. *Communications in Statistics - Theory and Methods, 42*, 2543–2566.
Randles, H. R., & Wolfe, D. A. (1991). Introduction to the theory of nonparametric statistics. Malabar, FL: Krieger Publishing Company.
Sun, Y. (2007). *Rank-sum test for two-sample location problem under order restricted randomized design*. Ph.D. dissertation, Department of Statistics, The Ohio State University.
Wolfe, D. A. (2012). Ranked set sampling, its relevance and impact on statistical inference. *ISRN Probability and Statistics, 2012*(1), 1–32.

Chapter 9
On a Partially Sequential Ranked Set Sampling Paradigm

Douglas A. Wolfe

Abstract In a two-sample setting it is important to design statistical procedures that can take advantage of additional information to minimize the sample sizes required to reach reliable inferences about possible differences between the two populations. This is particularly true when it is difficult and/or costly to obtain sample observations from one or both of the populations. One class of procedures designed with this goal in mind uses the partially sequential sampling (*PS*) approach, first introduced by Wolfe (Journal of the American Statistical Association 72:202–205, 1977a). The use of ranked set sampling (*RSS*), first introduced by McIntyre (Australian Journal of Agricultural Research 3:385–390, 1952, reprinted in 2005), offers another approach for minimizing required sample sizes through the mechanism of obtaining more representative samples than can be achieved using simple random samples. In this paper we provide a review of these two sampling techniques and discuss options for melding the two methodologies to obtain partially sequential ranked set sample (*PSRSS*) two-sample test procedures that take advantage of the sample saving properties of both the *PS* and *RSS* approaches. To illustrate this combination, we consider *PSRSS* procedures where the fixed (control) sample is obtained via simple random sampling and the sequential (treatment) sample is obtained via ranked set sampling. Properties of the associated tests are discussed, including the limiting distributions as the fixed sample size tends to infinity.

Keywords Distribution free tests • Judgment ranked order statistics • Minimizing sample sizes • Negative binomial distribution • Using auxiliary sampling information

9.1 Introduction

Minimizing the cost associated with collection of the sample data is a critical feature of most statistical analyses. As a result, it is important to develop statistical approaches to sampling that minimize the sample sizes necessary to achieve desired

D.A. Wolfe (✉)
Department of Statistics, Ohio State University, Columbus, OH 43210, USA
e-mail: daw@stat.osu.edu

© Springer International Publishing Switzerland 2016

R.Y. Liu, J.W. McKean (eds.), *Robust Rank-Based and Nonparametric Methods*, Springer Proceedings in Mathematics & Statistics 168,
DOI 10.1007/978-3-319-39065-9_9

properties, whether it be precision of estimators, length of confidence intervals, or power of statistical tests. One technique that has been shown to be useful in this regard is ranked set sampling (*RSS*), first introduced by McIntyre (1952, reprinted in 2005) in the context of sampling from pasture and crop plots. This sampling approach uses readily available auxiliary information from individual units in a population to aid in the selection of more representative units for measurement than are typically generated by simple random sampling (*SRS*). Development of statistical procedures using this *RSS* approach remains an active area of research. [See, for example, the recent survey article by Wolfe (2012).] A second approach to data collection designed to reduce the sample size in a treatment versus control two-sample setting is the partially sequential (*PS*) paradigm introduced by Wolfe (1977a,b). This approach uses a negative binomial sampling framework to minimize the number of treatment observations necessary for reaching satisfactory statistical conclusions regarding the treatment's efficacy.

In this paper we review the basic tenets of both the *RSS* and *PS* methodologies and discuss how to combine these approaches to develop partially sequential ranked set sample (*PSRSS*) two-sample test procedures. In Sect. 9.2 we present the *PS* two-sample framework and review previous work in this area. We describe the basic *RSS* approach in Sect. 9.3 and discuss a number of options available within this structure. We propose a class of melded *PSRSS* two-sample test procedures in Sect. 9.4 and develop their basic small sample and asymptotic properties as the control sample size becomes large. Section 9.5 is devoted to a general discussion of the opportunities presented by this new methodology as well as extensions for future research.

9.2 Partially Sequential Two-Sample Procedures

The partially sequential approach to data collection in the two-sample setting was first introduced by Wolfe (1977a). It is particularly appropriate for data collection settings such as the following:

1. A sample from the first population (e.g., control) has already been collected and we do not wish to collect any more observations from the second population (e.g., new treatment) than are necessary for reaching a decision.
2. Neither sample has been collected, but one of the samples (say the 'standard'procedure observations) is relatively easy and inexpensive to collect, while the other sample observations (corresponding to the 'new treatment') are costly and/or difficult to collect. In such situations our goal would be to collect a sample (usually large) of standard observations and then collect only enough difficult-to-obtain new treatment observations necessary to reach statistically valid conclusions about potential differences between the two populations.

We first describe a general *PS* procedure to test for differences between two distributions. Let $X_1, \cdots, X_m$ be a random sample from a continuous probability distribution with p.d.f. $f(x)$ and c.d.f. $F(x)$, where m is a fixed positive integer, and let $G(y)$ be a second continuous distribution function with associated p.d.f. $g(y)$. Let $(x_1, \cdots, x_m)$ be an arbitrary m-tuple of real numbers and let $A(x_1, \cdots, x_m)$ be a subset of the real line R depending on the m-tuple $(x_1, \cdots, x_m)$. For example, $A(\cdot)$ could be the portion of R below the minimum x-value or the portion of R above the maximum x-value. Define the indicator function $\Psi(\cdot)$ by

$$\Psi(y) = \begin{cases} 1, & \text{if } y \in A(x_1, \cdots, x_m), \\ 0, & \text{if } y \notin A(x_1, \cdots, x_m). \end{cases} \tag{9.1}$$

Now let Y be a random variable (independent of $X_1, \cdots, X_m$) from the second distribution $G(y)$. Applying $\Psi(y)$ to these random variables $X_1, \cdots, X_m$ and Y, we obtain the following

$$\Psi(Y) = \begin{cases} 1, & \text{if } Y \in A(X_1, \cdots, X_m), \\ 0, & \text{if } Y \notin A(X_1, \cdots, X_m). \end{cases} \tag{9.2}$$

Thus, $\Psi(Y)$ is the random indicator variable for the random set $A(X_1, \cdots, X_m)$.

With (9.2) in mind we sequentially sample mutually independent $Y's$ from the distribution $G(y)$ until a preset number, say r, of these $Y's$ are in the set $A(x_1, \cdots, x_m)$, where $(x_1, \cdots, x_m)$ is the observed value of the previously collected random vector $(X_1, \cdots, X_m)$. Define the statistic N_m (having random contributions from both the X and Y samples) by

$$N_m = \{\text{number of } Y \text{ observations required to get } r\ Y's \text{ in } A(x_1, \cdots, x_m)\}. \tag{9.3}$$

Wolfe (1977a) discussed how to use N_m to test the null hypothesis $H_0 : F(x) \equiv G(x)$ against appropriate alternatives [depending on the nature of the set $A(x_1, \cdots, x_m)$]. The decision rule he proposed is to reject H_0 when $N_m \leq N_0(\alpha, r, m, A)$, where $N_0(\alpha, r, m, A)$ is the lower αth percentile point for the null (H_0) distribution of N_m. Note that with this approach we will never need to collect more than $N_0(\alpha, r, m, A)$ Y observations. In fact, we would stop even sooner with an even smaller Y sample size and (1) reject H_0 as soon as we obtain r Y observations in $A(x_1, \cdots, x_m)$ or (2) fail to reject H_0 as soon as we obtain $\{N_0(\alpha, r, m, A) - r + 1\}$ Y observations not in $A(x_1, \cdots, x_m)$.

9.2.1 Properties of Partially Sequential Procedures

For given $X_1 = x_1, \cdots, X_m = x_m$, let

$$p_m = p_m(x_1, \cdots, x_m) = P_G\{Y \in A(x_1, \cdots, x_m)\}. \tag{9.4}$$

Thus, p_m is the conditional probability that an observation from the distribution G falls in the set $A(x_1, \cdots, x_m)$ prescribed by the observed values from the F distribution. Then, conditional on $X_1 = x_1, \cdots, X_m = x_m$, N_m has a negative binomial distribution with parameters r and p_m; that is,

$$P(N_m = n|X_1 = x_1, \cdots, X_m = x_m)$$
$$= \binom{n-1}{r-1}[p_m(x_1, \cdots, x_m)]^r[1 - p_m(x_1, \cdots, x_m)]^{n-r} I_{\{r,r+1,r+2,\cdots\}}(n). \tag{9.5}$$

The unconditional distribution of N_m is obtained from the result in (9.5) by integrating over the distribution of the $X's$, namely,

$$P(N_m{=}n){=}E_F\{\binom{n-1}{r-1}[p_m(X_1, \cdots, X_m)]^r[1{-}p_m(X_1, \cdots, X_m)]^{n-r}\}I_{\{r,r+1,r+2,\cdots\}}(n). \tag{9.6}$$

Since the investigator has flexibility in setting the sample size m for the X observations, it is of interest to know how N_m behaves as m becomes large, that is, as $m \to \infty$. If, for given F and G, $p_m(X_1, \cdots, X_m)$ converges in probability to a fixed number $p = p(F, G)$, $0 < p \leq 1$, as $m \to \infty$, then the limiting distribution $(m \to \infty)$ of N_m is negative binomial with parameters r and p; that is, the asymptotic distribution $(m \to \infty)$ of N_m is

$$P^*(N_m = n) = \binom{n-1}{r-1} p^r(1 - p)^{n-r} I_{\{r,r+1,r+2,\cdots\}}(n). \tag{9.7}$$

(Note: A limiting value of $p = p(F, G) = 0$ does not satisfy the conditions for this result. If a pair (F, G) produces a limiting value of $p = 0$, the statistic N_m does not possess a limiting distribution as $m \to \infty$, since in such cases N_m increases stochastically without limit as $m \to \infty$.)

When m is fixed and large we can use the limiting distribution in (9.7) to select r to guarantee asymptotic $(m \to \infty)$ power against an alternative to H_0 of interest. Let H_a be an alternative to H_0 against which we require an approximate power β, where $0 < \beta < 1$ is arbitrary. Let p^* be the value of p in (9.7) that corresponds to the alternative H_a. Then from the definition of $N_0(\alpha, r, \infty, A)$ (i.e., the approximate α-level critical value for the asymptotic, $m \to \infty$, distribution of N_m), this approximate power requirement corresponds to

$$\sum_{n=r}^{N_0(\alpha,r,\infty,A)} \binom{n-1}{r-1} (p^*)^r(1 - p^*)^{n-r} \geq \beta. \tag{9.8}$$

For many partially sequential procedures, the left side of the inequality in (9.8) is a non-decreasing function of r. In this case, to satisfy our asymptotic power requirements with the fewest Y observations, we can preset r to be $r = r^*$, where r^* is the smallest integer for which (9.8) is satisfied.

9.2.2 Examples

The *PS* approach can be used equally well in parametric or nonparametric settings. We briefly discuss two such examples.

Example 1: Parametric Setting

Let $F(x) = \Phi\{\frac{x-\mu_1}{\sigma}\}$ and $G(y) = \Phi\{\frac{y-\mu_2}{\sigma}\}$, where $\Phi(t)$ is the standard normal distribution function. The null hypothesis of interest is $H_0 : \mu_1 = \mu_2$ and we consider here the alternative $H_a : \mu_2 > \mu_1$.

One method for selecting the set $A(x_1, \cdots, x_m)$ would be to view the indicator $\Psi(\cdot)$ in (9.1) as a critical function for testing H_0 against H_a for random samples of sizes m and 1 from the F and G distributions, respectively. For example, we know that the uniformly most powerful level α^* test of H_0 against H_a for m X observations and a single Y observation has critical region

$$C(y, x_1, \cdots, x_m) = \left\{ (y, x_1, \cdots, x_m) : \frac{m^{1/2}}{(m+1)^{1/2}} \frac{(y-\bar{x})}{s} \geq t_{\alpha^*}(m-1) \right\},$$

where $\bar{x} = \sum_{i=1}^{m} x_i/m$, $s^2 = \sum_{i=1}^{m}(x_i - \bar{x})^2/(m-1)$ and $t_{\alpha^*}(m-1)$ is the upper α^* percentile point for the t distribution with $m-1$ degrees of freedom.

Thus, in this setting it is natural to take the set $A(x_1, \cdots, x_m)$ to be

$$A(x_1, \cdots, x_m) = \{y : y \geq \bar{x} + t_{\alpha^*}(m-1)(\{m+1\}s^2/m)^{\frac{1}{2}}\}. \tag{9.9}$$

In fact, Orban and Wolfe (1978) showed that this choice of $A(x_1, \cdots, x_m)$ leads to the asymptotically ($m \to \infty$) most powerful level α^* partially sequential procedure for testing H_0 against H_a.

With $A(x_1, \cdots, x_m)$ given by (9.9), we have

$$p_m = 1 - \Phi(\{\bar{x} + t_{\alpha^*}(m-1)(\{m+1\}s^2/m)^{\frac{1}{2}} - \mu_2\}/\sigma)$$

and the limiting distribution of N_m as $m \to \infty$ is negative binomial (9.7) with parameters r and $p = \lim_{m\to\infty} p_m = 1 - \Phi\left\{\frac{\mu_1-\mu_2}{\sigma} + z_{\alpha^*}\right\}$.

Example 2: Nonparametric Setting

Let F and G be arbitrary, continuous distribution functions. We wish to test $H_0 : F \equiv G$ against the alternative $H_a : \xi_2 > \xi_1$, where ξ_1 and ξ_2 are the medians of the F and G distributions, respectively. Assume that m is an odd integer (more complicated, but tractable for m even) and define $A(x_1, \cdots, x_m)$ by

$$A(x_1, \cdots, x_m) = \{y : y > m_x\}, \tag{9.10}$$

where $m_x = \text{median}(x_1, \cdots, x_m)$. Then the *PS* two-sample median test associated with N_m (9.3) has the following properties:

(a) $p_m = 1 - G(m_x)$ and the exact null (H_0) distribution of N_m is given by

$$P_0(N_m = n) = \begin{cases} \binom{n-1}{r-1} \frac{m!}{[\{(m-1)/2\}!]^2} \frac{\Gamma\left(\frac{m+2n-2r+1}{2}\right)\Gamma\left(\frac{2r+m+1}{2}\right)}{(m+n)!}, & n = r, r+1, \cdots \\ 0, & \text{elsewhere.} \end{cases} \tag{9.11}$$

(b) The limiting distribution of N_m as $m \to \infty$ is negative binomial (9.7) with parameters r and $p = \lim_{m\to\infty} p_m = 1 - G(\xi_1)$, with $p = 1/2$ or $> 1/2$ depending on whether H_0 or H_a is true, respectively.

Wolfe (1977b) initially proposed this *PS* two-sample median procedure and Orban and Wolfe (1982) studied its properties, including the expected number of Y observations required to conduct the test. They also provided the necessary tables for selecting r so that the approximate power requirement in (9.8) can be attained.

9.3 Ranked Set Sampling

The goal of *RSS* is to collect observations that are more likely to be representative of the full range of values in a population than the same number of observations obtained via *SRS*. To obtain a balanced *RSS* of k observations from a population, we proceed as follows. First, an initial *SRS* of k units is selected from the population and rank ordered on the attribute of interest. This ranking can be obtained through a variety of mechanisms, including visual comparisons, expert opinion, or through the use of correlated concomitant variables, but it cannot involve actual measurements of the attribute of interest on the selected units. The unit that is judged to be the smallest in this ranking is taken as the first item in the *RSS* and the attribute of interest is formally measured for the unit and denoted by $X_{[1]}$. Note that square brackets are used instead of the usual round brackets for the smallest order statistic since $X_{[1]}$ may or may not actually have the smallest attribute measurement among the k units in the *SRS*, even though our ranking judged it to be the smallest. The other remaining $k-1$ units in our initial *SRS* are not considered further in making inferences about the population—they were used solely to assist in the selection of the smallest ordered ranked unit for measurement.

Following the selection of $X_{[1]}$, a second *SRS* (independent of the first *SRS*) of size k is selected from the population and ranked in the same manner as the first *SRS*. From this second *SRS* we select the item ranked as the second smallest of the k units (i.e., the second judgment order statistic) and add its attribute measurement, $X_{[2]}$, to the RSS. From a third SRS (independent of both previous *SRS*'s) of size k we select the unit ranked to be the third smallest (i.e., the third judgment order statistic) and include its attribute measurement, $X_{[3]}$, in the *RSS*. This process continues until we have selected the unit ranked to be the largest of the k units in the kth independent *SRS* and included its attribute measurement, $X_{[k]}$, in our *RSS*.

This process results in the k measured observations $X_{[1]}, X_{[2]}, \cdots, X_{[k]}$ and is called a *cycle*. The number of units, k, in each *SRS* is called the set size. To complete a single ranked set cycle, we need to access a total of k^2 units from the population to separately rank k independent simple random samples of size k each. The measured observations, $X_{[1]}, X_{[2]}, \cdots, X_{[k]}$, constitute a *balanced ranked set sample of size* k, where the descriptor "balanced" refers to the fact that we have collected one judgment order statistic for each of the ranks $1, 2, \cdots, k$. To obtain a final balanced *RSS* with a desired total number of measured observations $n = qk$, we repeat the entire process for q independent cycles, yielding the balanced *RSS* of size n: $X_{[1]j}, X_{[2]j}, \cdots, X_{[k]j}$, for $j = 1, \cdots, q$.

Note that a balanced *RSS* of size n differs from an *SRS* of size n in a number of important ways. An *SRS* is designed so that the n observations in the sample are mutually independent and identically distributed. This means that, probabilistically speaking, each of the individual sample items can be viewed as representative of a typical value from the underlying population. That is certainly not the case for a balanced *RSS* of size n. While the individual observations in a balanced *RSS* are also mutually independent, they are clearly not identically distributed. As such, it is not the case that each of the individual observations in a balanced *RSS* represents a typical value from the underlying population. On the contrary, the individual judgment order statistics represent very distinctly different portions of the underlying population. It is, however, precisely this additional structure on the items in the balanced *RSS* that enables it to provide greater assurance that the entire range of population values are represented in the sample data.

There have been numerous papers in the literature demonstrating the advantages that balanced *RSS* provides relative to *SRS*, both in terms of precision accuracy and in terms of reducing required sample sizes. Dell and Clutter (1972) showed that the estimator of the population mean μ based on a balanced *RSS* is unbiased and it has a variance that is never larger than the variance of the estimator of μ based on a *SRS* of the same size. The remarkable thing is that this result is true even if the judgment ranking for the balanced *RSS* is not perfect. The better the judgment ranking, of course, the greater the improvement from using a balanced *RSS* instead of a *SRS*. Stokes and Sager (1988) obtained similar results for the *RSS* estimator of the distribution function of the population and Terpstra (2004) did the same for the *RSS* maximum likelihood estimator for a population proportion.

While a balanced *RSS* is the most commonly occurring form of ranked set sampling data, there are situations where it is not optimal to collect the same number of measured observations for each of the judgment order statistics. For example, suppose we are interested primarily in making inferences about the median ξ of a distribution based on an odd number of observations $k = 2d + 1$. It is well known that among all the order statistics the sample median, $X_{(d+1)}$, contains the most information about ξ when the underlying distribution is unimodal and symmetric.

Thus, to make inferences about ξ, it is natural to measure the same judgment order statistic, $X_{[d+1]}$, in each set so that it is measured all k times in each of the q cycles. The resulting *RSS* consists of qk measured observations, each of which is a judgment median from a set of size k. This is the most efficient *RSS* for making inferences about the population median ξ for a distribution that is both unimodal and symmetric, and it is clearly as unbalanced as possible. (A similar argument calls for a distinctly different unbalanced *RSS* for estimating the median of an asymmetric unimodal population or a multimodal population. See, for example, Ozturk and Wolfe 2000, and Chen et al. 2006.) We should point out, however, that such a median unbalanced *RSS* would not necessarily be a good idea if we wanted to make inference about other features of the population, such as its distribution function or the population variance.

RSS and related methodology has an active and rich literature. The interested reader is referred to the recent survey and review articles in Wolfe (2004) and Wolfe (2012) for more comprehensive discussions.

9.4 A Class of *PSRSS* Two-Sample Percentile Test Procedures

There are three approaches that can be taken to incorporate *RSS* into partially sequential procedures:

1. Use *RSS* for the X sample data and *SRS* for the sequentially obtained Y sample data.
2. Use *RSS* for both the X sample and Y sample data.
3. Use *SRS* for the X sample data and *RSS* for the Y sample data.

All three of these options are worthy of consideration, although the first approach is probably the least interesting in the context where partially sequential procedures would be most useful. In this paper we concentrate on the most natural third option to provide an illustration of how to introduce *RSS* into the partially sequential process. To facilitate the discussion we consider the particular unbalanced *RSS* corresponding to all of the observations being collected at a single judgment order statistic and we assume that the judgment ranking is perfect, so that the various judgment order statistics can be viewed as true order statistics.

As before, let $X_1, \cdots, X_m$ be a random sample from a probability distribution with p.d.f. $f(x)$ and c.d.f. $F(x)$, where m is an odd integer, and let $G(y)$ be a second distribution function with associated p.d.f. $g(y)$. Let M_X be the X sample median and let m_x be the observed value of M_X. Once again we wish to test $H_0 : F \equiv G$ against the alternative $H_a : \xi_2 > \xi_1$, where ξ_1 and ξ_2 are the medians of the F and G distributions, respectively.

For illustrative purposes, we consider collecting unbalanced *RSS* data from G using a single cycle ($q = 1$) with set size k and measuring the jth order statistic, $Y_{(j)}$, at each step of the sequential sampling, for fixed $j \in \{1, \cdots, k\}$. With this *RSS Y*-sampling scheme and the indicator set $A(x_1, \cdots, x_m) = \{y : y > m_x\}$, the associated *PSRSS* test of $H_0 : F \equiv G$ against the alternative $H_a : \xi_2 > \xi_1$ has the following properties:

(a) The unconditional exact distribution of N_m still has the form

$$P(N_m = n) = E_F\left\{\binom{n-1}{r-1}[p_m(X_1, \cdots, X_m)]^r[1 - p_m(X_1, \cdots, X_m)]^{n-r}\right\} I_{\{r,r+1,r+2,\cdots\}}(n), \tag{9.12}$$

but the parameter $p_m = p_m(x_1, \cdots, x_m)$ is now given by

$$p_m = P\{Y_{(j)} > m_x\} = 1 - Q_j(m_x),$$

where $Q_j(\cdot)$ is the c.d.f. for the jth order statistic for a random sample of size k from G, given by

$$Q_j(t) = \sum_{u=j}^{k} \binom{k}{u} [G(t)]^u [1 - G(t)]^{k-u}. \tag{9.13}$$

Combining (9.12) and (9.13), the unconditional distribution of N_m becomes

$$P(N_m = n) = E_{F_{M_X}}\left\{\binom{n-1}{r-1}[1 - Q_j(M_X)]^r[Q_j(M_X)]^{n-r}\right\} I_{\{r,r+1,r+2,\cdots\}}(n), \tag{9.14}$$

where F_{M_X} is the c.d.f. of the sample median for a random sample of size m from F. Using the standard form of F_{M_X} for an odd sample size m in expression (9.14), it follows that

$$P(N_m = n) = \int_{-\infty}^{\infty} \binom{n-1}{r-1} [1 - Q_j(t)]^r [Q_j(t)]^{n-r}$$
$$\times \frac{m!}{[(\frac{m-1}{2})!]^2} \{F(t)[1 - F(t)]\}^{\frac{m-1}{2}} f(t) dt I_{\{r,r+1,r+2,\cdots\}}(n), \tag{9.15}$$

Under $H_0 : F \equiv G$ it follows from the change of variable $v = F(t)$ in (9.15) that the null distribution for N_m "simplifies" to

$$P(N_m = n) = \int_0^1 \binom{n-1}{r-1} \left\{1 - \sum_{u=j}^{k} \binom{k}{u} [v]^u [1 - v]^{k-u}\right\}^r \left[\sum_{u=j}^{k} \binom{k}{u} [v]^u [1 - v]^{k-u}\right]^{n-r}$$
$$\times \frac{m!}{[(\frac{m-1}{2})!]^2} \{v(1 - v)\}^{\frac{m-1}{2}} dv I_{\{r,r+1,r+2,\cdots\}}(n), \tag{9.16}$$

This expression clearly does not depend on the form of the continuous F, so that the test based on N_m is distribution-free and the exact critical values for the test can be evaluated from (9.16) without knowledge of F.

(b) The limiting distribution of N_m as $m \to \infty$ is negative binomial with parameters r and $p_j^* = \lim_{m\to\infty} p_m = 1 - Q_j(\xi_1)$.

Using the expression for $Q_j(t)$ in (9.13), we see that

$$p_j^* = 1 - Q_j(\xi_1) = 1 - \sum_{u=j}^{k} \binom{k}{u} [G(\xi_1)]^u [1 - G(\xi_1)]^{k-u},$$

which simplifies under the null hypothesis to

$$p_{0j}^* = 1 - Q_j(\xi_1) = 1 - \sum_{u=j}^{k} \binom{k}{u} [F(\xi_1)]^u [1 - F(\xi_1)]^{k-u} = 1 - \sum_{u=j}^{k} \binom{k}{u} [0.5)]^u [0.5]^{k-u}, \quad (9.17)$$

9.4.1 *Special Cases*

1. $j = k$—here we are measuring the maximum judgment order statistic in each set and

$$p_{0k}^* = 1 - \sum_{u=k}^{k} \binom{k}{u} [0.5)]^u [0.5]^{k-u} = 1 - \binom{k}{k} [0.5)]^k [0.5]^{k-k} = 1 - (0.5)^k, \quad (9.18)$$

which converges to 1 as $k \to \infty$.

2. $j = 1$—here we are measuring the minimum judgment order statistic in each set and

$$\begin{aligned} p_{01}^* &= 1 - \sum_{u=1}^{k} \binom{k}{u} [0.5)]^u [0.5]^{k-u} \\ &= 1 - [\sum_{u=0}^{k} \binom{k}{u} [0.5)]^u [0.5]^{k-u} - \binom{k}{0} [0.5)]^0 [0.5]^{k-0}] \\ &= 1 - [1 - (0.5)^k] = (0.5)^k, \end{aligned} \quad (9.19)$$

which converges to 0 as $k \to \infty$. (Remember that this is not a viable option for *PSRSS*.)

3. $j = d + 1$, where $k = 2d + 1$ is an odd integer—here we are measuring the median judgment order statistic, $Y_{(d+1)}$, in each set and

$$
\begin{aligned}
p_{0(d+1)} &= 1 - \sum_{u=d+1}^{2d+1} \binom{2d+1}{u} [0.5)]^{u} [0.5]^{(2d+1)-u} \\
&= \sum_{u=0}^{d} \binom{2d+1}{u} [0.5)]^{u} [0.5]^{(2d+1)-u} \\
&= \frac{1}{2} \sum_{u=0}^{2d+1} \binom{2d+1}{u} [0.5)]^{u} [0.5]^{(2d+1)-u} = \frac{1}{2}(1) = 0.5. \quad (9.20)
\end{aligned}
$$

The fact that the limiting distribution under the general alternative $F \neq G$ depends on both the negative binomial stopping parameter r and the set size k provides us with even greater flexibility in designing a study with the idea of guaranteeing prescribed power against specific alternatives. Both increasing r and increasing k will lead to increased power for the *PSRSS* median procedure, but increasing r will also lead to a larger number of measured observations from the Y distribution, something that we are trying to avoid. Increasing k and/or increasing the initial sample size m from the X distribution can be used as effective alternatives for increasing the power without increasing the number of measured Y observations.

9.5 Discussion and Future Research

Small sample and asymptotic properties of the *PSRSS* two-sample median test procedure (corresponding to special case 3) based on measuring the Y sample median (for an odd set size k) in every ranked set have been investigated extensively by Matthews et al. (2016). They found that taking the *RSS* approach for collection of the Y sample observations leads to both increased power and decreased expected Y sample size relative to the *PSRSS* version studied by Orban and Wolfe (1982). This is due to both the intrinsic structure inherent in the partially sequential approach to the two-sample problem and the ranked set sampling methodology employed in obtaining the Y sample. As noted in Sect. 9.4, further improvements in both power and reduced Y sample size can likely be obtained by utilizing *RSS* to collect both the X and Y sample observations. The basic formulation of this dual *RSS* approach would be analogous to what we utilized in this paper using *SRS* to collect the X sample items, although the mathematical properties would be more complicated. Another intriguing possibility would be to develop *PSRSS* methodology that utilized a fully balanced *RSS* approach to the collection of the Y observations, rather than relying solely on the use of the medians of the ranked sets. This could also include a fully balanced *RSS* approach to collection of the initial X sample, leading to natural partially sequential analogues to the two- sample balanced *RSS* procedures considered by Bohn and Wolfe (1992) and Fligner and MacEachern (2006).

References

Bohn, L., & Wolfe, D. A. (1992). Nonparametric two-sample procedures for ranked-set samples data. *Journal of the American Statistical Association, 87*, 552–561.

Chen, H., Stasny, E. A., & Wolfe, D. A. (2006). Unbalanced ranked set sampling for estimating a population proportion. *Biometrics, 62*, 150–158.

Dell, T. R., & Clutter, J. L. (1972). Ranked set sampling theory with order statistics background. *Biometrics, 28*, 545–555.

Fligner, M. A., & MacEachern, S. N. (2006). Nonparametric two-sample methods for ranked-set sample data. *Journal of the American Statistical Association, 101*, 1107–1118.

Matthews, M. J., Stasny, E. A., & Wolfe, D. A. (2016). Partially sequential median test procedures using ranked set sample data. *Special Issue of Journal of Applied Statistical Sciences* (to appear).

McIntyre, G. A. (1952). A method for unbiased selective sampling, using ranked sets. *Australian Journal of Agricultural Research, 3*, 385–390.

McIntyre, G. A. (2005). A method for unbiased selective sampling, using ranked sets. *The American Statistician, 59*, 230–232 (Reprinted).

Orban, J., & Wolfe, D. A. (1978). Optimality criteria for the selection of a partially sequential indicator set. *Biometrika, 65*, 357–362.

Orban, J., & Wolfe, D. A. (1982). Properties of a distribution-free two-stage two-sample median test. *Statistica Neerlandica, 36*, 15–22.

Özturk, Ö., & Wolfe, D. A. (2000). Optimal allocation procedure in ranked set sampling for unimodal and multi-modal distributions. *Environmental and Ecological Statistics, 7*, 343–356.

Stokes, S. L., & Sager, T. W. (1988). Characterization of a ranked-set sample with application to estimating distribution functions. *Journal of the American Statistical Association, 83*, 374–381.

Terpstra, J. T. (2004). On estimating a population proportion via ranked set sampling. *Biometrical Journal, 46*, 264–272.

Wolfe, D. A. (1977a). On a class of partially sequential two-sample test procedures. *Journal of the American Statistical Association, 72*, 202–205.

Wolfe, D. A. (1977b). Two-stage two-sample median test. *Technometrics, 19*, 495–501.

Wolfe, D. A. (2004). Ranked set sampling: an approach to more efficient data collection. *Statistical Science, 19*, 636–643.

Wolfe, D. A. (2012). Ranked set sampling—its relevance and impact on statistical inference. *ISRN Probability and Statistics, 2012*, 32 p. (Spotlight Article).

Chapter 10
A New Scale-Invariant Nonparametric Test for Two-Sample Bivariate Location Problem with Application

Sunil Mathur, Deepak M. Sakate, and Sujay Datta

Abstract Diagnostic testing in medicine is crucial in determining interventions and treatment plans. It is important to analyze diagnostic tests accurately so that the right decision can be made by clinicians. A scale-invariant test is proposed for when treatment and control samples are available and a change in condition between the treatment and control groups is investigated. The proposed test statistic is shown to have an asymptotically normal distribution. The power of the proposed test is compared with that of several existing tests using Monte Carlo simulation techniques under different bivariate population set-ups. The power study shows that the proposed test statistic performs very well as compared to its competitors for almost all the changes in location and for almost all the distributions considered in this study. The computation of proposed test statistic is shown using a real-life data set.

Keywords Location test • Power • Bivariate • Wilcoxon's rank sum test • Mardia's test

10.1 Introduction

Diagnostic testing plays an important role in medicine, helping physicians determine treatment plans, interventions, and the need for new diagnostic tests (Deeks 2001). The patients' health and well-being are dependent on the decisions made based on these diagnostic tests. Thus, it becomes critical to analyze diagnostic tests

S. Mathur (✉) • D.M. Sakate
Department of Biostatistics and Epidemiology, Medical College of Georgia, Georgia Regents University, Augusta, GA 30912-4900, USA
e-mail: smathur@gru.edu; dsakate@gru.edu

S. Datta
Department of Statistics, Buchtel College of Arts and Sciences, University of Akron, Akron, OH 44325-1901, USA
e-mail: sd85@uakron.edu

© Springer International Publishing Switzerland 2016

R.Y. Liu, J.W. McKean (eds.), *Robust Rank-Based and Nonparametric Methods*, Springer Proceedings in Mathematics & Statistics 168,
DOI 10.1007/978-3-319-39065-9_10

accurately in order to select the correct treatment plans and interventions. Every year several new drugs are introduced to treat a variety of illnesses. In the case of epilepsy, for example, ten new antiepileptic drugs were introduced in a single year (Beghi 2004), which expanded the options available to physicians. The new drugs may be better in treating the illness in terms of efficacy, safety, and tolerability. The choice of a new drug is generally considered when there is no benefit or a negative effect from the old drug (Beghi 2004). In the area of drug development, it is essential to detect even the smallest change produced by a new drug in the treatment group's outcome (Pepe 2003). In most cases, the distributional form of the population remains unknown, which can make statistical tests based on a parametric model ineffective or may even lead to false conclusions (Woolson and Clarke 2011). The problem becomes more complicated if the sample size available is small and involves more than one outcome variable (Armitage et al. 2008). Our aim is to compare the clinical outcomes of two drugs in order to make a better treatment strategy. In other words, we would like to compare two clinical outcomes under two different conditions (placebo, treatment). In this paper, first we outline the problem in a statistical framework and then propose a test which can be applied when the population distribution is unknown, two outcome variables are available, and treatment and control samples are available.

We consider two independent random samples, denoted by $(X_{1i}, Y_{1i}), i = 1, \ldots, m$ and $(X_{2j}, Y_{2j}), j = 1, \ldots, n$, from bivariate populations with continuous distribution functions (cdfs) $F(x, y)$ and $G(x, y)$ respectively. It is assumed that the populations are elliptically symmetric about their respective medians. Our aim is to test

$$H_0 : F(x, y) = G(x, y) \tag{10.1}$$

against

$$H_A : F(x, y) = G(x + \delta_1, y + \delta_2) \tag{10.2}$$

where, $(\delta_1, \delta_2) \neq (0, 0)$.

There are several procedures available in the literature for the problem (Baringhaus and Franz 2004; Chung and Romano 2013; Davis and McKean 1993; Dietz 1982; García et al. 2010; Jurečková and Kalina 2012; Oja 1999; Peters and Randles 1990, 1991; Randles and Peters 1990; Wilcox 2012). The generalized multivariate median of Oja (1999) was used in Brown and Hettmansperger (1987) to define the multivariate notion of quantile or rank. Peddada et al. (2006) proposed the 'Dunnett-type' test procedures to test for simple tree order restrictions on the means of p independent normal populations and also suggested nonparametric versions based on ranked data for non-normal data. Mathur and Smith (2008) proposed a test statistic that had a U-statistic representation with a degenerate kernel for the two sample bivariate location problem. The limiting distribution for the proposed test statistic was *Gaussian chaos* (Van der Vaart 2000). However, some of these tests are complex and may require special conditions for applications.

In Sect. 10.2, we propose a test statistic U. The invariance property of the test is studied in Sect. 10.2.1. In Sect. 10.3, asymptotic properties are presented.

In Sect. 10.4, Monte Carlo results are presented. In Sect. 10.5, a bivariate two sample data set is used for applying the proposed test. In Sect. 10.6, the discussion is presented.

10.2 The Proposed Test Statistic

Let $(X^0_{1i}, Y^0_{1i}), i = 1, \ldots, m$ and $(X^0_{2j}, Y^0_{2j}), j = 1, \ldots, n$ be two independent random samples from the bivariate populations with cdfs $F(x, y)$ and $G(x, y)$, respectively. We wish to test the null hypothesis as given in Eq. (10.1).

Consider, $(\bar{X}_C, \bar{Y}_C,)$ as the mean of the combined sample which is obtained by pooling the two samples. Define $X_{1i} = X^0_{1i} - \bar{X}_C$, $Y_{1i} = Y^0_{1i} - \bar{Y}_C$, $i = 1, \ldots, m$ and $X_{2j} = X^0_{2j} - \bar{X}_C$, $Y_{2j} = Y^0_{2j} - \bar{Y}_C$, $j = 1, \ldots, n$. We compute the angles made by $Z_i = (X_{1i}, Y_{1i})$ and $W_j = (X_{2j}, Y_{2j})$ with the positive directions of X- axis, measured from 0 to 2π, and denote them by θ_{1ci} and θ_{2cj} respectively. The slopes are defined as $m_{1i} = \frac{Y_{1i}}{X_{1i}}$, and $m_{2j} = \frac{Y_{2j}}{X_{2j}}$. Looking at the combined samples, let $Rank(\theta_{1ci})$, $i = 1, \ldots, m$ and $Rank(\theta_{2cj}), j = 1, \ldots, n$ be the ranks of angles in first and second sample respectively. Blumen (1958) proposed a test statistic based on slopes for one sample location problem. Mardia (1967) used the slopes for the two-sample location problem. Motivated by the work in Blumen (1958) and Mardia (1967), we base our test statistic on the center of gravity of the unit circle. It is similar to that in Blumen (1958) and Mardia (1967) but we use the ranks of angles of two samples in different manner which we believe would lead to a more powerful test statistic. Similar approach was used in Peters and Randles (1991). A test statistic based on the direction of the vectors $X_i = (X_{1i}, Y_{1i})$ and $Y_j = (X_{2j}, Y_{2j})$ with the positive directions, measured from 0 to 2π, and slopes $m_{1i} = \frac{Y_{1i}}{X_{1i}}$, and $m_{2j} = \frac{Y_{2j}}{X_{2j}}$, will be independent of unit of measurements used and correlation between two variates. Moreover, the use of ranks of θ_{1ci} and θ_{2cj} provides a way to compare variables in two populations without using population distributions, thus leading to a nonparametric test statistic. Also, the direction of the slopes incorporated into building of a test statistic enables us to see whether vectors are lying in the upper half of the unit circle or lower half of the unit circle. The direction of slopes can be found by looking at the sign of y_i and under the null hypothesis, we should have probability of y lying in the upper half of a unit circle to be one half and that in negative half of the unit circle is also one half. Now, if we consider the slopes and directions of the unit vectors, we should see a random pattern if the null hypothesis is true otherwise there should be a preferred direction of these vectors. Thus, calculating the mean value of the vectors, which in our case is the center of the gravity of the unit circle, will enable us to see whether there is a random pattern of observations exists or not.

We define statistics T_1 and T_2 as

$$T_1 = \frac{1}{m} \sum_{i=1}^{m} A_i Rank(\theta_{1ci}) \tag{10.3}$$

and

$$T_2 = \frac{1}{n}\sum_{j=1}^{n} B_j Rank(\theta_{2cj}) \tag{10.4}$$

where,

$$A_i = \begin{cases} 1, & \text{if } y_{1i} \text{ is Positive in } i^{th} \text{ ordered slope } m_{1i} \\ 0, & \text{if } y_{1i} \text{ is Negative in } i^{th} \text{ ordered slope } m_{1i}, \end{cases}$$

$$B_j = \begin{cases} 1, & \text{if } y_{2j} \text{ is Positive in } j^{th} \text{ ordered slope } m_{2j} \\ 0, & \text{if } y_{2j} \text{ is Negative in } j^{th} \text{ ordered slope } m_{2j} \end{cases}$$

Here, A's and B's are not mutually independent as both depend on the commonly centered data values. Also, the test procedure will not be exactly distribution-free but only asymptotically.

Under H_0, we expect

$$P[A_i = 1] = \frac{1}{2} = P[A_i = 0],$$

$$P[B_j = 1] = \frac{1}{2} = P[B_j = 0].$$

We combine both T_1 and T_2 and propose the statistic

$$U = T'\Sigma^{-1}T, \tag{10.5}$$

where, $T'_{1\times 2} = (T_1, T_2)'$ and Σ is a 2×2 covariance matrix of T. For very small or large values of U, H_0 is rejected.

10.2.1 Invariance

A statistic T is said to be scale-invariant if

$$T(cX_1, cX_2, \ldots, cX_m; cY_1, cY_2, \ldots, cY_n) = T(X_1, X_2, \ldots, X_m; Y_1, Y_2, \ldots, Y_n),$$

where, c is a constant.

In our case, $X_i = (X_{1i}, Y_{1i}), i = 1, 2, \ldots, m$, and $Y_j = (X_{2j}, Y_{2j}), j = 1, 2, \ldots, n$. We find that m_{1i}'s hence θ_{1i}'s are invariant under above transformation. Similarly, θ_{2j}'s are invariant under above transformation. Hence, the proposed statistic U is scale-invariant.

10.3 Asymptotic Properties

In order to study asymptotic properties of the proposed test statistic, we denote the probability density function (pdf) of Y_1 and Y_2 by $f_{Y_1}(y_1)$ and $h_{Y_2}(y_2)$ in the ordered slopes m_1 and m_2, respectively and corresponding cumulative distribution function (cdf) for Y_1 and Y_2 by $F_{Y_1}(y_1)$ and $H_{Y_2}(y_2)$ respectively.

We define $p_1 = P(Y_1 > 0)$, and $p_2 = P(Y_2 > 0)$, under H_0, $p_1 = 1/2 = p_2$. Define $p_{11} = P[Y_{1j} < Y_{1i}]$, $p_{22} = P[Y_{2j} < Y_{2i}]$, and $p_{12} = P[Y_{1j} < Y_{2i}]$.

We find that $E(T_1) = \frac{m+1}{2} p_1$, $E(T_2) = \frac{n+1}{2} p_2$,

$$Cov(T_1, T_2) = \sum_{i \neq j}^{m} \sum^{n} (E(A_i B_j) - E(A_i)E(B_j)) Rank(\theta_{1ci}) Rank(\theta_{2cj})$$

$$= \sum_{i \neq j}^{m} \sum^{n} (p_{12} - p_{11} p_{22}) Rank(\theta_{1ci}) Rank(\theta_{2cj}), \quad (10.6)$$

$Var(T_1) = \sum_{i=1}^{m} Var(A_i)[Rank(\theta_{1ci})]^2$, and $Var(T_2) = \sum_{j=1}^{n} Var(B_j)[Rank(\theta_{2cj})]^2$.

Under H_0, $E(T_1) = \frac{m+1}{4}$ and $E(T_2) = \frac{n+1}{4}$.

Let $\delta = (\delta_1, \delta_2)$ and $N = n + m$. Denote $F_{M_1}(m_1)$ and $H_{M_2}(m_2)$ the cdf of m_1 and m_2, respectively.

Theorem 10.1. *Suppose that $n/N \to r \in (0, 1)$. Then*

$$\frac{\sqrt{N}[\bar{U} - u(\delta)]}{\sigma(\delta)} \xrightarrow{D} N(0, 1) \quad (10.7)$$

where,

$$\bar{U} = \frac{U}{mn}$$

$$u(\delta) = E(\bar{U}) \quad (10.8)$$

and

$$\sigma^2(\delta) = Var(\bar{U}) \quad (10.9)$$

Remark 10.1. The convergence is asymptotic as both m and n tend to infinity such that n/N converges to r.

Remark 10.2. If F_{M_1} and H_{M_2} are continuous under H_0, then $u(0) = 1/2$ and $\sigma^2(0) = 1/12r(1-r)$.

Under the large sample, the asymptotic power function of the test yields

$$\begin{aligned}
\pi_N(\delta) &= P_\delta(\sqrt{N}(\bar{U}-u(0)) > Z_{1-\alpha}\sigma(0)) \\
&= P_\delta(\sqrt{N}(\bar{U}-u(\delta)) > Z_{1-\alpha}\sigma(0) - \sqrt{N}(u(\delta)-u(0)) \\
&= P_\delta\left(\frac{\sqrt{N}(\bar{U}-u(\delta))}{\sigma(\delta)} > \frac{Z_{1-\alpha}\sigma(0)-\sqrt{N}(u(\delta)-u(0))}{\sigma(\delta)}\right) \\
&= 1-\Phi\left(Z_{1-\alpha} - \frac{\sqrt{N}(u(\delta)-u(0))}{\sigma(\delta)}\right) \\
&= 1-\Phi\left(Z_{1-\alpha} - \frac{\sqrt{N}\nabla u(0)\delta'}{\sigma(\delta)}\right) + o(1)
\end{aligned}$$

Proof (Proof of Theorem 10.1). The test statistic is $U = T'\Sigma^{-1}T$.

Let $\bar{U} = U/nm$. Considering an equivalent form of U as $X'\Sigma^{-1}X$, where $X' = (X_1, X_2)$, and Σ is variance-covariance matrix, using the kernel as $h(x_1, \ldots, x_m) = X'\Sigma^{-1}X$, the U-statistic can be defined as $U_n(h) = \frac{1}{\binom{n}{m}} \sum_{C_{m,n}} h(x_1, \ldots, x_m)$ (Lee 1990). Similarly, taking $h(x_1, \ldots, x_m) = T'\Sigma^{-1}T$ as kernel function, we find that $\bar{U}$ is a two-sample U-statistic. By the two-sample U-statistic theorem,

$$\begin{aligned}
\sqrt{N}(\bar{U} - P_\delta(m_2 < m_1)) = &\frac{\sqrt{N}}{n}\sum_{i=1}^{n}(H_{M_2}(m_{1i}) - E[H_{M_2}(m_{1i})]) \\
&-\frac{\sqrt{N}}{m}\sum_{j=1}^{m}\left(F_{M_2}(m_{2j}) - E[F_{M_2}(m_{2j})]\right) + o_{P_\delta}(1).
\end{aligned}$$

This leads to the asymptotic normality in Eq. (10.7) with $u(\delta)$, $\sigma^2(\delta)$ defined in Eqs. (10.8) and (10.9) respectively. □

Remark 10.3. We notice that U has a form similar to Mahalanobis' $d^2 = (X-\mu)'\Sigma^{-1}(X-\mu)$, which has a Chi-square distribution with p degrees of freedom. This leads to the conclusion that U has Chi-square distribution with 2 degrees of freedom asymptotically. As an anonymous reviewer pointed out, the Monte Carlo results in Sect. 10.4 tend to support the normal approximation for the standardized test statistic.

Special Case.

Suppose that $(X_1, Y_1) \sim N\left(\begin{pmatrix}0\\0\end{pmatrix}, \begin{pmatrix}1&0\\0&1\end{pmatrix}\right)$ and $(X_2, Y_2) \sim N\left(\begin{pmatrix}\delta_1\\\delta_2\end{pmatrix}, \begin{pmatrix}1&0\\0&1\end{pmatrix}\right)$. Therefore, the asymptotic power function in this case is

$$\pi_N(\delta) = 1-\Phi\left(Z_{1-\alpha} - \sqrt{\frac{12nm}{n+m}}\left(\frac{1}{2} - \frac{exp(-||\delta||^2/2)}{2}\right)\right) + o(1) \qquad (10.10)$$

10.4 Power Study

Power is compared using Monte Carlo simulation method. We compared the powers of the proposed test with those of Hotelling's T^2 test, Mardia's test (Mardia 1967), and Wilcoxon's rank sum test (Peters and Randles 1991). We denote proposed statistic by U, Mardia's statistic by M, Wilcoxon's rank sum test statistic by WRS and Hotelling's T^2 by T^2. Four test statistics U, M, WRS and T^2 are compared using samples from the following seven distributions: Bivariate normal, Bivariate normal mixtures with probability $P = 0.5$ and $P = 0.9$ respectively, Pearson type II (light tailed), Pearson type VII (heavy tailed), populations 6 (heavy tailed) and 7 (light tailed). Population 6 and 7 are from the class of densities in Randles and Peters (1990). We have chosen $\upsilon = 0.10$ and $\upsilon = 30$ for population 6 and 7 respectively. The proportion of times that each test statistic exceeded its critical value when the sample size is kept fixed and location parameters are changed provides the measure of power. The results are presented in Appendix in Tables 10.2 and 10.3. Results in Table 10.2 are based on 2000 samples of size $n = 15$ from a bivariate population with location (δ_1, δ_2) where, $\delta_1, \delta_2 > 0$ and 2000 samples of size $m = 18$ from population with location (0,0). This process is repeated for samples of sizes $n = 25$ and $m = 28$ respectively for producing Table 10.3 given in Appendix. A nominal significance level of 0.05 was used. The variance-covariance matrix used in the simulation is $\begin{pmatrix} 1 & 0.5 \\ 0.5 & 1 \end{pmatrix}$.

For sample size $m = 15$, and $n = 18$, proposed test statistic performs very well as compared to Mardia's, and Wilcoxon's test statistics when the underlying distribution is normal. We find that for all shifts in location parameters considered in the study, Hotelling's T^2 dominates when the underlying population is normal. When the underlying distribution is non-normal, the proposed test statistic works very well as compared to its competitors including Hotelling's T^2. For example, in case of bivariate normal mixtures, the proposed test statistic U dominates the rest of test statistics for small as well as large changes in the location. In case of light-tailed distributions, the proposed test statistic U performs very well for almost all the values of location parameters considered here. For heavy tailed distributions, we notice that the power of the proposed test statistic is very high for all values of the location parameters as compared to all the statistics considered here. The proposed test statistic U approaches the power value 1 at a very faster rate as compared to its competitors. Also, for $m = 25$, and $n = 28$, similar results are seen.

Thus, the power study shows that the proposed test statistic will work very well when the underlying population is non-normal and is able to detect the smallest shift in the location which is a highly desirable property of any test statistic. It is highly desirable that a nonparametric test's performance be as good as its parametric counterpart under normal distribution conditions (Dietz 1982; Larocque et al. 2003; Peters and Randles 1991) which is achieved by the proposed test.

10.5 Real-Life Application

The proposed test statistic is applied to bivariate data, Table 10.1, taken from Peterson's Data (Belle et al. 2004) which describes the inverse relation of sudden infant death syndrome (SIDS) rate with birthweight. The data consists of the birthweight (in grams) of 22 dizygous twins and 19 monozygous twins.

Let $\delta = (\delta_1, \delta_2)$ be the difference in the medians of two populations from which the samples were drawn. We state our null hypothesis as

$$H_0 : \delta = 0$$

against

$$H_A : \delta \neq 0$$

We find that $T_1 = 8.219$, $T_2 = 9.125$, and $U = 1.214$. The approximate p-value of the test statistic is 0.002. Therefore, the null hypothesis is rejected at 5 % level of significance.

Table 10.1 Birthweight and SID

Dizygous twins		Monozygous twins	
SID	Non-SID	SID	Non-SID
1474	2098	1701	1956
3657	3119	2580	2438
3005	3515	2750	2807
2041	2126	1956	1843
2325	2211	1871	2041
2296	2750	2296	2183
3430	3402	2268	2495
3515	3232	2070	1673
1956	1701	1786	1843
2098	2410	3175	3572
3204	2892	2495	2778
2381	2608	1956	1588
2892	2693	2296	2183
2920	3232	3232	2778
3005	3005	1446	2268
2268	2325	1559	1304
3260	3686	2835	2892
3260	2778	2495	2353
2155	2552	1559	2466
2835	2693		
2466	1899		
3232	3714		

10.6 Discussion

One of the important applications of nonparametric testing is in the field of diagnostic testing in medical science. Physicians base their future treatment plans based on diagnostic testing. Since, the nature of population is generally unknown, parametric testing procedures cannot be applied with certainty. Nonparametric tests present a useful alternative to parametric procedures under these conditions. We developed a strictly nonparametric test. The proposed test is also scale-invariant, which is a desirable property of any test statistic (Kowalski and Tu 2008; Sugiura 1965). The proposed test statistic takes into account the information contained in the samples, including the correlation structure of the population through Σ. This makes the test statistic more sensitive to location changes, thus making it a highly powerful test statistic. The power study shows that under normal distribution it is more powerful than Mardia's test, and more powerful than Wilcoxon's test for almost all the values of the location parameters considered here. For non-normal populations, the proposed test statistic dominates the test statistics considered here for almost all the values of location parameters. It approaches power level 1 very quickly for very small shifts of location parameters. The power simulation study shows that the proposed test statistic works extremely well when an underlying population is non-normal. The proposed test statistic is able to detect the smallest shift in the location parameter. Since, the distribution of an underlying population is hardly ever known in real-life situations, the proposed test statistic may work very well in such situations. Thus, in many fields, such as the medical field where the smallest change in a drug's effectiveness can be critical (Schneeweiss et al. 2011) for a clinician to determine a treatment plan or intervention for a patient (McDonald et al. 2002), this test can provide highly desirable accuracy and efficiency. Several medical diagnosis require more than one characteristic of a population to be considered for making the decision on future course of treatment. The proposed test fits very well with those requirements. The proposed test can take into account two characteristics of the population at a time and compare the control group and treatment group based on these two characteristics. The higher power of the proposed test as compared to some of its competitors makes it a more attractive option than its competitors.

Acknowledgements Authors would like to thank the anonymous referee for his/her useful comments which enhanced the clarity of the paper and led to significant improvements in the paper.

Appendix

See Tables 10.2 and 10.3.

Table 10.2 Monte Carlo rejection proportion, sample size $m = 15, n = 18$

Distribution	δ_1, δ_2	U	M	WRS	T^2
Bivariate normal	0.00, 0.00	0.0545	0.046	0.0445	0.0485
	0.10, 0.07	0.176	0.065	0.098	0.3875
	0.30, 0.10	0.3885	0.0905	0.145	0.7405
	0.70, 0.50	0.8995	0.367	0.81	1
	1.20, 1.00	0.99	0.659	0.9995	1
	2.40, 3.00	1	0.9375	1	1
BVN mixture P = 0.5	0.00, 0.00	0.055	0.0505	0.0475	0.0505
	0.10, 0.07	0.2115	0.068	0.066	0.112
	0.30, 0.10	0.4735	0.152	0.1725	0.213
	0.70, 0.50	0.8975	0.523	0.845	0.8705
	1.20, 1.00	1	0.855	1	1
	2.40, 3.00	1	0.9995	1	1
BVN mixture P = 0.9	0.00, 0.00	0.047	0.0495	0.0475	0.044
	0.10, 0.07	0.295	0.067	0.061	0.0615
	0.30, 0.10	0.546	0.152	0.125	0.135
	0.70, 0.50	0.9575	0.447	0.687	0.682
	1.20, 1.00	0.9985	0.798	0.9975	0.995
	2.40, 3.00	1	0.9995	1	1
Type VII	0.00, 0.00	0.0515	0.0515	0.0445	0.0495
	0.10, 0.07	0.266	0.0725	0.154	0.098
	0.30, 0.10	0.699	0.193	0.49	0.5965
	0.70, 0.50	1	0.8855	0.9985	1
	1.20, 1.00	1	0.9885	1	1
	2.40, 3.00	1	0.999	1	1
Type II	0.00, 0.00	0.0455	0.035	0.051	0.039
	0.10, 0.07	0.393	0.058	0.1505	0.1515
	0.30, 0.10	0.785	0.2105	0.5135	0.716
	0.70, 0.50	1	0.685	1	1
	1.20, 1.00	1	0.7655	1	1
	2.40, 3.00	1	0.8655	1	1
Population 6	0.00, 0.00	0.05	0.0535	0.0585	0.044
	0.10, 0.07	0.5935	0.069	0.519	0.1825
	0.30, 0.10	0.891	0.0705	0.8635	0.714
	0.70, 0.50	1	0.2755	1	1
	1.20, 1.00	1	0.4225	1	1
	2.40, 3.00	1	0.5110	1	1
Population 7	0.00, 0.00	0.053	0.0535	0.049	0.0495
	0.10, 0.07	0.2135	0.071	0.062	0.0575
	0.30, 0.10	0.368	0.1165	0.1115	0.105
	0.70, 0.50	0.7785	0.3495	0.422	0.4645
	1.20, 1.00	0.9735	0.803	0.927	0.9605
	2.40, 3.00	1	1	1	1

Table 10.3 Monte Carlo rejection proportion, sample size $m = 25$, $n = 28$

Distribution	δ_1, δ_2	U	M	WRS	T^2
Bivariate normal	0.00, 0.00	0.0525	0.046	0.0445	0.0485
	0.10, 0.07	0.0905	0.0585	0.058	0.183
	0.30, 0.10	0.3855	0.0905	0.145	0.7405
	0.70, 0.50	0.888	0.367	0.81	1
	1.20, 1.00	1	0.659	0.9995	1
	2.40, 3.00	1	0.9375	1	1
BVN mixture P = 0.5	0.00, 0.00	0.0535	0.0505	0.0445	0.05
	0.10, 0.07	0.1715	0.0725	0.075	0.0815
	0.30, 0.10	0.387	0.2325	0.2395	0.347
	0.70, 050	1	0.736	0.948	0.991
	1.20, 1.00	1	0.993	1	1
	2.40, 3.00	1	1	1	1
BVN mixture P = 0.9	0.00, 0.00	0.045	0.051	0.058	0.0515
	0.10, 0.07	0.1655	0.0785	0.0795	0.0775
	0.30, 0.10	0.381	0.2085	0.1965	0.2375
	0.70, 0.50	0.995	0.5915	0.833	0.9165
	1.20, 1.00	1	0.9105	1	1
	2.40, 3.00	1	1	1	1
Type VII	0.00, 0.00	0.0445	0.0425	0.047	0.0525
	0.10, 0.07	0.4095	0.0725	0.3285	0.2525
	0.30, 0.10	1	0.353	0.6985	0.9445
	0.70, 0.50	1	0.9995	1	1
	1.20, 1.00	1	1	1	1
	2.40, 3.00	1	1	1	1
Type II	0.00, 0.00	0.048	0.046	0.043	0.046
	0.10, 0.07	0.3205	0.0835	0.1925	0.219
	0.30, 0.10	1	0.416	0.611	0.914
	0.70, 0.50	1	0.825	1	1
	1.20, 1.00	1	0.858	1	1
	2.40, 3.00	1	0.9985	1	1
Population 6	0.00, 0.00	0.0506	0.057	0.049	0.051
	0.10, 0.07	0.899	0.0685	0.7225	0.275
	0.30, 0.10	1	0.0975	0.921	0.8975
	0.70, 0.50	1	0.2345	1	1
	1.20, 1.00	1	0.4775	1	1
	2.40, 3.00	1	0.5655	1	1
Population 7	0.00, 0.00	0.051	0.0465	0.047	0.0585
	0.10, 0.07	0.323	0.0715	0.0655	0.0725
	0.30, 0.10	0.536	0.2005	0.125	0.145
	0.70, 0.50	0.941	0.604	0.6025	0.762
	1.20, 1.00	0.999	0.9815	0.99	0.9995
	2.40, 3.00	1	1	1	1

References

Armitage, P., Berry, G., & Matthews, J. N. S. (2008). *Statistical methods in medical research*. Wiley-Blackwell, Massachusetts, USA.

Baringhaus, L., & Franz, C. (2004). On a new multivariate two-sample test. *Journal of Multivariate Analysis, 88*, 190–206.

Beghi, E. (2004). Efficacy and tolerability of the new antiepileptic drugs: Comparison of two recent guidelines. *The Lancet Neurology, 3*, 618–621.

Belle, G. V., Fisher, L. D., Heaerty, P. J., & Lumley, T. (2004). *Biostatistics: A methodology for the health sciences* (2nd ed.). New York: Wiley.

Blumen, I. (1958). A new bivariate sign test. *Journal of the American Statistical Association, 53*, 448–456.

Brown, B., & Hettmansperger, T. (1987). Affine invariant rank methods in the bivariate location model. *Journal of the Royal Statistical Society Series B (Methodological), 49*, 301–310.

Chung, E., & Romano, J. P. (2013). Exact and asymptotically robust permutation tests. *The Annals of Statistics, 41*, 484–507.

Davis, J. B., & McKean, J. W. (1993). Rank-based methods for multivariate linear models. *Journal of the American Statistical Association, 88*, 245–251.

Deeks, J. J. (2001). Systematic reviews in health care: Systematic reviews of evaluations of diagnostic and screening tests. *British Medical Journal, 323*, 157.

Dietz, E. J. (1982). Bivariate nonparametric tests for the one-sample location problem. *Journal of the American Statistical Association, 77*, 163–169.

García, S., Fernández, A., Luengo, J., & Herrera, F. (2010). Advanced nonparametric tests for multiple comparisons in the design of experiments in computational intelligence and data mining: Experimental analysis of power. *Information Sciences, 180*, 2044–2064.

Jurečková, J., & Kalina, J. (2012). Nonparametric multivariate rank tests and their unbiasedness. *Bernoulli, 18*, 229–251.

Kowalski, J., & Tu, X. M. (2008). *Modern applied U-statistics* (Vol. 714). Wiley.com.

Larocque, D., Tardif, S., & Eeden, C. V. (2003). An affine-invariant generalization of the Wilcoxon signed-rank test for the bivariate location problem. *Australian and New Zealand Journal of Statistics, 45*, 153–165.

Lee, A. J. (1990). *U-Statistics: Theory and practice*. Boca Raton, FL: CRC Press.

Mardia, K. (1967). A non-parametric test for the bivariate two-sample location problem. *Journal of the Royal Statistical Society. Series B (Methodological), 29*, 320–342.

Mathur, S. K., & Smith, P. F. (2008). An efficient nonparametric test for bivariate two-sample location problem. *Statistical Methodology, 5*, 142–159.

McDonald, H. P., Garg, A. X., & Haynes, R. B. (2002). Interventions to enhance patient adherence to medication prescriptions. *The Journal of the American Medical Association, 288*, 2868–2879.

Oja, H. (1999). Affine invariant multivariate sign and rank tests and corresponding estimates: A review. *Scandinavian Journal of Statistics, 26*, 319–343.

Peddada, S. D., Haseman, J. K., Tan, X., & Travlos, G. (2006). Tests for a simple tree order restriction with application to dose–response studies. *Journal of the Royal Statistical Society: Series C (Applied Statistics), 55*, 493–506.

Pepe, M. S. (2003). *The statistical evaluation of medical tests for classification and prediction*. Oxford: Oxford University Press.

Peters, D., & Randles, R. H. (1990). A multivariate signed-rank test for the one-sample location problem. *Journal of the American Statistical Association, 85*, 552–557.

Peters, D., & Randles, R. H. (1991). A bivariate signed rank test for the two-sample location problem. *Journal of the Royal Statistical Society. Series B (Methodological), 53*, 493–504.

Randles, R. H., & Peters, D. (1990). Multivariate rank tests for the two-sample location problem. *Communications in Statistics-Theory and Methods, 19*, 4225–4238.

Schneeweiss, S., Gagne, J., Glynn, R., Ruhl, M., & Rassen, J. (2011). Assessing the comparative effectiveness of newly marketed medications: Methodological challenges and implications for drug development. *Clinical Pharmacology and Therapeutics, 90*, 777–790.

Sugiura, N. (1965). *Multisample and multivariate nonparametric tests based on U statistics and their asymptotic efficiencies. Osaka Journal of Mathematics, 2*, 385–426.

Van der Vaart, A. W. (2000). *Asymptotic statistics* (Vol. 3). Cambridge: Cambridge University Press.

Wilcox, R. R. (2012). *Introduction to robust estimation and hypothesis testing*. Academic Press, USA.

Woolson, R. F., & Clarke, W. R. (2011). *Statistical methods for the analysis of biomedical data.* (Vol. 371). John Wiley and Sons, New York.

Chapter 11
Influence Functions and Efficiencies of k-Step Hettmansperger–Randles Estimators for Multivariate Location and Regression

Sara Taskinen and Hannu Oja

Abstract In Hettmansperger and Randles (Biometrika 89:851–860, 2002) spatial sign vectors were used to derive simultaneous estimators of multivariate location and shape. Oja (Multivariate nonparametric methods with R. Springer, New York, 2010) proposed a similar approach for the multivariate linear regression case. These estimators are highly robust and have under general assumptions a joint limiting multinormal distribution. The estimates are easy to compute using fixed-point algorithms. There are however no exact proofs for the convergence of these algorithms. The existence and uniqueness of the solutions also still remain unproven although we believe that they hold under general conditions. To circumvent these problems, we consider in this paper k-step versions of Hettmansperger and Randles (HR) location and shape estimators and their extensions to the linear regression problem. The influence functions, limiting distributions and asymptotical efficiencies of the estimators are derived at the multivariate elliptical case.

Keywords Affine equivariance • Efficiency • Influence function • Location vector • Multivariate regression • Shape matrix • Spatial sign

11.1 Introduction

In this paper we consider so called Hettmansperger–Randles (HR) estimators for multivariate location-scatter and regression-scatter models. In the case of the regular least square estimates with simultaneous covariance matrix estimation, the residuals are made mutually uncorrelated as well as uncorrelated with the explaining

---

S. Taskinen
Department of Mathematics and Statistics, University of Jyvaskyla, Jyvaskyla, Finland
e-mail: sara.l.taskinen@jyu.fi

H. Oja (✉)
Department of Mathematics and Statistics, University of Turku, Turku, Finland
e-mail: hannu.oja@utu.fi

© Springer International Publishing Switzerland 2016
R.Y. Liu, J.W. McKean (eds.), *Robust Rank-Based and Nonparametric Methods*, Springer Proceedings in Mathematics & Statistics 168,
DOI 10.1007/978-3-319-39065-9_11

variables. The HR estimators are obtained in a similar way, the residuals are just replaced by the multivariate spatial signs of the residuals. The spatial sign of the p-vector $\mathbf{y}$ is a unit vector in the direction of $\mathbf{y}$, that is,

$$\mathbf{S}(\mathbf{y}) = \begin{cases} \|\mathbf{y}\|^{-1}\mathbf{y}, & \mathbf{y} \neq \mathbf{0} \\ \mathbf{0}, & \mathbf{y} = \mathbf{0}, \end{cases} \tag{11.1}$$

where $\|\mathbf{y}\| = (\mathbf{y}^T\mathbf{y})^{1/2}$ is the Euclidean length of the vector $\mathbf{y}$. Spatial sign function thus projects the residuals on the unit sphere making the estimators obtained in this way highly robust. For statistical inference based on spatial signs and ranks, see Oja (2010). For other concepts of multivariate signs and ranks, see also Puri and Sen (1971), Oja (1999), and Hettmansperger and McKean (2011).

Simple location and scatter estimators based on spatial sign vectors, that is, the spatial median and the spatial sign covariance matrix, were studied in Brown (1983), Locantore et al. (1999), Marden (1999) and Visuri et al. (2000) among others. These estimators are easy to compute and have some nice statistical properties e.g. bounded influence functions, positive breakdown points and high efficiency at heavy-tailed distributions. Unfortunately, the estimators are only rotational equivariant therefore their usage in multivariate analysis is limited. The affine equivariant shape matrix of Tyler (1987) is based on an affine transformation such that the spatial signs of transformed data points are uncorrelated. In Hettmansperger and Randles (2002) simultaneous estimators of location and scatter were found such that the spatial signs of the estimated residuals were standardized with mean vector zero and covariance matrix proportional to the identity matrix. In multivariate regression setting, the estimators based on spatial sign vectors were developed and studied in Bai et al. (1990) and Arcones (1998). These estimators are not affine equivariant either, but as in the case of location and scatter estimation, fully equivariant estimators can be derived by applying spatial sign function to the residual points, Oja (2010).

The HR estimators are computed via simple iterative algorithms, but so far the proofs for the convergence of these algorithms are missing. Also the existence and uniqueness of the solutions remain unproven. Notice that conditions for the existence and uniqueness of simultaneous M-estimators of multivariate location and scatter were derived in Maronna (1976), but HR estimators do not satisfy such conditions. Due to these shortcomings, we study in this paper so called k-step versions of the estimators. In Sect. 11.2, we consider estimators for the location-scatter model and in Sect. 11.3, the multivariate regression model is considered. The paper is concluded with some final comments in Sect. 11.4.

11.2 Hettmansperger–Randles Estimators of Location and Shape

11.2.1 Elliptic Model

Assume that the p-variate observations $\mathbf{y}_i$ follow the location-scatter model

$$\mathbf{y}_i = \boldsymbol{\Omega}\mathbf{e}_i + \boldsymbol{\mu},$$

where $\boldsymbol{\mu}$ is the p-variate symmetry center and $\boldsymbol{\Omega}$ is a full-rank $p \times p$ mixing matrix The standardized observations (residuals) $\mathbf{e}_i = \boldsymbol{\Omega}^{-1}(\mathbf{y}_i - \boldsymbol{\mu})$ are assumed to have a spherical distribution, that is, its density function is of the form

$$f_0(\mathbf{e}) = \exp\{-\rho(||\mathbf{e}||)\}$$

for a suitable function ρ. Notice that $\mathbf{e}_i$ can be decomposed as $\mathbf{e}_i = r_i\mathbf{u}_i$, where the modulus $r_i = ||\mathbf{e}_i||$ and the direction vector $\mathbf{u}_i = ||\mathbf{e}_i||^{-1}\mathbf{e}_i$ are independent with $\mathbf{u}_i$ being uniformly distributed on the unit sphere, Fang et al. (1990). To fix $\boldsymbol{\Omega}$, we assume that ρ is chosen so that $E[r_i^2] = p$, that is, $Cov(\mathbf{e}_i) = \mathbf{I}_p$. (If one wish to avoid moment assumptions, one could for example assume that $Med[r_i^2] = p$.) Then $Cov(\mathbf{y}_i) = \boldsymbol{\Sigma} = \boldsymbol{\Omega}\boldsymbol{\Omega}'$ and we can without loss of generality choose $\boldsymbol{\Omega} = \boldsymbol{\Sigma}^{1/2}$ (positive definite and symmetric (PDS) version). We further assume that the density function f_0 is continuous and finite in an open neighborhood of the origin implying that $E(||\mathbf{e}_i||^{-1}||)$ is finite. Under these assumptions, $\mathbf{y}_i$ comes from an elliptically symmetric distribution with probability density function

$$f(\mathbf{y}) = |\boldsymbol{\Omega}|^{-1} f_0(\boldsymbol{\Omega}^{-1}(\mathbf{y} - \boldsymbol{\mu})). \tag{11.2}$$

The scatter parameter $\boldsymbol{\Sigma} = \boldsymbol{\Omega}^2$ can be decomposed as $\boldsymbol{\Sigma} = \sigma\boldsymbol{\Lambda}$, where $\sigma = \sigma(\boldsymbol{\Sigma})$ is a scale parameter and $\boldsymbol{\Lambda} = \sigma^{-1}\boldsymbol{\Sigma}$ is a shape matrix. Natural choices for the scale parameter are given in Paindaveine (2008) and Frahm (2009). In this paper we choose $\sigma = tr(\boldsymbol{\Sigma})/p$, that is, $\boldsymbol{\Lambda}$ is standarised so that $Tr(\boldsymbol{\Lambda}) = p$. Notice that in several applications it is enough to estimate the shape matrix only as it provides all information of the shape and orientation of the multivariate distribution.

Beyond the elliptic model, the properties of multivariate distributions can be described using location and shape functionals. If $F_{\mathbf{y}}$ denotes a cumulative distribution function of $\mathbf{y}$, then a p-variate vector valued functional $\boldsymbol{\mu}(F_{\mathbf{y}})$ is a location vector if it is affine equivariant in the sense that

$$\boldsymbol{\mu}(F_{\mathbf{Ay}+b}) = \mathbf{A}\boldsymbol{\mu}(F_{\mathbf{y}}) + \mathbf{b}, \tag{11.3}$$

for any nonsingular $p \times p$ matrix $\mathbf{A}$ and p-vector $\mathbf{b}$. Further, we call $\mathbf{V}(F_{\mathbf{y}})$ a $p \times p$ shape matrix functional if it is *PDS* and affine equivariant in the sense that

$$\mathbf{V}(F_{\mathbf{Ay}+b}) = \frac{p}{Tr(\mathbf{AV}(F_{\mathbf{y}})\mathbf{A}^T)}\mathbf{AV}(F_{\mathbf{y}})\mathbf{A}^T.$$

Affine equivariance properties of the functionals imply that in the elliptic case, $\mu(F_{\mathbf{y}}) = \mu$ and $\mathbf{V}(F_{\mathbf{y}}) = p\,\Sigma/Tr(\Sigma) = \Lambda$.

When location and shape functionals are applied to empirical distribution function based on the sample $\mathbf{Y} = (\mathbf{y}_1, \ldots, \mathbf{y}_n)^T$, we obtain estimators that we denote from now on by $\hat{\mu} = \mu(\mathbf{Y})$ and $\hat{\mathbf{V}} = \mathbf{V}(\mathbf{Y})$. The estimators are then naturally affine equivariant as well and, in the elliptic model, all location and shape estimates then estimate the same population quantities μ and Λ and are directly comparable without any modifications.

11.2.2 k-Step Location and Shape Estimators

The location estimator $\hat{\mu}$ based on a chosen location score function $T(\mathbf{y})$ solves the estimating equation

$$ave\{\mathbf{T}(\mathbf{y}_i - \hat{\mu})\} = \mathbf{0}.$$

The corresponding location functional $\mu(F_{\mathbf{y}})$ is then defined by $E\left[\mathbf{T}(\mathbf{y}-\mu(F_{\mathbf{y}}))\right] = \mathbf{0}$. If the identity score, $T(\mathbf{y}) = \mathbf{y}$, were used, the classical sample mean vector is obtained that is optimal in the case of multivariate normality. Optimal location score function in the spherical case is $\mathbf{T}(\mathbf{y}) = \nabla\rho(||\mathbf{y}||)$. The spatial median, Brown (1983), is obtained by using the spatial sign score function $\mathbf{S}(\mathbf{y})$ defined in (11.1). The spatial median is highly robust estimator of symmetry center having 50 % breakdown point and bounded influence function. It can be computed using a simple iteration steps

$$\hat{\mu}_k = \hat{\mu}_{k-1} + \frac{ave\{\mathbf{S}(\mathbf{y}_i - \hat{\mu}_{k-1})\}}{ave\{||\mathbf{y}_i - \hat{\mu}_{k-1}||^{-1}\}}.$$

The estimator is however only rotation equivariant, that is, it satisfies (11.3) only for orthogonal $p \times p$ matrices $\mathbf{A}$.

The affine equivariant spatial median can be obtained using so called transformation-retransformation technique. In that case, the observations are first standardized, the spatial median is then found for the standardized observations, and the estimate is then transformed back to the coordinate system of the original observations. See Chakraborty et al. (1998), Tyler et al. (2009) and Ilmonen et al. (2012) and the references therein. In Hettmansperger and Randles (2002), location and shape estimators are estimated simultaneously: $\hat{\mu}$ and $\hat{\mathbf{V}}$ are chosen to satisfy

$$ave\{\mathbf{S}(\hat{\mathbf{e}}_i)\} = \mathbf{0} \quad \text{and} \quad p\,ave\{\mathbf{S}(\hat{\mathbf{e}}_i)\mathbf{S}(\hat{\mathbf{e}}_i)^T\} = \mathbf{I}_p, \tag{11.4}$$

where $\hat{\mathbf{e}}_i = \hat{\mathbf{V}}^{-1/2}(\mathbf{y}_i - \hat{\mu})$ and $\hat{\mathbf{V}}$ is standardized so that $Tr(\hat{\mathbf{V}}) = p$. The resulting location estimate $\hat{\mu}$ is an affine equivariant spatial median and the shape matrix estimate $\hat{\mathbf{V}}$ is the Tyler's M-estimate, Tyler (1987), with respect to the spatial median.

The statistical properties of HR estimators were studied in Hettmansperger and Randles (2002), Tyler (1987), and Dümbgen and Tyler (2005). They showed that the location and shape estimators have bounded influence functions, positive breakdown points and limiting multivariate normal distributions. The computation is very simple. As in general M-estimation case, the estimating equations (11.4) can rewritten in a way that provides the following iteration steps.

Iteration Steps 1 *The HR location-scatter estimate is obtained using the following steps*

1. $\hat{\mathbf{e}}_i = \hat{\mathbf{V}}_{k-1}^{-1/2}(\mathbf{y}_i - \hat{\boldsymbol{\mu}}_{k-1}),\ i = 1, \ldots, n,$
2. $\hat{\boldsymbol{\mu}}_k = \hat{\boldsymbol{\mu}}_{k-1} + \hat{\mathbf{V}}_{k-1}^{1/2}[ave\{||\hat{\mathbf{e}}_i||^{-1}\}]^{-1} ave\{\mathbf{S}(\hat{\mathbf{e}}_i)\},$
3. $\hat{\mathbf{V}}_k = \hat{\mathbf{V}}_{k-1}^{1/2}\ ave\{\mathbf{S}(\hat{\mathbf{e}}_i)\mathbf{S}(\hat{\mathbf{e}}_i)^T\}\ \hat{\mathbf{V}}_{k-1}^{1/2}.$

and $\hat{\mathbf{V}}_k$ *is standardized so that* $Tr(\hat{\mathbf{V}}_k) = p$.

Unfortunately there is no proof for the convergence of the above algorithm nor the existence and uniqueness of the HR estimates. It is however well known that the convergence is attained if one repeats the steps 1 and 2 alone (spatial median) or the steps 1 and 3 alone (Tyler's scatter matrix). In the paper we proceed with the same practical solution as in Taskinen et al. (2010), that is, we start the iteration with some $\sqrt{n}$-consistent estimates and stop iterating after k steps. The estimate then inherits some properties of the initial estimate but, with large k, the behavior is almost as that of the regular HR estimate. We then give the following.

Definition 11.1. Let $\hat{\boldsymbol{\mu}}_0$ and $\hat{\mathbf{V}}_0$ be initial location and shape estimators. The k-step HR estimators $\hat{\boldsymbol{\mu}}_k$ and $\hat{\mathbf{V}}_k$ for location and shape are obtained by starting with $\hat{\boldsymbol{\mu}}_0$ and $\hat{\mathbf{V}}_0$ and repeating Iteration Steps 1 k times.

Notice that the k-step estimators are affine equivariant if the initial estimators are affine equivariant. In Croux et al. (2010), the robustness and efficiency properties of k-step Tyler's shape estimator were studied. They for example showed that the breakdown property of the k-step estimator is inherited from the initial estimator. The approach used in Croux et al. (2010) differs from ours in that the location center is assumed to be fixed. In the following sections, we derive influence functions and asymptotic properties for the simultaneous k-step HR location and shape estimators.

11.2.3 Influence Functions

The robustness of a functional T against a single outlier $\mathbf{y}$ can be measured using the influence functions Hampel et al. (1986). Let

$$F_\epsilon = (1-\epsilon)F + \epsilon\Delta_{\mathbf{y}},$$

denote the contaminated distribution, where $\Delta_{\mathbf{y}}$ is the cdf of a distribution with probability mass one at point $\mathbf{y}$. The influence function of T is then given by

$$IF(\mathbf{y}; T, F) = \lim_{\epsilon \to 0} \frac{T(F_\epsilon) - T(F)}{\epsilon}.$$

A continuous and bounded influence function indicates good local robustness properties of an estimator. We now find the influence functions of the k-step HR estimators in the elliptic case.

Due to affine equivariance properties of our estimators, it suffices to derive influence functions at a spherical distribution F_0 of $\mathbf{e}$. Hampel et al. (1986) and Ollila et al. (2004) showed that in that case, the influence functions of all location and shape functionals, $\boldsymbol{\mu}(F)$ and $\mathbf{V}(F)$, are of the form

$$IF(\mathbf{y}; \boldsymbol{\mu}, F_0) = \gamma(r)\,\mathbf{u}, \tag{11.5}$$

and

$$IF(\mathbf{y}; \mathbf{V}, F_0) = \alpha(r) \left[\mathbf{u}\mathbf{u}^T - \frac{1}{p}\mathbf{I}_p \right], \tag{11.6}$$

where $r = ||\mathbf{y}||$, $\mathbf{u} = ||\mathbf{y}||^{-1}\mathbf{y}$ and real-valued functions $\gamma(r)$ and $\alpha(r)$ depend both on the functionals and on the underlying distribution F_0. When comparing robustness properties of different estimators, it is enough to compare weight functions γ and α only. In the following we will derive these functions for k-step HR-estimators.

Let now $\boldsymbol{\mu}_k = \boldsymbol{\mu}_k(F_\mathbf{y})$ and $\mathbf{V}_k = \mathbf{V}_k(F_\mathbf{y})$ be the functionals corresponding to k-step HR-estimators $\hat{\boldsymbol{\mu}}_k$ and $\hat{\mathbf{V}}_k$, that is,

$$\boldsymbol{\mu}_k = \boldsymbol{\mu}_{k-1} + \frac{\mathbf{V}_{k-1}^{1/2}\, E[\mathbf{S}(\mathbf{e})]}{E[||\mathbf{e}||^{-1}]} \tag{11.7}$$

and

$$\mathbf{V}_k = p\,[Tr(\mathbf{V}_{k-1}^{1/2}\, E_F\left[\mathbf{S}(\mathbf{e})\mathbf{S}(\mathbf{e})^T\right] \mathbf{V}_{k-1}^{1/2})]^{-1}\, \mathbf{V}_{k-1}^{1/2}\, E_F\left[\mathbf{S}(\mathbf{e})\mathbf{S}(\mathbf{e})^T\right] \mathbf{V}_{k-1}^{1/2}, \tag{11.8}$$

where $\mathbf{e} = \mathbf{V}_{k-1}^{-1/2}(\mathbf{y} - \boldsymbol{\mu}_{\mathbf{k-1}})$. We prove the following result in Appendix.

Theorem 11.1. *The influence functions of k-step HR location and scatter functionals $\boldsymbol{\mu}_k$ and $\mathbf{V}_k$ with initial functionals $\boldsymbol{\mu}_0$ and $\mathbf{V}_0$ at F_0, the distribution of spherical $\mathbf{e}$ with $Cov(\mathbf{e}) = \mathbf{I}_p$, is given by (11.5) and by (11.6), respectively, with*

$$\gamma_k(r) = \left(\frac{1}{p}\right)^k \gamma_0(r) + \left[1 - \left(\frac{1}{p}\right)^k\right] p\,[(p-1)E(||\mathbf{e}||^{-1})]^{-1},$$

and

$$\alpha_k(r) = \left(\frac{p}{p+2}\right)^k \alpha_0(r) + \left[1 - \left(\frac{p}{p+2}\right)^k\right] (p+2).$$

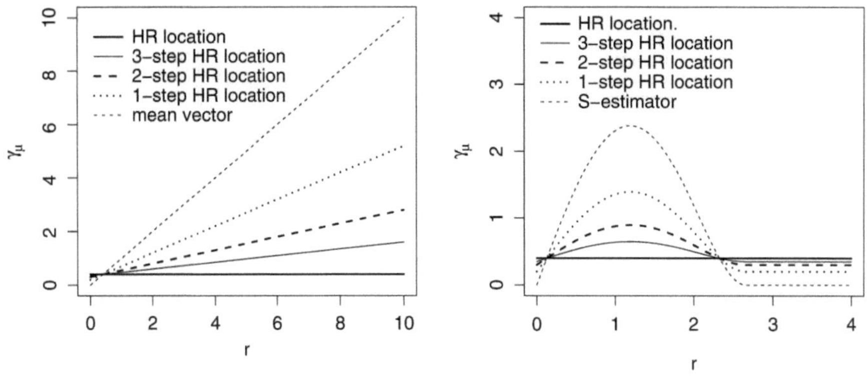

Fig. 11.1 Functions γ_k for the k-step HR location functionals with $k = 0, 1, 2, 3$ and ∞ when the regular mean vector (*left figure*) and 50 % BP S-estimator with biweight loss-function (*right figure*) are used as starting functionals. The functions are computed at the bivariate standard normal distribution case

First note that, the influence functions of the regular HR location-scatter estimate is obtained when $k \to \infty$, that is,

$$\gamma(r) = p[(p-1)E(||\mathbf{e}||^{-1})]^{-1} \text{ and } \alpha(r) = p + 2.$$

The above influence functions are clearly bounded if those of the initial estimators are bounded. In Fig. 11.1 we illustrate the behaviour of the function γ_k at bivariate standard normal case using two different initial estimators. When the sample mean vector is used as a starting value, resulting k-step estimators have naturally unbounded influence functions, although after few steps the influence function is very close to that of the affine equivariant spatial median. When highly robust 50 % breakdown point S-estimator with biweight loss-function, Davies (1987), is used as an initial estimator, bounded influence functions are obtained. Notice that after few steps the influence function does not differ much from that of the location HR estimator.

In Fig. 11.2, the influence functions for k-step HR shape estimators are illustrated at bivariate standard normal case. As initial estimators we use the sample covariance matrix as well as the 50 % breakdown point S-estimator with biweight loss-function. When we start with non-robust sample covariance matrix, unbounded influence functions are obtained, but the estimator with better robustness properties is again obtained after few steps. By using S-estimator as a starting value, the influence functions of resulting k-step estimators are naturally bounded.

In the following section, we will compare the efficiency properties of k-step HR estimators with different initial estimators. We will show that, after only few steps, the initial estimator has very little influence on the resulting efficiencies.

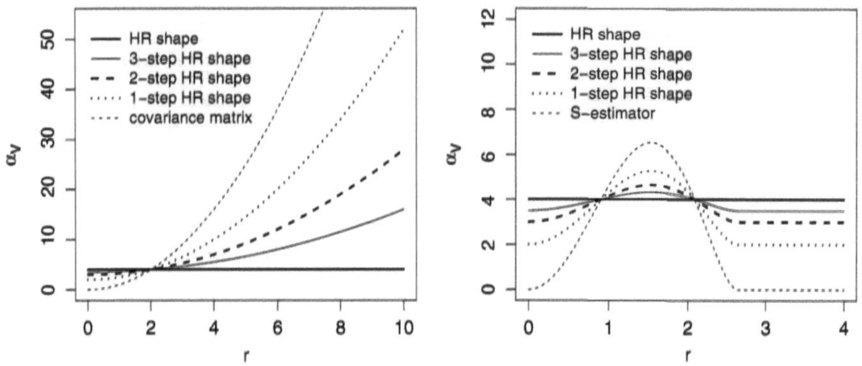

Fig. 11.2 Functions α_k for the k-step HR shape functionals with $k = 0, 1, 2, 3$ and ∞ when the regular covariance matrix (*left figure*) and 50 % BP S-estimator with biweight loss-function (*right figure*) are used as starting functionals. The functions are computed at the bivariate standard normal distribution case

11.2.4 Limiting Distributions and Asymptotic Relative Efficiencies

The asymptotic normality of k-step HR estimators follows if the initial estimators are $\sqrt{n}$-consistent and have limiting multinormal distributions. In the following, we write $vec(\mathbf{V})$ for the vectorization of a matrix $\mathbf{V}$, obtained by stacking the columns of $\mathbf{V}$ on top of each other. We also denote

$$\mathbf{C}_{p,p}(\mathbf{V}) = (\mathbf{I}_{p^2} + \mathbf{K}_{p,p})(\mathbf{V} \otimes \mathbf{V}) - \frac{2}{p} vec(\mathbf{V}) vec^T(\mathbf{V}),$$

where $\mathbf{K}_{p,p}$ is the commutation matrix, that is, a $p^2 \times p^2$ block matrix with (i,j)-block being equal to a $p \times p$ matrix that has 1 at entry (j,i) and zero elsewhere.

Theorem 11.2. *Let $\mathbf{y}_1, \ldots, \mathbf{y}_n$ be a random sample from F_0, the distribution of spherical $\mathbf{e}$ with $Cov(\mathbf{e}) = \mathbf{I}_p$. Assume that $\sqrt{n}\hat{\boldsymbol{\mu}}_0$ and $\sqrt{n}\, vec(\hat{\mathbf{V}}_0 - \mathbf{I}_p)$ have a joint limiting multivariate normal distribution. Then*

$$\sqrt{n}\,\hat{\boldsymbol{\mu}}_k \xrightarrow{d} N(\mathbf{0}, \tau_{1k}\mathbf{I}_p) \quad \text{and} \quad \sqrt{n}\, vec(\hat{\mathbf{V}}_k - \mathbf{I}_p) \xrightarrow{d} N(\mathbf{0}, \tau_{2k}\, \mathbf{C}_{p,p}(\mathbf{I}_p)),$$

where

$$\tau_{1k} = \frac{E[\gamma_k^2(||\mathbf{e}||)]}{p} \quad \text{and} \quad \tau_{2k} = \frac{E[\alpha_k^2(||\mathbf{e}||)]}{p(p+2)}.$$

Functions γ_k and α_k are given in Theorem 11.1.

The limiting distributions at elliptical distribution follow from the affine equivariance properties of the estimators. See for example Ollila et al. (2004) and Taskinen et al. (2010).

Corollary 11.1. *Let $\mathbf{y}_1, \ldots, \mathbf{y}_n$ be a random sample from F, an elliptical distribution of $\boldsymbol{\Sigma}^{1/2}\mathbf{e} + \boldsymbol{\mu}$ where $\mathbf{e}$ is spherical with $Cov(\mathbf{e}) = \mathbf{I}_p$. Write $\boldsymbol{\Lambda} = (p/tr(\boldsymbol{\Sigma}))\boldsymbol{\Sigma}$. Then*

$$\sqrt{n}\,(\hat{\boldsymbol{\mu}}_k - \boldsymbol{\mu}) \xrightarrow{d} N(\mathbf{0}, \tau_{1k}\boldsymbol{\Sigma}) \quad and \quad \sqrt{n}\,vec(\hat{\mathbf{V}}_k - \boldsymbol{\Lambda}) \xrightarrow{d} N(\mathbf{0}, \tau_{2k}\,\mathbf{W}\mathbf{C}_{p,p}(\boldsymbol{\Lambda})\mathbf{W}^T),$$

where τ_{1k} and τ_{2k} are given in Theorem 11.2 and $\mathbf{W} = \mathbf{I}_{p^2} - p^{-1}\,vec(\boldsymbol{\Lambda})vec(\mathbf{I}_p)^T$.

In order to compare asymptotic relative efficiencies of different estimators, one only has to compare scalars τ_{1k} and τ_{2k}. In Table 11.1 we list the asymptotic relative efficiencies of k-step HR location estimators as compared to the sample mean at different p-variate t-distributions with selected values of dimension p and degrees of freedom ν, where $\nu = \infty$ refers to the multinormal case. As in previous section, we use the sample mean and 50 % BP S-estimator as starting values.

Table 11.1 Asymptotic relative efficiencies of k-step HR location estimators as compared to the sample mean at different p-variate t-distributions with selected values of dimension p and degrees of freedom ν

		(a)			(b)		
	k	$\nu = 3$	$\nu = 6$	$\nu = \infty$	$\nu = 3$	$\nu = 6$	$\nu = \infty$
$p = 2$	1	1.600	1.135	0.936	2.025	1.035	0.697
	2	1.882	1.135	0.867	2.045	1.074	0.747
	3	1.969	1.115	0.827	2.031	1.083	0.768
	4	1.992	1.101	0.806	2.017	1.085	0.777
	5	1.998	1.093	0.796	2.009	1.085	0.781
	∞	2.000	1.084	0.785	2.000	1.084	0.785
$p = 5$	1	2.094	1.238	0.937	2.359	1.259	0.898
	2	2.274	1.250	0.912	2.318	1.253	0.904
	3	2.300	1.250	0.907	2.308	1.251	0.905
	4	2.304	1.250	0.906	2.306	1.250	0.905
	5	2.305	1.250	0.905	2.306	1.250	0.905
	∞	2.306	1.250	0.905	2.306	1.250	0.905
$p = 10$	1	2.302	1.297	0.960	2.451	1.320	0.950
	2	2.412	1.312	0.952	2.426	1.314	0.951
	3	2.421	1.313	0.951	2.423	1.313	0.951
	4	2.422	1.313	0.951	2.423	1.313	0.951
	5	2.422	1.313	0.951	2.422	1.313	0.951
	∞	2.422	1.313	0.951	2.422	1.313	0.951

The sample mean (a) and the 50 % BP S-estimator (b) are used as a starting values

Consider first the efficiency results for the simple k-step estimator that uses sample mean vector as a starting value. In case of high-dimensional data, the k-step estimators are very efficient even in the multinormal case, and after few steps the efficiencies are already very close to those of regular HR estimators. As seen in previous section, in case of low-dimensional data, several steps are needed to obtain estimator with reasonable robustness properties. When multinormal data in considered, such estimator seems to lack efficiency. To study the effect of an initial estimator to efficiencies, 50 % BP S-estimator was also used as a starting value. When k is large enough, the initial estimator has very little influence on the efficiencies. For example when different 5-step estimators are compared, regardless of the distribution, the efficiencies are almost alike.

In Table 11.2 the asymptotic relative efficiencies of k-step HR shape estimators are given as compared to the sample covariance matrix based shape estimator. We again use the sample covariance matrix as well as the 50 % BP S-estimator as starting values. As seen in Table 11.2, in multinormal case, the k-step shape estimators are very inefficient no matter which estimator is used as a starting value. For heavy-tailed distributions, the k-step estimators outperform the initial sample covariance matrix. Again after five steps, the efficiencies are very close to those of the limiting estimators, and the efficiencies of sample covariance matrix based estimators are very similar to those of the S-estimator based estimators.

11.3 Hettmansperger–Randles Estimators of Regression

11.3.1 k-Step Regression Estimators

Assume next the linear regression model

$$\mathbf{y}_i = \mathbf{B}^T\mathbf{x}_i + \boldsymbol{\Sigma}^{1/2}\mathbf{e}_i, \quad i = 1, \ldots, n,$$

where $\mathbf{y}_i$ are the p-variate response vectors, $\mathbf{B}$ is the $q \times p$ matrix of unknown regression parameters and $\boldsymbol{\Sigma}$ is the covariance matrix of the residuals. The q-vector of explaining variables $\mathbf{x}_i$ and the standardized p-variate residuals $\mathbf{e}_i$ are independent and $\mathbf{e}_i$ is spherical around zero with $Cov(\mathbf{e}_i) = \mathbf{I}_p$. Finally $(\mathbf{x}_i, \mathbf{e}_i)$, $i = 1, \ldots, n$, are iid. We may then also write

$$\mathbf{Y} = \mathbf{X}\mathbf{B} + \mathbf{E}\boldsymbol{\Sigma}^{1/2}, \tag{11.9}$$

where $\mathbf{Y} = (\mathbf{y}_1, \ldots, \mathbf{y}_n)^T$ and $\mathbf{E} = (\mathbf{e}_1, \ldots, \mathbf{e}_n)^T$ are $n \times p$ matrices, and $\mathbf{X} = (\mathbf{x}_1, \ldots, \mathbf{x}_n)^T$ is an $n \times q$ matrix.

The regression estimator $\hat{\mathbf{B}}$ based on the location score function $\mathbf{T}(\mathbf{y})$ solves

$$ave\{\mathbf{T}(\mathbf{y}_i - \mathbf{B}^T\mathbf{x}_i)\mathbf{x}_i^T\} = \mathbf{0} \tag{11.10}$$

Table 11.2 Asymptotic relative efficiencies of k-step HR shape estimators as compared to the sample covariance matrix based shape estimator at different p-variate t-distributions with selected values of dimension p and degrees of freedom ν

		(a)			(b)		
	k	$\nu = 5$	$\nu = 8$	$\nu = \infty$	$\nu = 5$	$\nu = 8$	$\nu = \infty$
$p = 2$	1	1.714	1.091	0.800	1.377	0.687	0.458
	2	1.778	0.941	0.640	1.472	0.733	0.489
	3	1.670	0.846	0.566	1.495	0.746	0.496
	4	1.590	0.796	0.532	1.500	0.749	0.498
	5	1.546	0.774	0.516	1.500	0.750	0.499
	∞	1.500	0.750	0.500	1.500	0.750	0.500
$p = 5$	1	2.194	1.205	0.831	2.173	1.094	0.744
	2	2.221	1.119	0.748	2.159	1.081	0.723
	3	2.170	1.086	0.724	2.149	1.074	0.717
	4	2.151	1.075	0.717	2.144	1.072	0.715
	5	2.145	1.073	0.715	2.143	1.072	0.715
	∞	2.143	1.071	0.714	2.143	1.071	0.714
$p = 10$	1	2.512	1.301	0.878	2.521	1.245	0.851
	2	2.520	1.261	0.841	2.505	1.253	0.836
	3	2.504	1.252	0.835	2.501	1.251	0.834
	4	2.501	1.250	0.834	2.500	1.250	0.834
	5	2.500	1.250	0.833	2.500	1.250	0.833
	∞	2.500	1.250	0.833	2.500	1.250	0.833

The sample covariance matrix (a) and the 50 % BP S-estimator (b) are used as starting values

With the identity score $\mathbf{T}(\mathbf{y}) = \mathbf{y}$, the classical least squares (LS) estimator for model (11.9) is obtained. The solution $\hat{\mathbf{B}} = \hat{\mathbf{B}}(\mathbf{X}, \mathbf{Y}) = (\mathbf{X}^T\mathbf{X})^{-1}\mathbf{X}^T\mathbf{Y}$ is then fully equivariant, that is, it satisfies

$$\hat{\mathbf{B}}(\mathbf{X}, \mathbf{X}\mathbf{H} + \mathbf{Y}) = \hat{\mathbf{B}}(\mathbf{X}, \mathbf{Y}) + \mathbf{H},$$

for all $q \times p$ matrices $\mathbf{H}$ (regression equivariance). Further,

$$\hat{\mathbf{B}}(\mathbf{X}, \mathbf{Y}\mathbf{W}) = \hat{\mathbf{B}}(\mathbf{X}, \mathbf{Y})\mathbf{W},$$

for all nonsingular $p \times p$ matrices $\mathbf{W}$ ($\mathbf{Y}$-equivariance) and

$$\hat{\mathbf{B}}(\mathbf{X}\mathbf{V}, \mathbf{Y}) = \mathbf{V}^{-1}\hat{\mathbf{B}}(\mathbf{X}, \mathbf{Y}),$$

for all nonsingular $q \times q$ matrices $\mathbf{V}$ ($\mathbf{X}$-equivariance).

As in case of location estimation, robust regression estimator is obtained by replacing identity scores used in (11.10) with spatial sign scores $\mathbf{S}(\mathbf{y})$. This choice yields to the multivariate least absolute deviation (LAD) estimator, Bai et al. (1990). The solution $\hat{\mathbf{B}}$ cannot be given in a closed form, but may be obtained using simple iterative algorithm.

1. $\hat{\mathbf{e}}_i = \mathbf{y}_i - \hat{\mathbf{B}}_{k-1}^T\mathbf{x}_i$, for $i = 1, \ldots, n$,
2. $\hat{\mathbf{B}}_k = \hat{\mathbf{B}}_{k-1} + [ave\{||\hat{\mathbf{e}}_i||^{-1}\mathbf{x}_i\mathbf{x}_i^T\}]^{-1}\, ave\{\mathbf{x}_i\mathbf{S}(\hat{\mathbf{e}}_i)^T\}$.

LAD-estimator is regression and **X**-equivariant, but **Y**-equivariant only with respect to orthogonal transformations. As in the case of location and shape estimation, a fully equivariant estimator is obtained using a similar inner standardisation. The regression estimator $\hat{\mathbf{B}}$ and the residual scatter matrix $\hat{\mathbf{V}}$ then solve

$$ave\{\mathbf{S}(\hat{\mathbf{e}}_i)\mathbf{x}_i^T\} = \mathbf{0} \text{ and } p\,ave\{\mathbf{S}(\hat{\mathbf{e}}_i)\mathbf{S}(\hat{\mathbf{e}}_i)^T\} = \mathbf{I}_p$$

where $\hat{\mathbf{e}}_i = \mathbf{V}^{-1/2}(\mathbf{y}_i - \mathbf{B}^T\mathbf{x}_i)$ and $\hat{\mathbf{V}}$ is standardized so that $Tr(\hat{\mathbf{V}}) = p$. As in the case of regular LAD-estimator, the solution $\hat{\mathbf{B}}$ cannot be given in a closed form, and the estimate is obtained using a fixed-point algorithm with the following steps.

Iteration Steps 2 *The HR regression-scatter estimate is obtained using the following steps*

1. $\hat{\mathbf{e}}_i = \hat{\mathbf{V}}_{k-1}^{-1/2}(\mathbf{y}_i - \hat{\mathbf{B}}_{k-1}^T\mathbf{x}_i)$, *for* $i = 1, \ldots, n$,
2. $\hat{\mathbf{B}}_k = \hat{\mathbf{B}}_{k-1} + [ave\{||\hat{\mathbf{e}}_i||^{-1}\mathbf{x}_i\mathbf{x}_i^T\}]^{-1}\, ave\{\mathbf{x}_i\mathbf{S}(\hat{\mathbf{e}}_i)^T\}\mathbf{V}_{k-1}^{1/2}$,
3. $\hat{\mathbf{V}}_k = p\,\hat{\mathbf{V}}_{k-1}^{1/2}\, ave\{\mathbf{S}(\hat{\mathbf{e}}_i)\mathbf{S}(\hat{\mathbf{e}}_i)^T\}\,\hat{\mathbf{V}}_{k-1}^{1/2}$.

$\hat{\mathbf{V}}$ *is scaled so that* $Tr(\hat{\mathbf{V}}) = p$.

Again there is no proof for the convergence of the above algorithm. We therefore proceed as in the case of location and shape estimation and define k-step HR regression estimators as follows.

Definition 11.2. Let $\hat{\mathbf{B}}_0$ and $\hat{\mathbf{V}}_0$ be initial regression and scatter matrix estimates. The k-step HR estimators $\hat{\mathbf{B}}_k$ and $\hat{\mathbf{V}}_k$ are then the estimators obtained by starting the iteration with $\hat{\mathbf{B}}_0$ and $\hat{\mathbf{V}}_0$ and repeating Iteration Steps 2 k times.

11.3.2 Influence Functions and Limiting Distributions

Let $\mathbf{B}_k = \mathbf{B}_k(F_{\mathbf{x},\mathbf{y}})$ and $\mathbf{V}_k = \mathbf{V}_k(F_{\mathbf{x},\mathbf{y}})$ be the functionals corresponding to k-step HR-estimators $\hat{\mathbf{B}}_k$ and $\hat{\mathbf{V}}_k$, that is,

$$\mathbf{B}_k = \mathbf{B}_{k-1} + \{E[||\mathbf{e}||^{-1}\mathbf{x}\mathbf{x}^T]\}^{-1}\, E[\mathbf{x}\mathbf{S}(\mathbf{e})^T]\mathbf{V}_{k-1}^{1/2} \tag{11.11}$$

and

$$\mathbf{V}_k = p\,\mathbf{V}_{k-1}^{1/2}\, E_F\left[\mathbf{S}(\mathbf{e})\mathbf{S}^T(\mathbf{e})\right]\, \mathbf{V}_{k-1}^{1/2}, \tag{11.12}$$

where $\mathbf{e} = \mathbf{V}_{k-1}^{-1/2}(\mathbf{y} - \mathbf{B}_{k-1}^T\mathbf{x})$.

If $\mathbf{B}_0$ and $\mathbf{V}_0$ are affine equivariant functionals then so are $\mathbf{B}_k$ and $\mathbf{V}_k$, $k = 1, 2, \ldots$ Due to this equivariance, we may consider without loss of generality the spherical case with $\mathbf{B} = \mathbf{0}$ and $Cov(\mathbf{y}) = Cov(\mathbf{e}) = \mathbf{I}_p$. The influence function of $\mathbf{V}_k$ is then as in Theorem 11.1, and therefore bounded if the influence functional of $\mathbf{V}_k$ is bounded. The influence function of $\mathbf{B}_k$ in spherical case is given in the following theorem.

Theorem 11.3. *For $F_{\mathbf{x},\mathbf{y}}$ with $\mathbf{B} = \mathbf{0}$, $\boldsymbol{\Sigma} = \mathbf{I}_p$ and spherical $\mathbf{e}$ with $Cov(\mathbf{e}) = \mathbf{I}_p$, the influence function of k-step HR regression functional $\mathbf{B}_k$ with initial estimator $\mathbf{B}_0$ is at $\mathbf{z} = (\mathbf{x}, \mathbf{y})$ given by*

$$IF(\mathbf{z}; \mathbf{B}_k, F_{\mathbf{x},\mathbf{y}}) = \left(\frac{1}{p}\right)^k IF(\mathbf{z}; \mathbf{B}_0, F_{\mathbf{x},\mathbf{y}}) + \left[1 - \left(\frac{1}{p}\right)^k\right] \left[E(\mathbf{x}\mathbf{x}^T)\right]^{-1} p\,[(p-1)E(||\mathbf{e}||^{-1})]^{-1}\mathbf{x}\mathbf{S}(\mathbf{y})^T.$$

The latter part of the influence function is bounded in $\mathbf{y}$, but unbounded in $\mathbf{x}$, therefore even if the initial estimator has bounded influence function, the HR estimator is sensitive to bad leverage points.

Assume next that the influence function of an initial estimator is of the type

$$IF(\mathbf{z}; \mathbf{B}_0, F_{\mathbf{x},\mathbf{y}}) = \eta_0(r) \left[E(\mathbf{x}\mathbf{x}^T)\right]^{-1} \mathbf{x}\mathbf{S}(\mathbf{y})^T, \tag{11.13}$$

where the weight function η_0 depends on the functional $\mathbf{B}_0$ and the underlying spherical distribution of $\mathbf{e}$. Then

$$IF(\mathbf{z}; \mathbf{B}_k, F_{\mathbf{x},\mathbf{y}}) = \eta_k(r) \left[E(\mathbf{x}\mathbf{x}^T)\right]^{-1} \mathbf{x}\mathbf{S}(\mathbf{y})^T,$$

where

$$\eta_k(r) = \left(\frac{1}{p}\right)^k \eta_0(r) + \left[1 - \left(\frac{1}{p}\right)^k\right] p\,[(p-1)E(r^{-1})]^{-1}. \tag{11.14}$$

See Fig. 11.1 for an illustration of $\eta_k(r)$; in the left figure the initial regression estimator is the LS estimator with $\eta_0(r) = r$. The LS-estimator is highly non-robust estimator as it is sensitive to leverage points as well as vertical outliers. By taking just few steps in our estimation procedure, the effect of y-outliers is reduced. However, the estimator stays sensitive to leverage points through the term $\mathbf{x}\mathbf{S}(\mathbf{y})^T$.

The joint asymptotic normality of $\hat{\mathbf{B}}_k$ and $\hat{\mathbf{V}}_k$ follows if the initial estimators are $\sqrt{n}$-consistent with joint limiting multinormal distributions (see the proof of Theorem 11.4 in Appendix). The limiting distribution for $\hat{\mathbf{V}}_k$ is given in Theorem 11.2. If $\hat{\mathbf{B}}_0$ is an regression estimator with influence function as given in (11.13), the limiting distribution of $\hat{\mathbf{B}}_k$ reduces to a following simple form.

Theorem 11.4. *Let* $(\mathbf{x}_1, \mathbf{y}_1), \ldots, (\mathbf{x}_n, \mathbf{y}_n)$ *be a random sample from a distribution of* $(\mathbf{x}, \mathbf{y})$ *with* $\mathbf{B} = \mathbf{0}$, $\boldsymbol{\Sigma} = \mathbf{I}_p$ *and* $\mathbf{y} = \mathbf{e}$ *spherical around zero with* $Cov(\mathbf{e}) = \mathbf{I}_p$. *Let* $\mathbf{B}_0$ *be an initial estimator with influence function as given in (11.13). Then*

$$\sqrt{n}\, vec(\hat{\mathbf{B}}_k) \xrightarrow{d} N(\mathbf{0}, \tau_{3k}(\mathbf{I}_p \otimes \left[E(\mathbf{x}\mathbf{x}^T)\right]^{-1})),$$

where $\tau_{3k} = E[\eta_k^2(||\mathbf{e}||)]/p$, *and* $\eta_k(r)$ *as given in (11.14).*

The limiting distribution at elliptical case follows from the affine equivariance properties of $\hat{\mathbf{B}}_k$:

Corollary 11.2. *Let* $(\mathbf{x}_1, \mathbf{y}_1), \ldots, (\mathbf{x}_n, \mathbf{y}_n)$ *be a random sample from a distribution of* $(\mathbf{x}, \mathbf{y})$ *with* $Cov(\mathbf{e}) = \mathbf{I}_p$. *Let* $\mathbf{B}_0$ *be an initial estimator with influence function as given in (11.13) Then*

$$\sqrt{n}\, vec(\hat{\mathbf{B}}_k - \mathbf{B}) \xrightarrow{d} N\left(\mathbf{0}, \tau_{3k}\left(\boldsymbol{\Sigma} \otimes \left[E(\mathbf{x}\mathbf{x}^T)\right]^{-1}\right)\right),$$

where τ_{3k} *is as in Theorem 11.4.*

The asymptotic relative efficiencies of k-step HR regression estimators relative to LS-estimator equal to those obtained in the location estimation case. The efficiencies at p-variate t-distribution cases with selected values of ν and p are listed in Table 11.1.

11.4 Discussion

In this paper, the location, shape and regression estimators based on the spatial sign score function were considered. It was shown how the problems encountered in simultaneous location and shape estimation of Hettmansperger and Randles (2002) as well as in regression estimation of Oja (2010) can be circumvented by using corresponding k-step estimators.

The influence functions and asymptotic properties of k-step HR estimators were derived. In our example we used both robust and non-robust initial estimators. The use of the sample mean, the sample covariance matrix as well as the least squares estimator as initial estimators yields to estimators with unbounded influence functions. The robustness studies however indicate that already after few steps, estimators with better robustness properties are obtained, as the influence functions of k-step estimators are very close those of the limiting estimators. The efficiency studies demonstrate that when k is large enough, the use of non-robust initial estimators yields to efficiencies that are similar to those obtained using robust initial estimators. Based on these studies, we conclude that to obtain simple and practical estimators for location, shape and regression, one could use as initial estimators the sample mean, the sample covariance matrix and the least squares estimator, respectively. One has, however, to keep in mind that the breakdown properties of k-step estimators are inherited from the initial estimators, Croux et al. (2010).

Acknowledgements The authors wish to thank a referee for several helpful comments and suggestions. The Research was funded by the Academy of Finland (grants 251965 and 268703).

Appendix

Proof (Theorem 11.1).
The functional (11.7) solves

$$E_F\left[\frac{\mathbf{y}-\boldsymbol{\mu}_k(F)}{||\mathbf{z}||}\right]=\mathbf{0}, \tag{11.15}$$

where $\mathbf{z}=\mathbf{V}_{k-1}^{-1/2}(F)(\mathbf{y}-\boldsymbol{\mu}_{k-1}(F))$. Write $F_\epsilon=(1-\epsilon)F_0+\epsilon\Delta_{\mathbf{y}_0}$. Then

$\boldsymbol{\mu}_{k-1}(F_\epsilon)=\epsilon IF(\mathbf{y}_0;\boldsymbol{\mu}_{k-1},F_0)+o(\epsilon)$ and $\mathbf{V}_{k-1}(F_\epsilon)=\mathbf{I}_p+\epsilon IF(\mathbf{y}_0;\mathbf{V}_{k-1},F_0)+o(\epsilon)$

and, further,

$$||\mathbf{z}||^{-1}=\frac{1}{r}\left[1+\frac{\epsilon}{r}\mathbf{u}^T IF(\mathbf{y}_0;\boldsymbol{\mu}_{k-1},F_0)+\frac{\epsilon}{2}\mathbf{u}^T IF(\mathbf{y}_0;\mathbf{V}_{k-1},F_0)\mathbf{u}+o(\epsilon)\right].$$

Substituting these in (11.15) and having the expectation at F_ϵ gives

$$IF(\mathbf{y}_0;\boldsymbol{\mu}_k,F_0)=[E(r^{-1})]^{-1}\mathbf{u}_0+\frac{1}{p}IF(\mathbf{y}_0;\boldsymbol{\mu}_{k-1},F_0).$$

Find next the influence function of $\mathbf{V}_k(F)$. Write (11.8) as

$$\mathbf{V}_k(F)\,E_F\left[\frac{(\mathbf{y}-\boldsymbol{\mu}_{k-1}(F))^T(\mathbf{y}-\boldsymbol{\mu}_{k-1}(F))}{||\mathbf{z}||^2}\right]-p\,\mathbf{V}_{k-1}^{1/2}(F)\,E_F\left[\mathbf{S}(\mathbf{z})\mathbf{S}^T(\mathbf{z})\right]\,\mathbf{V}_{k-1}^{1/2}(F)=0,$$

where again $\mathbf{z}=\mathbf{V}_{k-1}^{-1/2}(F)(\mathbf{y}-\boldsymbol{\mu}_{k-1}(F))$. Proceeding then as in the proof for $\boldsymbol{\mu}_k(F)$, we get

$$IF(\mathbf{y}_0;\mathbf{V}_k,F_0)=\frac{2}{p+2}IF(\mathbf{y}_0;\mathbf{V}_{k-1},F_0)+p\left[\mathbf{u}_0\mathbf{u}_0^T-\frac{1}{p}\mathbf{I}_p\right].$$

The result then follows from the above recursive formulas for $IF(\mathbf{y};\boldsymbol{\mu}_k,F_0)$ and $IF(\mathbf{y};\mathbf{V}_k,F_0)$.

□

Proof (Theorem 11.2). Consider first the limiting distribution of 1-step HR location estimator. Let $\mathbf{y}_i, \ldots, \mathbf{y}_n$ be a sample from a spherically symmetric distribution F_0 and write $r_i = \|\mathbf{y}_i\|$ and $\mathbf{u}_i = r_i^{-1}\mathbf{y}_i$. Further, as we assume that $\hat{\boldsymbol{\mu}}_0$ and $\hat{\mathbf{V}}_0$ are $\sqrt{n}$-consistent, we write $\boldsymbol{\mu}_0^* := \sqrt{n}\hat{\boldsymbol{\mu}}_0$ and $\mathbf{V}_0^* := \sqrt{n}(\hat{\mathbf{V}}_0 - \mathbf{I}_p)$, where $\boldsymbol{\mu}_0^* = O_p(1)$ and $\mathbf{V}_0^* = O_p(1)$. Now using the delta-method as in Taskinen et al. (2010), we get

$$\sqrt{n}\hat{\boldsymbol{\mu}}_1 = \boldsymbol{\mu}_0^* + \left[ave\left\{\frac{1}{r_i}\left(1 + \frac{1}{r_i\sqrt{n}}\mathbf{u}_i^T\boldsymbol{\mu}_0^* + \frac{1}{2\sqrt{n}}\mathbf{u}_i^T\mathbf{V}_0^*\mathbf{u}_i\right)\right\}\right]^{-1}$$
$$\cdot\sqrt{n}\,ave\left\{\mathbf{u}_i - \frac{1}{\sqrt{n}r_i}\boldsymbol{\mu}_0^* + \frac{1}{\sqrt{n}r_i}\mathbf{u}_i\mathbf{u}_i^T\boldsymbol{\mu}_0^* + \frac{1}{2\sqrt{n}}\mathbf{u}_i\mathbf{u}_i^T\mathbf{V}_0^*\mathbf{u}_i\right\} + o_p(1). \tag{11.16}$$

As $\sqrt{n}\hat{\boldsymbol{\mu}}_0 = \sqrt{n}\,ave\{\gamma_0(r_i)\mathbf{u}_i\} + o_p(1)$, the asymptotic normality of $\sqrt{n}\hat{\boldsymbol{\mu}}_1$ follows from the Slutsky's theorem and joint limiting multivariate normality of $\sqrt{n}\,ave\{\mathbf{u}_i\}$ and $\boldsymbol{\mu}_0^* = \sqrt{n}\hat{\boldsymbol{\mu}}_0$ (and $E[\mathbf{u}_i^T\mathbf{V}^*\mathbf{u}_i] = Tr(\mathbf{V}^*) = 0$). Equation (11.16) reduces to

$$\sqrt{n}\hat{\boldsymbol{\mu}}_1 = p^{-1}\sqrt{n}\,\hat{\boldsymbol{\mu}}_0 + [E(r_i^{-1})]^{-1}\sqrt{n}\,ave\{\mathbf{u}_i\} + o_p(1).$$

Continuing in a similar way with $\sqrt{n}\hat{\boldsymbol{\mu}}_2$, $\sqrt{n}\hat{\boldsymbol{\mu}}_3$, and so on, we finally get

$$\sqrt{n}\hat{\boldsymbol{\mu}}_k = \left(\frac{1}{p}\right)^k\sqrt{n}\hat{\boldsymbol{\mu}}_0 + \left[1 - \left(\frac{1}{p}\right)^k\right]p[(p-1)E(r_i^{-1})]^{-1}\sqrt{n}\,ave\{\mathbf{u}_i\} + o_p(1).$$

Thus $\sqrt{n}\,\hat{\boldsymbol{\mu}}_k = \sqrt{n}\,ave\{\gamma_k(r_i)\mathbf{u}_i\} + o_p(1)$. and the limiting covariance matrix of $\sqrt{n}\hat{\boldsymbol{\mu}}_k$ equals to $E[\gamma_k^2(r)\mathbf{u}\mathbf{u}^T] = p^{-1}E[\gamma_k^2(r)]\mathbf{I}_p$.

The limiting distribution for k-step HR shape estimator can be computed as above starting from 1-step estimator

$$\hat{\mathbf{V}}_1 = p\left[ave\left\{\frac{(\mathbf{y}_i - \hat{\boldsymbol{\mu}}_0)^T(\mathbf{y}_i - \hat{\boldsymbol{\mu}}_0)}{||\hat{\mathbf{V}}_0^{-1/2}(\mathbf{y}_i - \hat{\boldsymbol{\mu}}_0)||^2}\right\}\right]^{-1} ave\left\{\frac{(\mathbf{y}_i - \hat{\boldsymbol{\mu}}_0)(\mathbf{y}_i - \hat{\boldsymbol{\mu}}_0)^T}{||\hat{\mathbf{V}}_0^{-1/2}(\mathbf{y}_i - \hat{\boldsymbol{\mu}}_0)||^2}\right\}.$$

Note that the estimator is scaled so that $Tr(\hat{\mathbf{V}}_1) = p$. After some straightforward derivations,

$$\sqrt{n}(\hat{\mathbf{V}}_1 - \mathbf{I}_p) = \left[1 + \frac{1}{\sqrt{n}}ave\{\mathbf{u}_i^T\mathbf{V}_0^*\mathbf{u}_i\}\right]^{-1}\left[p\sqrt{n}\left(ave\{\mathbf{u}_i\mathbf{u}_i^T\} - \frac{1}{p}\mathbf{I}_p\right)\right.$$
$$\left. + p\,ave\left\{\mathbf{u}_i^T\mathbf{V}_0^*\mathbf{u}_i\mathbf{u}_i\mathbf{u}_i^T + \frac{2}{r_i\sqrt{n}}\mathbf{u}_i\mathbf{u}_i^T\mathbf{u}_i^T\boldsymbol{\mu}_0^* - \frac{2}{r_i\sqrt{n}}\boldsymbol{\mu}_0^*\mathbf{u}_i^T - \mathbf{u}_i^T\mathbf{V}_0^*\mathbf{u}_i\mathbf{I}_p\right\}\right] + o_p(1).$$

As the joint limiting distribution of $\sqrt{n}\,(ave\{\mathbf{u}_i\mathbf{u}_i^T\} - p^{-1}\mathbf{I}_p)$ and $\sqrt{n}(\hat{\mathbf{V}}_0 - \mathbf{I}_p) = \sqrt{n}ave\{\alpha_0(r_i)(\mathbf{u}_i\mathbf{u}_i^T - p^{-1}\mathbf{I}_p)\} + o_p(1)$ is multivariate normal, the asymptotic normality of $\sqrt{n}(\hat{\mathbf{V}}_1 - \mathbf{I}_p)$ follows and

$$\sqrt{n}(\hat{\mathbf{V}}_1 - \mathbf{I}_p) = 2(p+2)^{-1}\sqrt{n}(\hat{\mathbf{V}}_0 - \mathbf{I}_p) + p\sqrt{n}(ave\{\mathbf{u}_i\mathbf{u}_i^T\} - p^{-1}\mathbf{I}_p) + o_p(1)$$
$$= \sqrt{n}\,ave\{\alpha_k(r_i)(\mathbf{u}_i\mathbf{u}_i^T - p^{-1}\mathbf{I}_p)\} + o_p(1).$$

Continuing in the same way, we obtain

$$\sqrt{n}(\hat{\mathbf{V}}_k - \mathbf{I}_p) = \sqrt{n}\,ave\{\alpha_k(r_i)(\mathbf{u}_i\mathbf{u}_i^T - p^{-1}\mathbf{I}_p)\} + o_p(1).$$

The limiting covariance matrix of $\sqrt{n}\,vec(\hat{\mathbf{V}}_k - \mathbf{I}_p)$ is then

$$E\left[\alpha_k^2(r)\,vec(\mathbf{u}\mathbf{u}^T - p^{-1}\mathbf{I}_p)vec^T(\mathbf{u}\mathbf{u}^T - p^{-1}\mathbf{I}_p)\right]$$
$$= \frac{E[\alpha_k^2(r)]}{p(p+2)}(\mathbf{I}_{p^2} + \mathbf{K}_{p,p} - 2p^{-1}\mathbf{J}_p) = \frac{E[\alpha_k^2(r)]}{p(p+2)}\mathbf{C}_{p,p}(\mathbf{I}_p).$$

□

Proof (Theorem 11.3). First note that (11.11) is equivalent to

$$E[||\mathbf{e}||^{-1}\mathbf{x}\mathbf{x}^T]\,\mathbf{B}_k - E[||\mathbf{e}||^{-1}\mathbf{x}\mathbf{x}^T]\,\mathbf{B}_{k-1} - E[\mathbf{x}\mathbf{S}(\mathbf{e})^T]\mathbf{V}_{k-1}^{1/2} = \mathbf{0},$$

where $\mathbf{e} = \mathbf{V}_{k-1}^{-1/2}(\mathbf{y} - \mathbf{B}_k^T\mathbf{x})$. Proceeding as in the Proof of Theorem 11.1, and assuming (without loss of generality) the spherical case with $\mathbf{B} = \mathbf{0}$ and $\boldsymbol{\Sigma} = \mathbf{I}_p$, we end up after some tedious derivations to

$$E\left[\frac{\mathbf{x}\mathbf{x}^T}{r}\right]IF(\mathbf{z};\mathbf{B}_k,F_0) - E\left[\frac{\mathbf{x}\mathbf{x}^T IF(\mathbf{z};\mathbf{B}_{k-1},F_0)\mathbf{u}\mathbf{u}^T}{r}\right] - \mathbf{x}\mathbf{u}^T = \mathbf{0},$$

where $\mathbf{y} = r\mathbf{u}$ with $r = ||\mathbf{y}||$ and $\mathbf{u} = \mathbf{y}/r$. As $E[\mathbf{u}\mathbf{u}^T] = p^{-1}\mathbf{I}_p$, this simplifies to

$$IF(\mathbf{z};\mathbf{B}_k,F_0) = \frac{1}{p}IF(\mathbf{z};\mathbf{B}_{k-1},F_0) + E[\mathbf{x}\mathbf{x}^T]^{-1}\frac{\mathbf{x}\mathbf{u}^T}{E[r^{-1}]},$$

and as the influence functions for all k are of the same type, we get

$$IF(\mathbf{z};\mathbf{B}_k,F_0) = \left(\frac{1}{p}\right)^k IF(\mathbf{z};\mathbf{B}_0,F_0) + \left[1 - \left(\frac{1}{p}\right)^k\right] E[\mathbf{x}\mathbf{x}^T]^{-1}p\,[(p-1)E(r^{-1})]^{-1}\,\mathbf{x}\mathbf{u}^T.$$

□

Proof (Theorem 11.4). Consider first the general case, where $\hat{\mathbf{B}}_0$ and $\hat{\mathbf{V}}_0$ are assumed to be any $\sqrt{n}$-consistent estimators and write $\mathbf{B}_0^* = \sqrt{n}\hat{\mathbf{B}}_0$ and $\mathbf{V}_0^* = \sqrt{n}(\hat{\mathbf{V}}_0 - \mathbf{I}_p)$, where $\mathbf{B}_0^* = O_p(1)$ and $\mathbf{V}_0^* = O_p(1)$.

Without loss of generality, assume that $\mathbf{B} = \mathbf{0}$ and $\boldsymbol{\Sigma} = \mathbf{I}_p$ so that $\mathbf{y}_1,\ldots,\mathbf{y}_n$ is a random sample from a spherical distribution with zero mean vector zero and identity covariance matrix. Write $r_i = \|\mathbf{y}_i\|$ and $\mathbf{u}_i = \mathbf{y}_i/r_i$. Now as in the proof of Theorem 11.2, the 1-step HR regression estimator may be written as

$$\sqrt{n}\hat{\mathbf{B}}_1 = \mathbf{B}_0^* + \left[ave \left\{ \frac{1}{r_i} \left(1 + \frac{1}{\sqrt{n} r_i} \mathbf{x}_i^T \mathbf{B}_0^* \mathbf{u}_i + \frac{1}{2\sqrt{n} r_i} \mathbf{u}_i^T \mathbf{V}_0^* \mathbf{u}_i \right) \mathbf{x}_i \mathbf{x}_i^T \right\} \right]^{-1}$$

$$\cdot \sqrt{n}\, ave \left\{ \mathbf{x}_i \left(\mathbf{u}_i^T - \frac{1}{\sqrt{n} r_i} \mathbf{x}_i^T \mathbf{B}_0^* + \frac{1}{\sqrt{n} r_i} \mathbf{x}_i^T \mathbf{B}_0^* \mathbf{u}_i \mathbf{u}_i^T + \frac{1}{2\sqrt{n}} \mathbf{u}_i^T \mathbf{V}_0^* \mathbf{u}_i \mathbf{u}_i^T \right) \right\} + o_p(1).$$

The limiting multivariate normality of $\sqrt{n}\hat{\mathbf{B}}_1$ then follows from the joint limiting multivariate normality of $\mathbf{B}_0^* = \sqrt{n}\hat{\mathbf{B}}_0$ and $\sqrt{n} ave\{\mathbf{x}_i \mathbf{u}_i^T\}$ and the Slutsky's theorem. The above equation then reduces to

$$\sqrt{n}\hat{\mathbf{B}}_1 = p^{-1} \sqrt{n}\hat{\mathbf{B}}_0 + [E(r_i^{-1})]^{-1} \mathbf{D}^{-1} \sqrt{n}\, ave\{\mathbf{x}_i \mathbf{u}_i^T\} + o_p(1),$$

where $\mathbf{D} = E[\mathbf{x}\mathbf{x}^T]$, and for the k-step HR regression estimator we get

$$\sqrt{n}\hat{\mathbf{B}}_k = \left(\frac{1}{p}\right)^k \sqrt{n}\hat{\mathbf{B}}_0 + \left[1 - \left(\frac{1}{p}\right)^k\right] p[(p-1)E(r_i^{-1})]^{-1} \mathbf{D}^{-1} \sqrt{n}\, ave\{\mathbf{x}_i \mathbf{u}_i^T\} + o_p(1)$$

again with a limiting multivariate normality.

Let us next consider the simple case, where the initial estimator is of the type

$$\sqrt{n}\, \hat{\mathbf{B}}_k = \mathbf{D}^{-1} \sqrt{n}\, ave\{\eta_k(r_i) \mathbf{x}_i \mathbf{u}_i^T\} + o_p(1),$$

where η_k is given in (11.14). The covariance matrix of $\sqrt{n}\, vec(\hat{\mathbf{B}}_k)$ then equals to

$$\begin{aligned} & E[\eta_k^2(r)\, vec(\mathbf{D}^{-1}\mathbf{x}\mathbf{u}^T) vec^T(\mathbf{D}^{-1}\mathbf{x}\mathbf{u}^T)] \\ &= E[\eta_k^2(r)](\mathbf{I}_p \otimes \mathbf{D}^{-1}) vec(\mathbf{x}\mathbf{u}^T) vec^T(\mathbf{x}\mathbf{u}^T)](\mathbf{I}_p \otimes \mathbf{D}^{-1}) \\ &= E[\eta_k^2(r)](\mathbf{I}_p \otimes \mathbf{D}^{-1})(E(\mathbf{u}\mathbf{u}^T) \otimes \mathbf{D})(\mathbf{I}_p \otimes \mathbf{D}^{-1}) \\ &= p^{-1} E[\eta_k^2(r)](\mathbf{I}_p \otimes \mathbf{D}^{-1}). \end{aligned}$$

□

References

Arcones, M. A. (1998). Asymptotic theory for M-estimators over a convex kernel. *Economic Theory, 14*, 387–422.

Bai, Z. D., Chen, R., Miao, B. Q., & Rao, C. R. (1990). Asymptotic theory of least distances estimate in the multivariate linear models. *Statistics, 21*, 503–519.

Brown, B. M. (1983). Statistical uses of the spatial median. *Journal of the Royal Statistical Society, Series B, 45*, 25–30.

Chakraborty, B., Chaudhuri, P., & Oja, H. (1998). Operating transformation retransformation on spatial median and angle test. *Statistica Sinica, 8*, 767–784.

Croux, C., Dehon, C., & Yadine, A. (2010). The k-step spatial sign covariance matrix. *Advances in Data Analysis and Classification, 4*, 137–150.

Davies, P. L. (1987). Asymptotic behaviour of S-estimates of multivariate location and dispersion matrices. *Annals of Statistics, 15*, 1269–1292.

Dümbgen, L., & Tyler, D. (2005). On the breakdown properties of some multivariate M-functionals. *Scandinavian Journal of Statistics, 32*, 247–264.

Fang, K. T., Kotz, S., & Ng, K. W. (1990). *Symmetric multivariate and related distributions*. London: Chapman and Hall.

Frahm, G. (2009). Asymptotic distributions of robust shape matrices and scales. *Journal of Multivariate Analysis, 100*, 1329–1337.

Hampel, F. R., Ronchetti, E. M., Rousseeuw, P. J., & Stahel, W. J. (1986). *Robust statistics: The approach based on influence functions*. New York: Wiley.

Hettmansperger, T. P., & McKean, J. W. (2011). *Robust nonparametric statistical methods* (2nd ed.). London: Arnold.

Hettmansperger, T. P., & Randles, R. H. (2002). A practical affine equivariant multivariate median. *Biometrika, 89*, 851–860.

Ilmonen, P., Serfling, R., & Oja, H. (2012). Invariant coordinate selection (ICS) functionals. *International Statistical Review, 80*, 93–110.

Locantore, N., Marron, J. S., Simpson, D. G., Tripoli, N., Zhang, J. T., Cohen, K. L. (1999). Robust principal components for functional data. *Test, 8*, 1–28.

Marden, J. I. (1999). Some robust estimates of principal components. *Statistics & Probability Letters, 43*, 349–359.

Maronna, R. A. (1976). Robust M-estimators of multivariate location and scatter. *Annals of Statistics, 4*, 51–67.

Oja, H. (1999). Affine invariant multivariate sign and rank tests and corresponding estimates: A review. *Scandinavian Journal of Statistics, 26*, 319–343.

Oja, H. (2010). *Multivariate nonparametric methods with R*. New York: Springer.

Ollila, E., Hettmansperger, T. P., Oja, H. (2004). *Affine equivariant multivariate sign methods*. University of Jyvaskyla. Technical report.

Paindaveine, D. (2008). A canonical definition of shape. *Statistics & Probability Letters, 78*, 2240–2247.

Puri, M. L., & Sen, P. K. (1971). *Nonparametric methods in multivariate analysis*. New York: Wiley.

Taskinen, S., Sirkiä, S., & Oja, H. (2010). k-Step estimators of shape based on spatial signs and ranks. *Journal of Statistical Planning and Inference, 140*, 3376–3388.

Tyler, D., Critchley, F., Dümbgen, L., & Oja, H. (2009). Invariant coordinate selection. *Journal of Royal Statistical Society B, 71*, 549–592.

Tyler, D. E. (1987). A distribution-free M-estimator of multivariate scatter. *Annals of Statistics, 15*, 234–251.

Visuri, S., Oja, H., & Koivunen, V. (2000). Sign and rank covariance matrices. *Journal of Statistical Planning and Inference, 91*, 557–575.

Chapter 12
New Nonparametric Tests for Comparing Multivariate Scales Using Data Depth

Jun Li and Regina Y. Liu

Abstract In this paper, we introduce several nonparametric tests for testing scale differences in the two- and multiple-sample cases based on the concept of data depth. The tests are motivated by the so-called DD-plot (depth versus depth plot) and are implemented through a permutation test. Our proposed tests are completely nonparametric. An extensive power comparison study indicates that our tests are as powerful as the parametric test in the normal setting but significantly outperform the parametric one in the non-normal settings. As an illustration, the proposed tests are applied to analyze an airline performance dataset collected by the FAA in the context of comparing the performance stability of airlines.

Keywords Data depth • DD-plot • Multivariate scale difference • Permutation test

12.1 Introduction

Advanced computing and data acquisition technologies have made possible the gathering of large multivariate data sets in many fields. The demand for efficient multivariate analysis has never been greater. However, most existing multivariate analysis still relies on the assumption of normality which is often difficult to justify in practice. A nonparametric method which does not have such a restriction is more desirable in practical situations. The goal of this paper is to introduce several nonparametric tests for comparing the scales (or dispersions) of multivariate samples. These tests are completely nonparametric. Therefore, they have broader applicability than most of the existing tests in the literature.

J. Li
University of California, Riverside, Riverside, CA 92521, USA
e-mail: jun.li@ucr.edu

R.Y. Liu (✉)
Department of Statistics, Rutgers University, New Brunswick, NJ 08854, USA
e-mail: rliu@stat.rutgers.edu

© Springer International Publishing Switzerland 2016

R.Y. Liu, J.W. McKean (eds.), *Robust Rank-Based and Nonparametric Methods*, Springer Proceedings in Mathematics & Statistics 168,
DOI 10.1007/978-3-319-39065-9_12

We first consider two distributions which are identical except for a possible scale difference. If two random samples are drawn from the two distributions, any point would be relatively more central with respect to the sample with the larger scale and relatively more outlying with respect to the sample with the smaller scale. This phenomenon results in a particular pattern in the so-called DD-plot (depth versus depth plot). Based on this particular pattern in the DD-plot, we propose a test for scale differences and carry out the test through a permutation test. We present a simulation study to compare power between our proposed test, a rank test and a parametric test. The performance of our test is comparable to the parametric one and slightly better than the rank test under the multivariate normal setting. Under the non-normal setting, such as the multivariate exponential or Cauchy case, our test significantly outperforms the parametric one and is as good as the rank test.

We further generalize the above nonparametric test to the multiple-sample case. The power comparison study shows the efficiency and robustness of our proposed test in both the normal and non-normal settings. Motivated by the proposed multiple-sample test, we also introduce a DD-plot for the visual detection of inhomogeneity across multiple samples.

The rest of the paper is organized as follows. In Sect. 12.2, we give a brief review of data depth, depth-induced multivariate rankings, and DD-plot. In Sect. 12.3, we describe the test for scale differences in the two-sample case. The results from a simulation study are presented. We devote Sect. 12.4 to the testing of scale homogeneity across multiple samples. In particular, it includes the description of our depth-based nonparametric test, a power comparison study between our test and a rank test, and a DD-plot for scale differences in the multiple-sample case. In Sect. 12.5, we apply our tests to compare the performance stability of airlines using the airlines performance data collected by the FAA. Finally, we provide some concluding remarks in Sect. 12.6.

12.2 Notation and Background Material

12.2.1 Data Depth and Center Outward Ranking of Multivariate Data

A *data depth* is a measure of how deep or central a given point is with respect to a multivariate data cloud or its underlying distribution. The word "*depth*" was first used in Tukey (1975) for picturing data. Liu (1990) observed the natural *center-outward ordering* of the sample points in a multivariate sample that data depth induces. Since then, many new and efficient nonparametric methods based on data depth have been developed to characterize complex features of multivariate distributions or make statistical inference for multivariate data (Liu et al. 1999). In the literature, many different notions of data depth have been proposed for capturing different probabilistic features of multivariate data. [See, for example, the lists in Liu et al. (1999) and Zuo and Serfling (2000)]. In the following, we use the simplicial depth proposed in Liu (1990) as an example of data depth to describe the general concept of data depth and its corresponding center-outward ordering.

Let $\{Y_1, \cdots, Y_m\}$ be a random sample from the distribution $G(\cdot)$ in $\mathbb{R}^d$, $d \geq 1$. We begin with the bivariate setting, $d = 2$. Let $\triangle(a, b, c)$ denote the triangle with vertices a, b and c, and $I(\cdot)$ be the indicator function with $I(A) = 1$ if A occurs and 0 otherwise. Given the sample $\{Y_1, \cdots, Y_m\}$, the sample simplicial depth of y is defined as

$$D_{G_m}(y) = \binom{m}{3}^{-1} \sum_{(*)} I(y \in \triangle(Y_{i_1}, Y_{i_2}, Y_{i_3})),$$

which is the fraction of the triangles generated from the sample that contain the point y. Here (*) runs over all possible triplets of $\{Y_1, \cdots, Y_m\}$. A large value of $D_{G_m}(y)$ indicates that y is contained in many triangles generated from the sample, and thus it lies deeper within the data cloud. On the other hand, a small $D_{G_m}(y)$ indicates an outlying position of y. Thus D_{G_m} is a measure of "depth" of y with respect to the data cloud $\{Y_1, \cdots, Y_m\}$.

The above simplicial depth can be generalized to any dimension d as follows:

$$D_{G_m}(y) = \binom{m}{d+1}^{-1} \sum_{(*)} I(y \in s[Y_{i_1}, \cdots, Y_{i_{d+1}}]),$$

where (*) runs over all possible subsets of $\{Y_1, \cdots, Y_m\}$ of size $(d + 1)$. Here $s[Y_{i_1}, \cdots, Y_{i_{d+1}}]$ is the closed simplex whose vertices are $\{Y_{i_1}, \cdots, Y_{i_{d+1}}\}$. When the distribution G is known, then the simplicial depth of y with respect to G is defined as

$$D_G(y) = P_G\{y \in s[Y_1, \cdots, Y_{d+1}]\},$$

where $Y_1, \cdots, Y_{d+1}$ are $(d + 1)$ random observations from G. D_G measures how "deep" y is with respect to G. In Liu (1990), it is shown that $D_G(\cdot)$ is affine invariant, and that $D_{G_m}(\cdot)$ converges uniformly to $D_G(\cdot)$. The affine invariance ensures that our proposed inference methods are coordinate free, and the convergence of D_{G_m} to D_G allows us to approximate $D_G(\cdot)$ by $D_{G_m}(\cdot)$ when G is unknown.

For the given sample $\{Y_1, Y_2, \cdots, Y_m\}$, we can calculate all the depth values $D_{G_m}(Y_i)$, $i = 1, \ldots, m$, and then order the Y_i according to their ascending depth value. Denote by $Y_{[j]}$ the sample point associated with the jth largest depth values. We then obtain $\{Y_{[1]}, Y_{[2]}, \cdots, Y_{[m]}\}$, which is the depth order statistics of the Y_i, with $Y_{[1]}$ being the *deepest* point, and $Y_{[m]}$ the most outlying point. Here, a larger rank is associated with a more outlying position with respect to the underlying distribution G. Note that the order statistics derived from depth are different from the usual order statistics in the univariate case, since the latter are ordered from the smallest sample point to the largest, while the former start from the most *central* sample point and move outwards in all directions. This property is illustrated in

Fig. 12.1 Depth contours for a bivariate normal sample

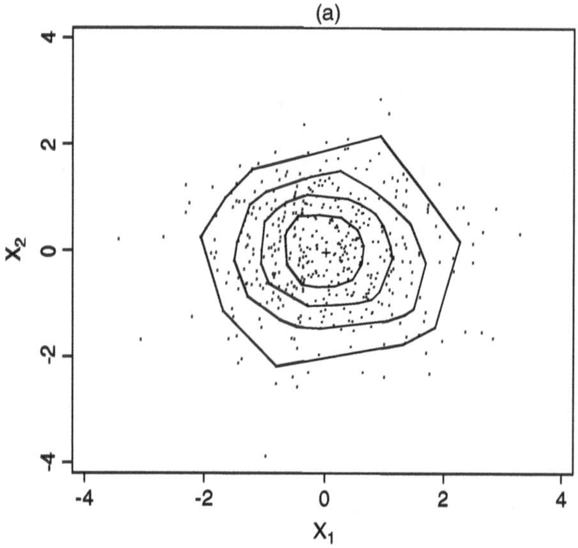

Fig. 12.1 which shows the depth ordering of a random sample of 500 points drawn from a bivariate normal distribution. The "+" marks the deepest point, and the most inner convex hull encloses the deepest 20 % of the sample points. The convex hull expands further to enclose the next deepest 20 % by each expansion. Such nested convex hulls determined by the decreasing depth value indicate that the depth ordering is from the center outward.

12.2.2 DD-Plot for Graphical Comparisons of Multivariate Samples

Suppose that $\{X_1, \ldots, X_n\}(= \mathbf{X})$ and $\{Y_1, \ldots, Y_m\}(= \mathbf{Y})$ are two random samples drawn respectively from F and G, where F and G are two continuous distributions in $\mathbb{R}^d$. Liu et al. (1999) proposed the so-called *depth versus depth (DD)-plot* for graphical comparisons of two multivariate samples. More specifically, the DD-plot is the plot of $DD(F_n, G_m)$, where

$$DD(F_n, G_m) = \left\{(D_{F_n}(x), D_{G_m}(x)),\ x \in \{\mathbf{X} \cup \mathbf{Y}\}\right\}.$$

This is the empirical version of

$$DD(F, G) = \left\{(D_F(x), D_G(x)),\ \text{for all } x \in \mathbb{R}^d\right\}.$$

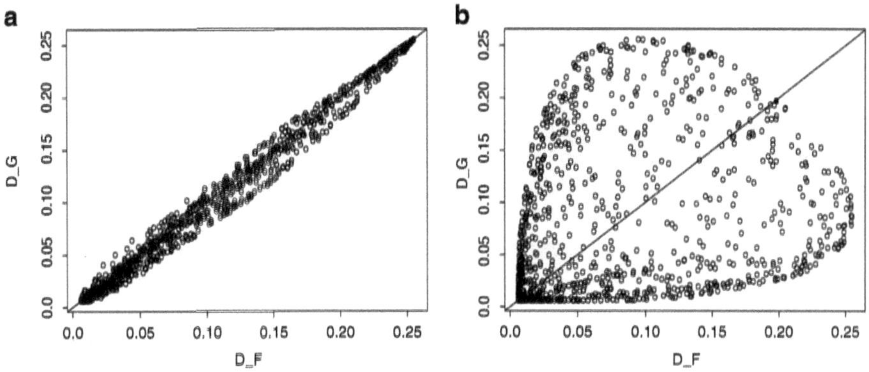

Fig. 12.2 (a) DD-plot: identical distributions (b) DD-plot: location shift

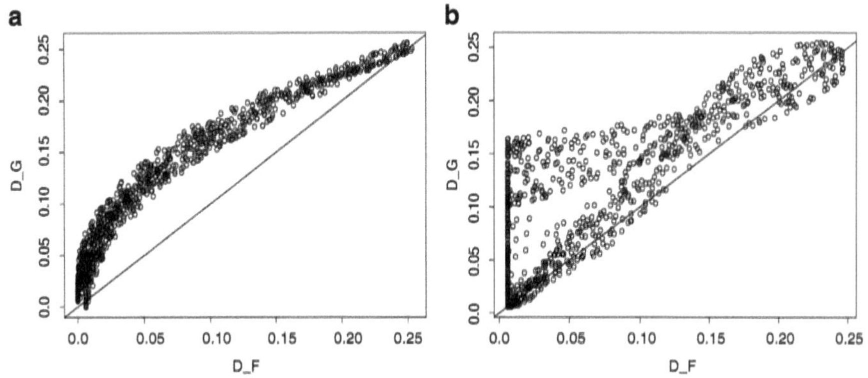

Fig. 12.3 (a) DD-plot: scale increase (b) DD-plot: skewness difference

If $F = G$, then $D_F(x) = D_G(x)$ for all $x \in \mathbb{R}^d$, and thus the resulting $DD(F, G)$ is simply a line segment on the diagonal line in the DD-plot. This is illustrated by the simulation result in Fig. 12.2a which is the DD-plot of two samples both drawn from the bivariate standard normal distribution. A deviation from the diagonal line segment in the DD-plot would suggest that there is a difference between the distributions F and G. As it turns out, each particular pattern of deviations from the diagonal line can be attributed to a specific type of differences between the two distributions. For example, as shown in Fig. 12.2b, in the presence of a location difference in the two samples, the DD-plot generally has a leaf-shaped figure. When there is a scale difference between the two samples, a half-moon pattern will appear in the DD-plot as shown in Fig. 12.3a. Figure 12.3b shows the wedge-like pattern of the DD-plot if there exists a skewness difference in the samples.

12.3 Nonparametric Test for Scale Differences between Two Multivariate Samples

As described above, the DD-plot can serve as a diagnostic tool for visually detecting the difference between two samples of any dimension. To make the visual diagnosis from the DD-plot statistically sound, we would need to establish a proper testing procedure to accompany the DD-plot to reach a decision. Several tests motivated by the DD-plot for location differences were proposed in Li and Liu (2004). In this paper, we focus specifically on developing proper procedures for testing scale differences.

Recall that $\mathbf{X} \equiv \{X_1, \cdots X_n\}$ and $\mathbf{Y} \equiv \{Y_1, \cdots, Y_m\}$ are two samples from F and G, respectively. Assume that F and G are identical except for a possible scale difference, i.e.,

$$g(\cdot - \theta) = f\left((\cdot - \theta)/\sigma\right),$$

where f and g are the density functions of F and G, respectively, and θ is the common location parameter for both distributions. For simplicity, we are interested in testing whether G has a larger scale. In other words, the hypotheses of interest are:

$$H_0 : \sigma = 1 \quad versus \quad H_a : \sigma > 1.$$

Under the null hypothesis, i.e., F and G are identical, the depth of any point with respect to F is equal to its depth with respect to G. Therefore, all the points in the DD-plot $DD(F_n, G_m)$ are clustered along the diagonal line as seen in Fig. 12.2a. If the alternative hypothesis is true, i.e., G has a larger scale, then the depth of any point with respect to G would be larger than its depth with respect to F. Figure 12.4a illustrates this phenomenon in a simple univariate setting. In this plot, it shows the depth curves from two symmetric functions with a scale difference. Due to the symmetry of the distribution function, their deepest points coincide with the centers of symmetry. Since G has a larger scale, its depth curve is more spread out than that of F. Thus, for any point, the depth value from G is larger than its corresponding value from F. In the DD-plot, this phenomenon is reflected by the following: y-coordinates of all the points, which represent the depth values with respect to G, are larger than their corresponding x-coordinates, which represent the corresponding depth values with respect to F. Therefore, in the DD-plot, there exist gaps between all the points and the diagonal line, which results in a half-moon shape as shown in Fig. 12.3a.

Figure 12.4b shows a DD-plot for two distributions with a larger scale difference than the one in Fig. 12.3a. It is clear that, as the scale difference between F and G increases, the differences between the y-coordinates and their corresponding x-coordinates in the DD-plot, i.e., the gaps between the points and the diagonal line, increase as well. Based on this observation, we propose to use the following sum of the gaps of all the points in the DD-plot,

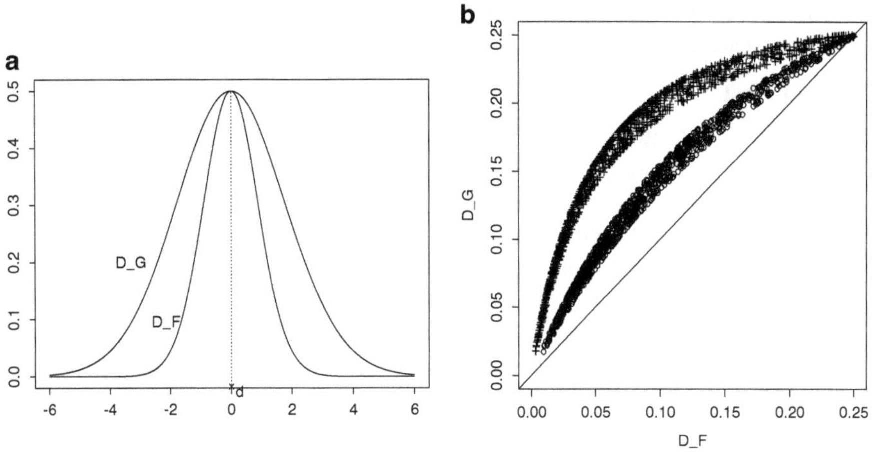

Fig. 12.4 (**a**) Two distributions with a scale difference (**b**) DD-plot of a large scale difference

$$S = \sum_{Z \in \mathbf{X} \bigcup \mathbf{Y}} (D_{G_m}(Z) - D_{F_n}(Z)) \tag{12.1}$$

as our test statistic for scale differences.

Intuitively, the larger the scale difference between the two distributions, the larger the S value and thus the stronger the evidence against H_0. The p-value based on this test statistic can then be determined by

$$p^S = P_{H_0}(S > S_{obs}), \tag{12.2}$$

where S_{obs} is the observed value of S based on the given sample $\mathbf{X} \cup \mathbf{Y}$.

Since the derivation of the distribution of S under the null hypothesis turns out to be quite challenging, we propose to use Fisher's permutation test to determine the above p-value. Fisher's permutation test is carried out as follows.

1. Permute the combined sample $\mathbf{X} \cup \mathbf{Y}$ B times. Here B is sufficiently large. For each permutation, we treat the first n elements as the X-sample and the remaining elements as the Y-sample. Denote the resulting samples of the ith permutation by $\mathbf{X}_i^* = \{X_{i1}^*, \cdots, X_{in}^*\}$, and $\mathbf{Y}_i^* = \{Y_{i1}^*, \cdots, Y_{in}^*\}$, for $i = 1, \ldots, B$.
2. Evaluate the corresponding S value [following (12.1)] for each $\mathbf{X}_i^* \cup \mathbf{Y}_i^*$, which is then denoted by S_i^*, $i = 1, \ldots, B$.

Then, the p^S defined in (12.2) can be estimated by

$$p_B^S = \sum_{i=1}^{B} I_{\{S_i^* > S_{obs}\}}/B. \tag{12.3}$$

To evaluate the performance of the proposed test, we compare it with the following tests in the literature.

- **Rank Test**

Let $\mathbf{W} \equiv \{W_1, W_2, \ldots, W_{n+m}\} \equiv \{X_1, \ldots, X_n, Y_1, \ldots, Y_m\}$. If G has a larger scale, then the X_i are more likely to cluster around the center of the pooled sample, while the Y_i are more likely to scatter at outlying positions. Based on this observation, Liu and Singh (2006) developed the following depth-based rank test for comparing scales between two multivariate samples.

Calculate the depth values of all the points with respect to the pooled sample $\mathbf{W}$. Then rank all the points according to their depth values. In other words, assign lower ranks to the points with lower depth values. Denote this depth-based rank of Y_i by $r(Y_i)$, and

$$r(Y_i) = \#\{W_j \in \mathbf{W} : D_{n+m}(W_j) \leq D_{n+m}(Y_i), j = 1, 2, \cdots, n+m\},$$

where $D_{n+m}(t)$ is the sample depth value of t with respect to $\{W_1, W_2, \ldots, W_{n+m}\}$. Then the rank test statistic R proposed in Liu and Singh (2006) is simply the sum of the depth ranks of the sample $\mathbf{Y}$, i.e.,

$$R = \sum_{i=1}^{m} r(Y_i).$$

Under H_0, if there are no ties, $\{r(Y_1), \ldots, r(Y_m)\}$ can be viewed as a random sample of size m drawn without replacement from the set $\{1, \ldots, n+m\}$. If H_a is true, then the Y_i tend to be more outlying, and assume smaller depth values and smaller ranks. Therefore, H_0 should be rejected if the rank-sum test statistic R is too small. The p-value of the above depth-based rank test can be obtained by using the Wilcoxon rank-sum table.

- **$F_{product}$ Test**

To compare the scales of two univariate normal samples, the standard level-α test is to reject the null hypothesis when $F = s_2^2/s_1^2 \geq F_\alpha(m-1, n-1)$, where s_1^2 and s_2^2 are the sample variances, and $F_\alpha(m-1, n-1)$ is the upper αth quantile of the F-distribution with degrees of freedom $(m-1, n-1)$. This F-test has been generalized to test scale differences in the case of two multivariate normal samples. More specifically, for testing $H_0 : |\Sigma_1| = |\Sigma_2|$ versus $H_a : |\Sigma_1| < |\Sigma_2|$, where $|\Sigma_i|$ is the determinant of the covariance matrix Σ_i, the test statistic is,

$$F_{\text{product}} = \frac{(m-1)^{d-1} \prod_{i=1}^{d-1}(n-i-1)}{(n-1)^{d-1} \prod_{i=1}^{d-1}(m-i-1)} \frac{|S_2|}{|S_1|}$$

where $|S_1|$ and $|S_2|$ are the sample covariance matrices for $\mathbf{X}$ and $\mathbf{Y}$, respectively [see Theorem 3.4.8 of Mardia et al. (1979)]. Under H_0, F_{product} follows the same

Table 12.1 Power comparison between the S test, the rank test and the parametric test

	Normal			Exponential			Cauchy		
σ	1	1.2	1.5	1	1.2	1.5	1	1.2	1.5
S test	**0.058**	**0.824**	1	**0.062**	**0.512**	**0.98**	**0.06**	**0.23**	**0.74**
Rank test	0.060	0.732	1	0.044	0.476	0.972	0.06	0.28	0.76
$F_{product}$	0.058	0.828	1	0.114	0.662	0.992	0.47	0.57	0.55

distribution as $V_1V_2\cdots V_d$, where V_i has the F-distribution with degrees of freedom $(m-i, n-i)$ and the V_i are independent of one another. Clearly, large values of $F_{product}$ indicate stronger support from the observed data for $H_a : |\Sigma_1| < |\Sigma_2|$. Therefore, H_0 is rejected at level α if $F_{product} > \tilde{F}_\alpha$, where $\tilde{F}$ denotes the distribution of $V_1V_2\cdots V_d$ and $\tilde{F}_\alpha$ denotes its upper αth quantile. The distribution $\tilde{F}$ can be approximated by using simulations. In our simulation study later, we draw 1000 random samples from $\tilde{F}$ and use the resulting empirical distribution, $\hat{\tilde{F}}$ as an approximation of $\tilde{F}$.

Table 12.1 shows the result of a simulation study which compares the power of our proposed S test with the rank test and the $F_{product}$ test for the bivariate normal, double exponential and Cauchy distributions. In this power comparison study, the sizes of both samples are 100. B, the number of permutations, is set to be 500. The power estimates are based on 500 replications. For the bivariate normal and double exponential distributions, the Mahalanobis depth (Mahalanobis 1936) is used to calculate the depth. For the bivariate Cauchy distribution, since it does not have any moment, we use the simplicial depth. The scale difference between F and G is denoted by σ (the same for both components) in the tables.

From the table, we can see that our proposed test is slightly better than the rank test and almost as powerful as the parametric test in the bivariate normal setting. In the non-normal setting, the type I errors of the parametric test far exceed the nominal level 0.05, especially in the Cauchy case. This clearly shows that the validity of the parametric test heavily depends on the normality assumption and is not appropriate for non-normal cases, while our S test is quite robust and its performance is as good as the rank test.

Remark. In the normal case, the power of our proposed S test is better than that of the rank test. This may be explained by the fact that the rank test is based on the rank induced by the depths of the sample points, while our S test is directly based on those depths. Some information in the data is lost when transforming the depths to their depth-based ranks, which may result in the minor loss of efficiency for the rank test.

12.4 Scale Comparisons for Multiple Multivariate Samples

12.4.1 Nonparametric Test of Scale Homogeneity

Let $\mathbf{X}_1 = \{X_{11}, \ldots, X_{1n_1}\}, \ldots, \mathbf{X}_k = \{X_{K1}, \ldots, X_{kn_k}\}$ be the k samples drawn respectively from the distributions $F_1, \ldots, F_k$. Let f_i denote the density function of F_i, $i = 1, \ldots, k$. We assume that the F_i only differ in scale. In other words, for $i = 1, \ldots, k$,

$$f_i(\cdot - \theta) = f_0((\cdot - \theta)/\sigma_i),$$

where θ is the common location parameter. The hypotheses of interest are

$$H_0 : \sigma_1 = \cdot = \sigma_k \quad versus \quad H_a : \exists\, i \neq j \quad such \quad that \quad \sigma_i \neq \sigma_j.$$

Similar to the two-sample case, we pool all the k samples together, and obtain the depth values of all the pooled sample points with respect to each individual sample. Denote S_i as the sum of the depths of all the pooled sample points with respect to the ith sample, $\mathbf{X}_i$. Under the null hypothesis of scale homogeneity, we would expect the S_i to be homogeneous. On the other hand, under the alternative hypothesis, we would expect these k values to be different. For example, if the sample $\mathbf{X}_\ell$ has a relatively small scale, the depth of any data point with respect to this sample would be relatively smaller than its depths with respect to the other samples. Then S_ℓ would be relatively smaller than the other S_i. If the sample $\mathbf{X}_\ell$ has a relatively large scale, the depth of any data point with respect to $\mathbf{X}_\ell$ would be relatively larger than its depths with respect to the other samples. Then S_ℓ would be relatively larger. Therefore, any scale inhomogeneity among the k samples would result in inhomogeneity among the S_i. To detect the inhomogeneity among the S_i, we can focus on the following statistic,

$$Q = \sum_{i=1}^{k} (S_i - \bar{S})^2,$$

where $\bar{S} = \sum_{i=1}^{k} S_i/k$. A larger Q value represents a stronger indication of inhomogeneity. Therefore, the corresponding p-value for the above test can be determined by

$$p^Q = P_{H_0}(Q > Q_{obs}),$$

where Q_{obs} is the observed value of Q based on the given samples $\mathbf{X}_i$, $i = 1, \ldots, k$.

Similar to the two-sample case, we turn to Fisher's permutation test to estimate the p-value, p^Q. We pool the k samples together and then randomly divide them into k subsamples with the sample sizes $n_1, n_2, \ldots, n_k$, and denote the resulting samples by $\mathbf{X}_1^*$, $\mathbf{X}_1^*$,...,$\mathbf{X}_k^*$. We then calculate the corresponding Q from the permuted samples $\mathbf{X}_1^*$,$\mathbf{X}_1^*$,...,$\mathbf{X}_k^*$. Repeat this random permutation procedure sufficiently many times, say B times, and denote by Q_i^* the Q value obtained in the ith permutation. The p-value p^Q is then approximated by

$$p_B^Q = \sum_{i=1}^{B} I_{\{Q_i^* > Q_{obs}\}}/B.$$

To evaluate the performance of our proposed Q test, we conduct a simulation study to compare it with a rank test proposed in Liu and Singh (2006) for testing scale homogeneity for multiple multivariate samples. Similar to the rank test (Liu and Singh 2006) proposed for the two-sample case, in this rank test, the depth of each sample point is calculated with respect to the pooled sample, and then all the depth values are ranked. Let $\bar{R}_i$ be the average of the ranks assigned to the ith sample, $i = 1, \ldots, K$. The test statistic proposed in Liu and Singh (2006) is then defined as,

$$T = \sum_{i=1}^{K} \left(1 - \frac{n_i}{N}\right) \left(\frac{\bar{R}_i - (N+1)/2}{\sqrt{(N-n_i)(N+1)/(12n_i)}} \right)^2,$$

where $N = n_1 + n_2 + \cdots + n_K$. The corresponding p-value is

$$p^T = P_{H_0}(T > T_{obs}).$$

Under the null hypothesis, T asymptotically follows a chi-square distribution with $k-1$ degrees of freedom, denoted by χ^2_{k-1}. Therefore, p^T can be approximated by

$$p^T = 1 - F_{\chi^2_{k-1}}(T_{obs}),$$

where $F_{\chi^2_{k-1}}(\cdot)$ is the cumulative distribution function of χ^2_{k-1}.

To compare our proposed Q test with the rank test, we simulate three random samples from bivariate distributions with some or no difference in their scales. Table 12.2 shows the powers of the two tests for detecting the scale inhomogeneity in the normal, double exponential and Cauchy setting, respectively. In each simulation, the sizes of the three samples, n_i, $i = 1, 2, 3$, are all set to be 100. For our proposed permutation test, B, the number of permutations, is chosen to be 500. All of the power estimates are based on 500 replications. The Mahalanobis depth is used for the bivariate normal and double exponential distributions, while the simplicial depth is applied to the bivariate Cauchy distribution.

From the table, we can observe the similar phenomenon as in the two-sample case. In the bivariate normal setting, our proposed Q test is more powerful than the rank test. In the non-normal settings, our test is still comparable to the rank test.

Table 12.2 Power comparison between the Q test and the rank test

	Normal		Exponential		Cauchy	
Σ	Q test	Rank test	Q test	Rank test	Q test	Rank test
$\Sigma_1 = \Sigma_2 = \Sigma_3 = \mathbf{I}$	**0.062**	0.06	**0.048**	0.052	**0.06**	0.06
$\Sigma_1 = \mathbf{I}, \Sigma_2 = \Sigma_3 = 1.1\mathbf{I}$	**0.292**	0.236	**0.131**	0.128	**0.07**	0.09
$\Sigma_1 = \mathbf{I}, \Sigma_2 = 1.1\mathbf{I}, \Sigma_3 = 1.2\mathbf{I}$	**0.624**	0.522	**0.262**	0.265	**0.12**	0.14
$\Sigma_1 = \mathbf{I}, \Sigma_2 = \Sigma_3 = 1.2\mathbf{I}$	**0.686**	0.608	**0.380**	0.398	**0.24**	0.21
$\Sigma_1 = \mathbf{I}, \Sigma_2 = 1.2\mathbf{I}, \Sigma_3 = 1.5\mathbf{I}$	**1**	0.994	**0.894**	0.898	**0.48**	0.53
$\Sigma_1 = \mathbf{I}, \Sigma_2 = \Sigma_3 = 1.5\mathbf{I}$	**1**	1	**0.968**	0.978	**0.60**	0.66

Table 12.3 Power comparison between the *Min* test and the Q test

	Normal		Exponential	
Σ	*Min* test	Q test	*Min* test	Q test
$\Sigma_1 = \Sigma_2 = \Sigma_3 = \mathbf{I}$	**0.038**	0.048	**0.058**	0.066
$\Sigma_1 = \mathbf{I}, \Sigma_2 = \Sigma_3 = 1.1\mathbf{I}$	**0.258**	0.256	**0.14**	0.13
$\Sigma_1 = \mathbf{I}, \Sigma_2 = \Sigma_3 = 1.2\mathbf{I}$	**0.754**	0.708	**0.416**	0.386
$\Sigma_1 = \mathbf{I}, \Sigma_2 = \Sigma_3 = 1.3\mathbf{I}$	**0.984**	0.98	**0.714**	0.68
$\Sigma_1 = \mathbf{I}, \Sigma_2 = \Sigma_3 = 1.5\mathbf{I}$	**1**	1	**0.986**	0.966

Remark. Our proposed test statistic takes into account all the information contained in the S_i. From some of our simulation results, we observe that, in some particular simulation setting, the permutation test based on other test statistics may yield higher power than the one based on the test statistic we propose here. For example, three samples are simulated with two having the same scales and one having a smaller scale. The resulting three S_i would have two similar values and one relatively smaller value. If we use the minimum of the S_i as the test statistic, the corresponding permutation test will be more powerful than the one we propose above. See the simulation results in Table 12.3. The test based on the minimum of the S_i is denoted as *Min* test in the tables. This result is not a surprise, since the only inhomogeneity among these three samples lies in the one with the smaller scale. The minimum of the S_i exactly captures this difference, while our proposed $\sum_{i=1}^{3}(S_i - \bar{S})^2$ dilutes this difference by taking the average across all the samples. This simulation study suggests that, if some extra information is available regarding the difference pattern of the multiple samples, we can improve the proposed test by choosing appropriate test statistics which well reflect that difference pattern. If there is no prior information about the difference pattern of the samples, the one based on $\sum_{i=1}^{3}(S_i - \bar{S})^2$ is always a reasonable choice.

12.4.2 DD-plot for Graphical Comparisons of Scales for Multiple Multivariate Samples

As described earlier, to carry out our Q test, we pool all the k samples together, and obtain the depth values of all the pooled sample points with respect to each individual sample. Thus, for each sample point, we have a set of k depth values corresponding to each of the k samples. If $k = 2$, we can plot the 2-dimensional vectors of all the sample points. This is essentially the DD-plot we mentioned earlier. This DD-plot is always a subset of $\Re^2$ no matter how large the dimension of the data is. The two-dimensional DD-plot is easy to visualize, and it provides a convenient tool for graphical comparisons of two multivariate samples as discussed in Sect. 12.2.2. For $k > 2$, plotting the k depth values of each point will result in a k-dimensional plot. It is not easy to visualize and find interesting patterns in a k-dimensional plot, especially when k is large. For the purpose of visually detecting the scale inhomogeneity among the k samples, we propose to construct a 2-dimensional plot as follows.

Recall that S_i is the sum of the depths of all the pooled sample points with respect to the sample $\mathbf{X}_i$. Since the smaller scale the sample, the smaller sum of those depths with respect to this sample, we rank the S_i and denote the sample associated with the largest S_i as the sample $\mathbf{X}_{max}$ and the one associate with the smallest S_i as the sample $\mathbf{X}_{min}$. Then the samples $\mathbf{X}_{max}$ and $\mathbf{X}_{min}$ represent the two samples with the largest and smallest scales, respectively. Since detecting the scale inhomogeneity of the k samples amounts to detecting the scale difference between these two extreme samples, a DD-plot can be defined as

$$DD(F_{min}, F_{max}) = \left\{(D_{F_{min}}(x), D_{F_{max}}(x)),\ x \in \{\mathbf{X}_1 \cup \cdots \cup \mathbf{X}_k\}\right\}, \qquad (12.4)$$

where $D_{F_{min}}(x)$ and $D_{F_{max}}(x)$ are the depths with respect to $\mathbf{X}_{min}$ and $\mathbf{X}_{max}$, respectively.

This plot can serve as a graphical tool for comparing the scales among multiple multivariate samples. If there is a scale difference among the k samples, then the sample $\mathbf{X}_{max}$ should have a larger scale than the sample $\mathbf{X}_{min}$. Similar to the two-sample case, the DD-plot defined in (12.4) will have a half-moon shape. When the scales of the k samples are homogeneous, there is no significant difference in scale between $\mathbf{X}_{max}$ and $\mathbf{X}_{min}$. Hence the points in the DD-plot will be clustered along the diagonal line. Figure 12.5 shows the corresponding plots.

12.5 Applications to Airlines Performance Data

We apply the tests described in Sects. 12.3 and 12.4 to an analysis of an airline performance dataset collected by the FAA. It consists of several monthly performance measures of the top ten air carriers from July 1993 to May 1998. To facilitate a visual

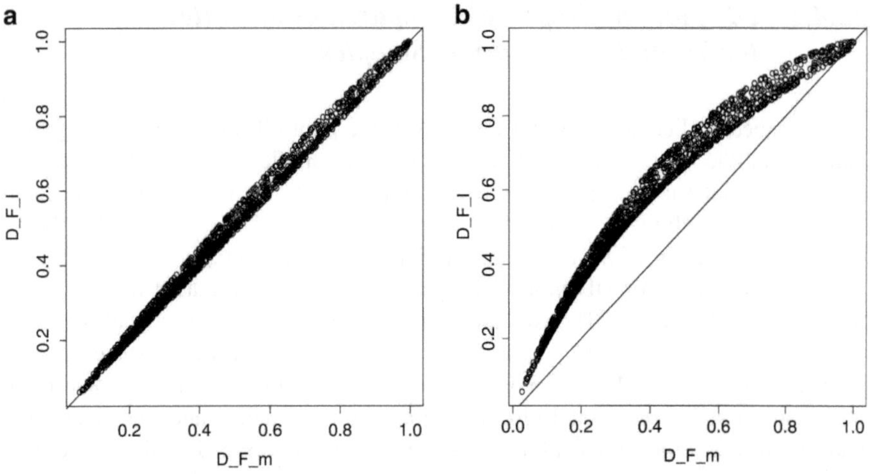

Fig. 12.5 DD-plots: (**a**) Three identical bivariate normal distributions (**b**) Three bivariate normal distributions with scale differences

confirmation of our test results with related bivariate scatter plots, we illustrate our proposed tests using only two performance measures, namely the fractions of nonconformity in airworthiness (AW) and operation (OP) surveillance. A smaller nonconformity fraction is a more desirable feature. Several depth based multivariate control charts (Cheng et al. 2000; Liu 1995) have been used to monitor and compare the performances of all the ten airlines. Using the DD-plot, Li and Liu (2004) propose two depth-based nonparametric tests to determine whether or not there are significant location differences in the underlying distributions of the air carriers. The location parameter here is referred to as the anticipated target performance in the aviation safety domain. In comparing Carriers 1 and 4, their scatter plots in Fig. 12.6a show that the location of Carrier 4 (i.e., the deepest point of Carrier 4, marked by a solid triangle) clearly shifts to the upper-right of that of Carrier 1 (marked by •). This observed location difference has been declared significant by the tests proposed in Li and Liu (2004).

In judging airline performance, in addition to examining the expected target performance (i.e. the location of the distribution) of the airlines, the stability of the performance within the airlines is also a major concern. When the performance of one air carrier is stable, all of its performance observations must cluster tightly and result in a small scale of its underlying distribution. On the other hand, when the performance of one air carrier is not very stable, all of its performance observations must scatter around and result in a large scale of its underlying distribution. Therefore, the measure of airline performance stability is simply the measure of scale or variation of the performance distribution. Thus, comparing performance stability amounts to comparing the scale of a distribution. Larger scales would mean less stable performance.

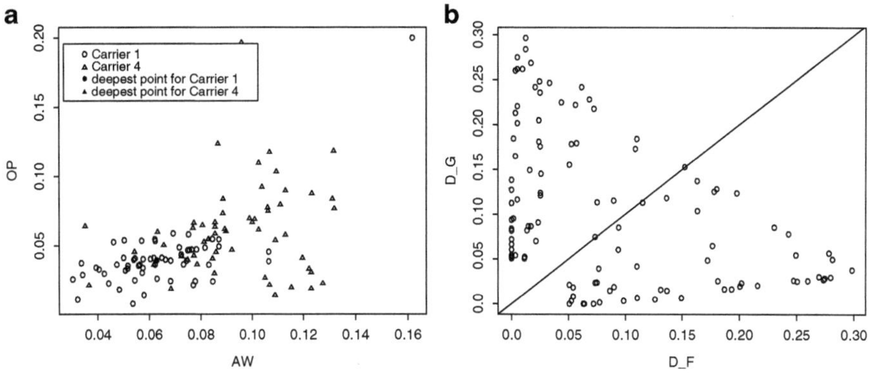

Fig. 12.6 (**a**) Scatter plot for Carriers 1 & 4 (**b**) DD-plot for Carriers 1 & 4

We next compare the scales of Carrier 1 and 4. Given that there exists a location difference between the two carriers, we first center both samples at their respective deepest points to ensure that they have the same location, and then apply our proposed S test to determine whether or not there is a significant difference in scale in the distributions underlying the two air carriers. The p-value is 0 from (12.3), which clearly supports the conclusion that Carrier 4 has a larger scale than Carrier 1. In other words, the performance of Carrier 4 is less stable. This same conclusion can also be reached by examining the two graphs in Fig. 12.7. Figure 12.7a is the DD-plot of Carriers 1 and 4 after centering the data respectively at their deepest points. It shows a pattern which combines Fig. 12.3a, b. This suggests that there are both scale and skewness differences between the two carriers. Figure 12.7b displays the scale curves of four carriers. The scale curve was first introduced in Liu et al. (1999) as another graphical tool to visualize scale differences between multivariate samples. The sample scale curve derived from a sample of size n is defined as,

$$S_n(p) = volume\{C_{n,p}\}, \quad \text{for } 0 \leq p \leq 1,$$

where $C_{n,p}$ is the convex hull containing the $\lceil np \rceil$ deepest points. Roughly speaking, the scale curve measures the volume of the nested depth contours, as seen in Fig. 12.1. The plot of $S_n(p)$ versus p shows the scale of the distribution as a simple curve. When comparing the scales of two samples, if the scale curve for the sample X is consistently above the scale curve for the sample Y, then the sample X is more spread out and thus has a larger scale than the sample Y. From Fig. 12.7b, it is obvious that the scale curve of Carrier 4 lies consistently above all others, including that of Carrier 1. The findings are also supported by the scatter plot in Fig. 12.6a which shows more scattered data for Carrier 4. In summary, the performance of Carrier 4 is inferior to that of Carrier 1, in that Carrier 4 has significantly higher target nonconformity ratios and it is also much less stable overall. Possible causes should be identified and corrective measures should be taken.

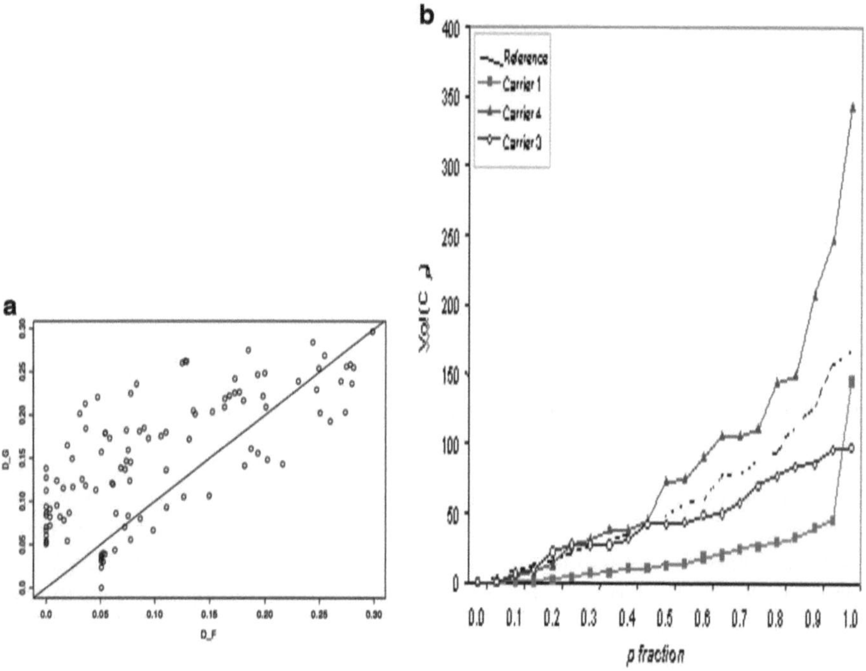

Fig. 12.7 (**a**) DD-plot for Carriers 1 & 4 after centering (**b**) Scale curves for air carriers

In testing the scale homogeneity among Carriers 1, 4, 9 and 10, we again first center all the samples at their respective deepest points, and then apply our Q test. The p-value we obtain is 0, which indicates scale inhomogeneity among these four carriers. This conclusion is also confirmed by the scale curve plot of the four carriers in Fig. 12.7b. Figure 12.8 shows the DD-plot of the four carriers defined in (12.4). It has a half-moon pattern, which also suggests scale inhomogeneity among those carriers. In summary, although most of the top ten carriers have roughly similar crew training, fleet size, and aircraft models, not all of them achieve the same level of performance stability.

12.6 Concluding Remarks

In this paper, we introduce several depth based tests for testing scale differences in the two- and multiple-sample cases. The tests are completely nonparametric, and they are easy to implement regardless of the dimensionality of the data. We also present several simulation and comparison studies to illustrate the properties of these tests. Although our illustrative examples are all in $\mathbb{R}^2$, all tests discussed in this paper apply to any dimension.

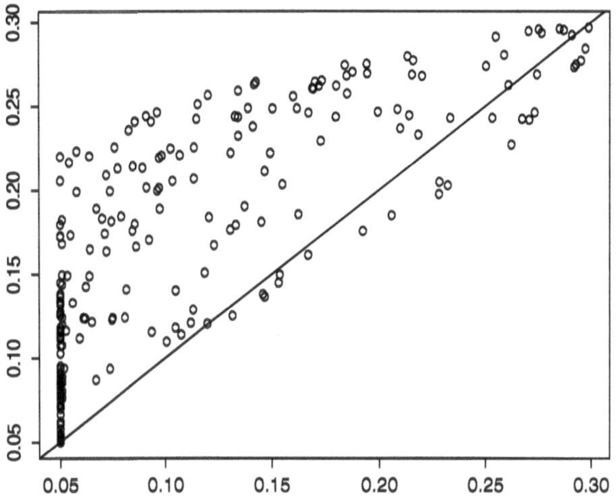

Fig. 12.8 DD-plot for Carriers 1, 4, 9 and 10

In principle, the tests proposed in this paper can be constructed using any notion of data depth which is affine invariant. Some notions of depth may be more suitable than others in capturing a certain feature of a distribution. For example, if the underlying distribution is close to elliptical, then it is more efficient to use the Mahalanobis depth. Otherwise, the more geometric depths such as the simplicial depth or the halfspace depth (Tukey 1975) may be more desirable since they do not require specific distributional structures or moment conditions. It may be worthwhile to see how the results of our proposed tests are affected by the different depths used in those tests. We plan to investigate this in a separate research project.

In this paper, we only focus on the case where the scale of one distribution is an expansion of that of the other one. It should be interesting to consider more general scale differences between two multivariate distributions. Combining proper statistics derived from graphical tools, such as DD-plots and scale curves, with the permutation test idea may prove to be a promising way in developing these tests.

Acknowledgements Regina Y. Liu gratefully acknowledges the support from the NSF grant DMS-1513483.

References

Cheng, A., Liu, R., & Luxhøj, J. (2000). Monitoring multivariate aviation safety data by data depth: control charts and threshold systems. *IIE Transactions on Operations Engineering, 32*, 861–872 (Best Paper Award for Feature Applications for 2000–2001 by Institute of Industrial Engineers).

Li, J., & Liu, R. (2004). New nonparametric tests of multivariate locations and scales using data depth. *Statistical Science, 19*, 686–696.

Liu, R. (1990). On a notion of data depth based on random simplices. *The Annals of Statistics, 18*, 405–414.

Liu, R. (1995). Control charts for multivariate processes. *Journal of the American Statistical Association, 90*, 1380–1388.

Liu, R., Parelius, J., & Singh, K. (1999). Multivariate analysis by data depth: descriptive statistics, graphics and inference (with discussions). *Annals of Statistics, 27*, 783–858.

Liu, R., & Singh, K. (2006). Rank tests for multivariate scale difference based on data depth. *Data depth: Robust multivariate analysis, computational geometry and applications, DIMACS series* (pp. 17–36). New York: AMS.

Mardia, K. V., Kent, J. T., & Bibby, J. M. (1979). *Multivariate analysis*. New York: Academic Press.

Mahalanobis, P. (1936). On the generalized distance in statistics. *Proceedings of the National Academy India, 12*, 49–55.

Tukey, J. (1975). Mathematics and picturing data. In *Proceedings of the 1975 International Congress of Mathematics* (Vol. 2, pp. 523–531).

Zuo, Y. and Serfling, R. (2000) Structural properties and convergence results for contours of sample statistical depth functions. *The Annals of Statistics, 28*, 483–499.

Chapter 13
Multivariate Autoregressive Time Series Using Schweppe Weighted Wilcoxon Estimates

Jaime Burgos and Jeff T. Terpstra

Abstract The vector autoregressive model in multivariate time series analysis is commonly used across different fields due to its simplicity in application. The traditional method for estimating the model parameters is the least squares minimization. However, since least squares estimates are sensitive to outliers, more robust techniques have become of interest. This paper investigates a robust alternative by obtaining the estimates using a weighted Wilcoxon dispersion with Schweppe weights. Under the so-called innovations outlier model where outliers are introduced through the error distribution, the proposed estimator is shown to be asymptotically multivariate normal, centered about the true model parameters, at a rate of $n^{-\frac{1}{2}}$. In addition, a Monte Carlo study is presented to evaluate the performance of various estimators. The study results suggest that the Schweppe-weighted Wilcoxon estimates will generally have best performance. This result is most noticeable under the presence of additive outliers or when the series is closer to non-stationarity.

Keywords Asymptotic normality • Outliers • U-statistics • Vector autoregressive • Wilcoxon estimates

13.1 Introduction

13.1.1 Time Series Model

A widely used model in multivariate time series analysis is the stationary m-variate vector autoregressive model of order p. Throughout this paper, we refer to it as the $\mathrm{VAR}_m(p)$. The (centered) model is generally expressed as

J. Burgos • J.T. Terpstra (✉)
Department of Statistics, Western Michigan University, Kalamazoo, MI 49008, USA
e-mail: jaime.burgosmariot@gmail.com; jeffrey.terpstra@wmich.edu

© Springer International Publishing Switzerland 2016

R.Y. Liu, J.W. McKean (eds.), *Robust Rank-Based and Nonparametric Methods*, Springer Proceedings in Mathematics & Statistics 168,
DOI 10.1007/978-3-319-39065-9_13

$$\begin{aligned} \boldsymbol{Y}_t &= \boldsymbol{\Phi}_1 \boldsymbol{Y}_{t-1} + \boldsymbol{\Phi}_2 \boldsymbol{Y}_{t-2} + \cdots + \boldsymbol{\Phi}_p \boldsymbol{Y}_{t-p} + \boldsymbol{\varepsilon}_t; \quad t = 1-p, \ldots, 0, 1, \ldots, T \\ &\stackrel{\text{def}}{=} \boldsymbol{\Phi} \boldsymbol{X}_{t-1} + \boldsymbol{\varepsilon}_t, \end{aligned} \tag{13.1}$$

where $p \in \{1, 2, \ldots\}$; $\boldsymbol{Y}_t, \boldsymbol{\varepsilon}_t \in \mathbb{R}^m$; $m \in \{2, 3, \ldots\}$; $\boldsymbol{X}_{t-1} = (\boldsymbol{Y}'_{t-1}, \boldsymbol{Y}'_{t-2}, \ldots, \boldsymbol{Y}'_{t-p})' \in \mathbb{R}^{mp}$; $\boldsymbol{\Phi}_i \in \mathbb{R}^{m \times m}$, $i = 1, 2, \ldots, p$; $\boldsymbol{\Phi} = (\boldsymbol{\Phi}_1, \ldots, \boldsymbol{\Phi}_p) \in \mathbb{R}^{m \times mp}$; and $p + T$ is the number of realizations in the time series. The process has a stationary solution if and only if

$$\det(x^p \mathbb{I}_m - x^{p-1} \boldsymbol{\Phi}_1 - \cdots - \boldsymbol{\Phi}_p) = 0 \;\Rightarrow\; |x| < 1, \tag{13.2}$$

where $\det(\cdot)$ is the determinant operator, $\mathbb{I}_m$ represents the dimension m identity matrix, x may be complex valued, and $|\cdot|$ is the modulus operator on the complex plane (Brockwell and Davis 2002). Furthermore, the $\boldsymbol{\varepsilon}_t$ are assumed to be independent with an identical continuous distribution function $\mathrm{F}(\cdot)$ that satisfies

$$\mathrm{V}[\boldsymbol{\varepsilon}_t] = \boldsymbol{\Omega} \;\; p.d., \tag{13.3}$$

where *p.d.* stands for positive definite. Under assumptions (13.1), (13.2) and (13.3), $\{\boldsymbol{Y}_t\}$ is causal (Brockwell and Davis 2002), ergodic (Krengel 1985), and geometrically absolutely regular (g.a.r.) (Terpstra and Rao 2002). Briefly, g.a.r. is a type of weak dependence property inherent in $\mathrm{VAR}_m(p)$ processes. It basically states that as observations become further apart, they become less dependent according to a geometric rate. As we shall see, the g.a.r. of the process plays a critical role in the theoretical development of this paper. For example, the covariance result found in Arcones (1998) and U-statistic theorems for g.a.r. processes form the theoretical paradigm of the paper.

13.1.2 Parameter Estimation

Estimates for the parameters in (13.1) are typically obtained by minimizing, with respect to $\boldsymbol{\Phi}$, a dispersion function. The most common selection, generalized with weights and for a multivariate setting, is the (conditional) L_2 dispersion function,

$$\mathrm{D}_2(\boldsymbol{\Phi}) = \sum_{t=1}^{T} b_t \| \boldsymbol{\varepsilon}_t(\boldsymbol{\Phi}) \|^2, \tag{13.4}$$

where $b_t \geq 0$ denotes a weight function that may depend on both the design $\boldsymbol{X}_{t-1}$ and the response $\boldsymbol{Y}_t$, $\boldsymbol{\varepsilon}_t(\boldsymbol{\Phi}) = \boldsymbol{Y}_t - \boldsymbol{\Phi} \boldsymbol{X}_{t-1}$, and $\| \cdot \|$ denotes the Euclidean norm. Typically $b_t \equiv 1$, which results in a component-wise ordinary least squares (OLS) estimation. The asymptotic theory for the OLS estimates can be found in many time series textbooks, including Lütkepohl (1993) and Fuller (1996).

As with most least squares estimates, outlying observations can yield unreliable estimates and predictions (Hettmansperger and McKean 2011; McKean and Sheather 2009; Rousseeuw and Leroy 1987). A number of robust estimators for the parameters in (13.1) have been studied. These include a functional least squares approach by Heathcote and Welsh (1988); an extension of the RA-estimates proposed by Bustos and Yohai (1986), Li and Hui (1989), and Ben et al. (1999); GR-estimates proposed by Terpstra and Rao (2002); weighted-L_1 estimates proposed by Reber et al. (2008); an extension of least trimmed squares by Croux and Joossens (2008); and an extension of MM-estimates by Muler and Yohai (2013).

Other robust estimators can be obtained by using different dispersion functions. For instance, a multivariate generalization of the weighted-L_1 dispersion function is given by

$$\mathrm{D}_1(\boldsymbol{\Phi}) = \sum_{t=1}^{T} b_t \| \boldsymbol{\varepsilon}_t(\boldsymbol{\Phi}) \|, \tag{13.5}$$

where $b_t \geq 0$ denotes a weight function that may depend on both the design $\boldsymbol{X}_{t-1}$ and the response $\boldsymbol{Y}_t$. Using Mallows weights (Mallows 1975), the asymptotic distribution of the weighted-L_1 estimates is described by Reber et al. (2008).

Another robust estimator can be obtained from a multivariate generalization of the weighted-Wilcoxon dispersion function

$$\mathrm{D}_{\mathrm{WW}}(\boldsymbol{\Phi}) = \sum_{i<j}^{T} b_{ij} \| \boldsymbol{\varepsilon}_j(\boldsymbol{\Phi}) - \boldsymbol{\varepsilon}_i(\boldsymbol{\Phi}) \|, \tag{13.6}$$

where $b_{ij} = b_{ji} \geq 0$ denotes a weight function that may depend on both the design and the response points at the ith and jth realizations. Using Mallows weights, the asymptotic theory of multivariate weighted-Wilcoxon (WW) estimates, known as Multivariate Generalized Rank (GR) estimates, was introduced by Terpstra and Rao (2002). This paper extends the research done by Terpstra and Rao (2002) by allowing the use of the more general class of Schweppe weights (Handschin et al. 1975).

The rest of the paper is outlined as follows. Section 13.2 discusses some commonly used weighting schemes. In Sect. 13.3, the proposed estimator is shown to be asymptotically multivariate normal, centered about the true model parameters, at a rate of $n^{-\frac{1}{2}}$. Section 13.4 presents details regarding the derivations of the main theoretical results. After that, in Sect. 13.5, a Monte Carlo study is presented to evaluate the performance of alternative estimators. Some final remarks are given in Sect. 13.6.

13.2 Weighting Schemes

The estimates obtained from (13.4), (13.5), and (13.6) are influenced by the selection of the weight functions. The weighting schemes used throughout this paper can be grouped into three classes: non-random weights, Mallows weights, and Schweppe weights.

13.2.1 Non-random Weights

Non-random weights do no explicitly depend on the time series realizations. The most common case is selecting the weights constant to one. Here $b_t \equiv 1$ and $b_{ij} \equiv 1$, thus weighing each realization equally.

13.2.2 Mallows Weights

A simple example of Mallows weights for (13.4), (13.5), and (13.6) are the Boldin (1994) and Theil (1950) weights. These can be generalized to the multivariate setting by defining them as

$$b_t = \|\boldsymbol{X}_{t-1}\|^{-1} \text{ if } \boldsymbol{X}_{t-1} \neq \boldsymbol{0} \quad \text{and} \quad b_{ij} = \|\boldsymbol{X}_{j-1} - \boldsymbol{X}_{i-1}\|^{-1} \text{ if } \boldsymbol{X}_{i-1} \neq \boldsymbol{X}_{j-1}. \tag{13.7}$$

In practice, when $\boldsymbol{X}_{t-1} = \boldsymbol{0}$ or $\boldsymbol{X}_{i-1} = \boldsymbol{X}_{j-1}$, b_t or b_{ij} can be defined arbitrarily since these points will not contribute to the dispersion function.

Another popular Mallows weighting scheme can be found in Chang et al. (1999). These can be generalized to the multivariate setting by defining them as

$$b_t = \min\left\{1, \left(\frac{c}{\mathrm{d}^2(\boldsymbol{X}_{t-1})}\right)^{k/2}\right\} \quad \text{and} \quad b_{ij} = b_i b_j, \tag{13.8}$$

where $\mathrm{d}(\cdot)$ denotes the Mahalanobis distance based on a robust measure of center and covariance and c and k are tuning constants. Typically, $c = \chi^2_{1-\alpha}(mp)$ is the $100(1-\alpha)$th percentile of a chi-squared distribution with mp degrees of freedom, k is set to 2, and the minimum covariance determinant (Rousseeuw and Leroy 1987) is used for the robust measures of center and covariance.

As can be seen from (13.7) and (13.8), Mallows weights depend only on the design point (i.e. $\boldsymbol{X}_{t-1}$ in the present context). Consequently, these types of weights do have some limitations. For example, consider an outlier which is introduced via the error distribution, F. Since observations in a VAR(p) model play a dual role as both a response and an explanatory variable, any resulting leverage point is likely to be a *good* leverage point in the sense that a small residual will be produced. Therefore, downweighting these so-called *good* leverage points (e.g. using Mallows weights) tends to decrease overall efficiency. That said, incorporating information from an initial fit into the weights (via the corresponding residuals) can regain some of this lost efficiency. This is the premise for considering Schweppe-type weighting schemes, which we now discuss.

13.2.3 Schweppe Weights

Schweppe weights depend on both design and response points. An example of Schweppe weights are the high breakdown weights defined in Chang et al. (1999). These can be generalized to the multivariate setting by defining them as

$$b_t = \psi\left(\frac{b}{a_t}\right) \quad \text{and} \quad b_{ij} = \psi\left(\frac{b^2}{a_i a_j}\right), \tag{13.9}$$

where $a_t = \frac{\mathrm{d}(\hat{\boldsymbol{\varepsilon}}_t)}{\psi(c/\mathrm{d}^2(\boldsymbol{X}_{t-1}))}$; $\hat{\boldsymbol{\varepsilon}}_t$ denotes the tth residual vector based on an initial robust estimate; and $\psi(x) = 1, x, -1$ when $x \geq 1, -1 < x < 1, x \leq -1$, respectively. The least trimmed squares estimate of Croux and Joossens (2008) or Mallows-based versions of (13.5) and (13.6) can be used for the initial fits while the tuning parameter b is typically set to $\mathrm{med_i}\{a_i\} + 3\,\mathrm{MAD_i}\{a_i\}$. Additional details regarding the selection of tuning constants can be found in Chang et al. (1999) and Hettmansperger and McKean (2011).

Another type of Schweppe weights are defined in Terpstra et al. (2001). These can be generalized to the multivariate setting by defining them as

$$b_t = 1 - \mathrm{I}(\mathrm{d}^2(\hat{\varepsilon}_t) > c_1)\,\mathrm{I}(\mathrm{d}^2(\boldsymbol{X}_{t-1}) > c_2)(1 - h_t) \quad \text{and} \quad b_{ij} = b_i b_j, \tag{13.10}$$

where h_t is the Mallows weight defined in (13.8) at the tth realization and $\mathrm{I}(\cdot)$ is an indicator function yielding 1 if the logical statement is true, and 0 otherwise. The tuning parameters c_1, c_2 are typically $\chi^2_{1-\alpha}(m)$ and $\chi^2_{1-\alpha}(mp)$, respectively. These weights make use of an indicator function to downweight only the *bad* leverage points.

13.2.4 The Proposed Estimate of $\boldsymbol{\Phi}$

In this paper, we propose the estimator of $\boldsymbol{\Phi}$ as a value that minimizes the weighted Wilcoxon dispersion function using Schweppe weights. Under the definition in (13.6), the dispersion function $\mathrm{D_{WW}}(\boldsymbol{\Phi})$ is non-negative, piecewise linear, and convex. Thus, $\mathrm{D_{WW}}(\boldsymbol{\Phi})$ has a minimum and an estimate can be obtained.

For notational convenience, the VAR$_m(p)$ model in (13.1) is rewritten in a notation where $\boldsymbol{\Phi}$ is vectorized. Let vec$(\cdot)$ denote a columns stacking operation that transforms a $r \times c$ matrix into a $rc \times 1$ vector. Also, let $\otimes$ denote the Kronecker product. Then,

$$\boldsymbol{Y}_t = \boldsymbol{\Phi}\boldsymbol{X}_{t-1} + \boldsymbol{\varepsilon}_t = (\boldsymbol{X}'_{t-1} \otimes \mathbb{I}_m)\,\mathrm{vec}(\boldsymbol{\Phi}) + \boldsymbol{\varepsilon}_t \stackrel{\text{def}}{=} \boldsymbol{x}'_{t-1}\boldsymbol{\beta} + \boldsymbol{\varepsilon}_t,$$

where $\boldsymbol{\beta} = \mathrm{vec}(\boldsymbol{\Phi})$ and $\boldsymbol{x}'_{t-1} = \boldsymbol{X}'_{t-1} \otimes \mathbb{I}_m$. More details on the use of vec$(\cdot)$ to rewrite models can be found in Lütkepohl (1993, Appendix A.12). Furthermore, we redefine $\boldsymbol{\varepsilon}_t(\boldsymbol{\beta}) = \boldsymbol{Y}_t - \boldsymbol{x}'_{t-1}\boldsymbol{\beta}$, so that the dispersion is now viewed as a function of $\boldsymbol{\beta}$ instead of $\boldsymbol{\Phi}$. Hence, our proposed estimator, denoted as $\hat{\boldsymbol{\beta}}_n$, is such that $\mathrm{D_{WW}}(\hat{\boldsymbol{\beta}}_n) = \min_{\boldsymbol{\beta}} \mathrm{D_{WW}}(\boldsymbol{\beta})$.

13.3 Theoretical Results

We start by stating a list of assumptions under which the asymptotic theory holds. As usual, asymptotic corresponds to the limit as $T = n$ goes to infinity. The results rely on the assumptions given below. In what follows, $\boldsymbol{\theta}$ denotes a vector of nuisance parameters that can be used in the calculation of the weights and needs to be estimated in practice (e.g. with $\hat{\boldsymbol{\theta}}$).

$$n^{\frac{1}{2}}(\hat{\boldsymbol{\theta}} - \boldsymbol{\theta}_0) = \boldsymbol{O}_p(1) \tag{W1}$$

$$\boldsymbol{D}_{\mathrm{ij}}(\boldsymbol{\theta}) \stackrel{\text{def}}{=} \nabla\, \mathrm{b_{ij}}(\boldsymbol{\theta}) \text{ is continuous } \forall\ (i,j,\boldsymbol{\theta}) \tag{W2}$$

$$\|\boldsymbol{D}_{\mathrm{ij}}(\boldsymbol{\theta})\| \le B_D < \infty\ \forall\ (i,j,\boldsymbol{\theta}) \tag{W3}$$

$$b_{ij} = \mathrm{b}(\boldsymbol{X}_{i-1}, \boldsymbol{\varepsilon}_i, \boldsymbol{X}_{j-1}, \boldsymbol{\varepsilon}_j) = \mathrm{b}(\boldsymbol{X}_{j-1}, \boldsymbol{\varepsilon}_j, \boldsymbol{X}_{i-1}, \boldsymbol{\varepsilon}_i) = b_{ji}\ \forall\ (i,j) \tag{W4}$$

$$\sup_{i<j} \mathrm{E}[\|\boldsymbol{X}_{j-1} - \boldsymbol{X}_{i-1}\|^2 \|\boldsymbol{\varepsilon}_j - \boldsymbol{\varepsilon}_i\|^{-1}]^{1+\delta} < \infty \text{ for some } \delta > 0 \tag{E1}$$

$$\sup_{i<j} \mathrm{E}[b_{ij}\|\boldsymbol{X}_{j-1} - \boldsymbol{X}_{i-1}\|^2 \|\boldsymbol{\varepsilon}_j - \boldsymbol{\varepsilon}_i\|^{-1}]^{1+\delta} < \infty \text{ for some } \delta > 0 \tag{E2}$$

$$\sup_{i<j} \mathrm{E}[\|\boldsymbol{X}_{j-1} - \boldsymbol{X}_{i-1}\|]^{1+\delta} < \infty \text{ for some } \delta > 0 \tag{E3}$$

$$\sup_{i<j} \mathrm{E}[b_{ij}\|\boldsymbol{X}_{j-1} - \boldsymbol{X}_{i-1}\|]^{2+\delta} < \infty \text{ for some } \delta > 0 \tag{E4}$$

We note that all of the above expectations are essentially taken with respect to the underlying error distribution, F. For instance, in the context of Sect. 13.5, the expectations are taken with respect to the core process (i.e. $\gamma = 0$). In addition, the expectations given in (E1), (E2), and (E3) must also be finite when taken with respect to the corresponding product distribution.

Since the main result to obtain is the asymptotic distribution of $\hat{\boldsymbol{\beta}}_n$, we start by denoting the true parameter vector for the VAR$_m(p)$ as $\boldsymbol{\beta}_0$. Following traditional methods of proof, we first need to establish asymptotic linearity (AL), asymptotic uniform linearity (AUL), and asymptotic uniform quadraticity (AUQ). Thus, we define

$$\begin{aligned}
\boldsymbol{S}(\boldsymbol{\beta}) &= -\nabla\, \mathrm{D_{WW}}(\boldsymbol{\beta}),\\
\mathrm{D_n}(\boldsymbol{\Delta}) &= n^{-1}\, \mathrm{D_{WW}}(\boldsymbol{\beta}_0 + n^{-\frac{1}{2}}\boldsymbol{\Delta}),\\
\boldsymbol{S}_{\mathrm{n}}(\boldsymbol{\Delta}) &= -\frac{\partial}{\partial \boldsymbol{\Delta}}\mathrm{D_n}(\boldsymbol{\Delta}) = n^{-\frac{3}{2}}\, \boldsymbol{S}(\boldsymbol{\beta}_0 + n^{-\frac{1}{2}}\boldsymbol{\Delta}), \text{ and}\\
\mathrm{Q_n}(\boldsymbol{\Delta}) &= \mathrm{D_n}(\mathbf{0}) - \boldsymbol{\Delta}'\boldsymbol{S}_{\mathrm{n}}(\mathbf{0}) + \boldsymbol{\Delta}'\boldsymbol{C}\boldsymbol{\Delta},
\end{aligned}$$

where ∇ is the gradient operator, $\boldsymbol{\Delta} \in \mathbb{R}^{m^2p}$ is arbitrary but fixed, and $\boldsymbol{C}$ is a fixed $m^2p \times m^2p$ positive definite matrix.

In the multivariate setting, AL, AUL and AUQ refer to

$$\begin{aligned}
\mathrm{AL}: &\quad \| \boldsymbol{S}_{\mathrm{n}}(\boldsymbol{\Delta}) - [\boldsymbol{S}_{\mathrm{n}}(\mathbf{0}) - 2\boldsymbol{C}\boldsymbol{\Delta}]\| = o_p(1),\\
\mathrm{AUL}: &\quad \sup_{\|\boldsymbol{\Delta}\|\le c} \| \boldsymbol{S}_{\mathrm{n}}(\boldsymbol{\Delta}) - [\boldsymbol{S}_{\mathrm{n}}(\mathbf{0}) - 2\boldsymbol{C}\boldsymbol{\Delta}]\| = o_p(1)\ \forall\, c>0, \text{ and}\\
\mathrm{AUQ}: &\quad \sup_{\|\boldsymbol{\Delta}\|\le c} |\mathrm{D_n}(\boldsymbol{\Delta}) - \mathrm{Q_n}(\boldsymbol{\Delta})| = o_p(1)\ \forall\, c>0.
\end{aligned}$$

Theorem 13.3.1 establishes these results and its proof is provided in Sect. 13.4.

Theorem 13.3.1. *Let* $\mathbb{E}_{\mathrm{i,j}}[\cdot]$ *denote the expectation with respect to the product distribution of* $(\boldsymbol{\varepsilon}_i', \boldsymbol{x}_{i-1}')'$ *and* $(\boldsymbol{\varepsilon}_j', \boldsymbol{x}_{j-1}')'$, $\boldsymbol{J}[\cdot]$ *the Jacobian operator,*

$$\begin{aligned}
\boldsymbol{u}_{\mathrm{ij}}(\boldsymbol{x}) &= \frac{\boldsymbol{\varepsilon}_j - \boldsymbol{\varepsilon}_i - \boldsymbol{x}}{\|\boldsymbol{\varepsilon}_j - \boldsymbol{\varepsilon}_i - \boldsymbol{x}\|}, \text{ and}\\
\boldsymbol{C} &= \frac{1}{4}\mathbb{E}_{\mathrm{i,j}}[b_{ij}(\boldsymbol{x}_{j-1} - \boldsymbol{x}_{i-1})\boldsymbol{J}[-\boldsymbol{u}_{\mathrm{ij}}(\mathbf{0})](\boldsymbol{x}_{j-1} - \boldsymbol{x}_{i-1})'].
\end{aligned}$$

Then, AL, AUL, *and* AUQ *hold under model assumptions* (13.1)–(13.3), (W1)–(W4), *and* (E1)–(E2).

With AUL and AUQ established, we proceed to derive the asymptotic distribution of $\boldsymbol{S}_{\mathrm{n}}(\mathbf{0})$. Theorem 13.3.2 establishes this result and its proof is provided in Sect. 13.4.

Theorem 13.3.2. *Let* $\mathbb{E}_{\rm i.j.k}[\cdot]$ *denote the expectation with respect to the product distribution of* $(\boldsymbol{\varepsilon}_i', \boldsymbol{x}_{i-1}')'$, $(\boldsymbol{\varepsilon}_j', \boldsymbol{x}_{j-1}')'$, *and* $(\boldsymbol{\varepsilon}_k', \boldsymbol{x}_{k-1}')'$,

$$\boldsymbol{u}(\boldsymbol{x}) = \frac{\boldsymbol{x}}{\|\boldsymbol{x}\|}, \text{ and}$$

$$\boldsymbol{\Sigma} = \mathbb{E}_{\rm i.j.k}[b_{ij}b_{ik}(\boldsymbol{x}_{j-1} - \boldsymbol{x}_{i-1})\,\boldsymbol{u}(\boldsymbol{\varepsilon}_j - \boldsymbol{\varepsilon}_i)\,\boldsymbol{u}(\boldsymbol{\varepsilon}_k - \boldsymbol{\varepsilon}_i)'(\boldsymbol{x}_{k-1} - \boldsymbol{x}_{i-1})'].$$

Then, $\boldsymbol{S}_{\rm n}(\boldsymbol{0}) \xrightarrow{D} \mathrm{N}_{\rm m^2p}(\boldsymbol{0}, \boldsymbol{\Sigma})$ *under model assumptions* (13.1)–(13.3), (W1)–(W4), *and* (E3)–(E4).

Now, by combining Theorems 13.3.1 and 13.3.2 with the Jaeckel (1972) convexity argument, we proceed to state our main result. Theorem 13.3.3 establishes this result and its proof is provided in Sect. 13.4.

Theorem 13.3.3. *Under the assumptions of Theorems* 13.3.1 *and* 13.3.2, *we have that*

$$n^{\frac{1}{2}}(\hat{\boldsymbol{\beta}}_n - \boldsymbol{\beta}_0) \xrightarrow{D} \mathrm{N}_{\rm m^2p}\left(\boldsymbol{0}, \frac{1}{4}\boldsymbol{C}^{-1}\boldsymbol{\Sigma}\boldsymbol{C}^{-1}\right).$$

For practicality of Theorem 13.3.3, estimates of $\boldsymbol{\Sigma}$ and $\boldsymbol{C}$ are needed. Note that $\boldsymbol{\Sigma}$ and $\boldsymbol{C}$ are actually functionals defined on the distribution function of $\boldsymbol{Z}_t = (\boldsymbol{\varepsilon}_t', \boldsymbol{X}_{t-1}')'$. Naturally then, we let $\hat{\boldsymbol{\Sigma}}_n$ and $\hat{\boldsymbol{C}}_n$ be the corresponding von Mises statistics where the $\boldsymbol{\varepsilon}_t$ are replaced with corresponding residuals.

13.4 Technical Details

The expression for $\boldsymbol{S}(\boldsymbol{\beta})$ is common to several proofs and is obtained by using rules of derivatives with respect to vectors and the definition of $\mathrm{D_{WW}}(\boldsymbol{\beta})$. Starting from the definition of $\boldsymbol{S}(\boldsymbol{\beta})$, we have that

$$\begin{aligned}
\boldsymbol{S}(\boldsymbol{\beta}) &= -\nabla\,\mathrm{D_{WW}}(\boldsymbol{\beta}) = -\frac{\partial}{\partial\boldsymbol{\beta}}\mathrm{D_{WW}}(\boldsymbol{\beta}) = -\frac{\partial}{\partial\boldsymbol{\beta}}\sum_{i<j}^{n} b_{ij}\|\,\boldsymbol{\varepsilon}_{\rm j}(\boldsymbol{\beta}) - \boldsymbol{\varepsilon}_{\rm i}(\boldsymbol{\beta})\| \\
&= -\frac{1}{2}\sum_{i<j}^{n} b_{ij}\|\,\boldsymbol{\varepsilon}_{\rm j}(\boldsymbol{\beta}) - \boldsymbol{\varepsilon}_{\rm i}(\boldsymbol{\beta})\|^{-1}\frac{\partial}{\partial\boldsymbol{\beta}}(\boldsymbol{\varepsilon}_{\rm j}(\boldsymbol{\beta}) - \boldsymbol{\varepsilon}_{\rm i}(\boldsymbol{\beta}))'(\boldsymbol{\varepsilon}_{\rm j}(\boldsymbol{\beta}) - \boldsymbol{\varepsilon}_{\rm i}(\boldsymbol{\beta})) \\
&= \sum_{i<j}^{n} b_{ij}\|\,\boldsymbol{\varepsilon}_{\rm j}(\boldsymbol{\beta}) - \boldsymbol{\varepsilon}_{\rm i}(\boldsymbol{\beta})\|^{-1}(\boldsymbol{x}_{j-1} - \boldsymbol{x}_{i-1})(\boldsymbol{\varepsilon}_{\rm j}(\boldsymbol{\beta}) - \boldsymbol{\varepsilon}_{\rm i}(\boldsymbol{\beta})) \\
&\stackrel{\text{def}}{=} \sum_{i<j}^{n} b_{ij}(\boldsymbol{x}_{j-1} - \boldsymbol{x}_{i-1})\,\boldsymbol{u}(\boldsymbol{\varepsilon}_{\rm j}(\boldsymbol{\beta}) - \boldsymbol{\varepsilon}_{\rm i}(\boldsymbol{\beta})),
\end{aligned}$$

where $\boldsymbol{u}(\boldsymbol{x}) = \boldsymbol{x}/\|\boldsymbol{x}\|$.

Establishing asymptotic results is simplified under the assumption that b_{ij} has no additional stochastic components besides $\boldsymbol{X}_{i-1}$, $\boldsymbol{\varepsilon}_i$, $\boldsymbol{X}_{j-1}$, and $\boldsymbol{\varepsilon}_j$. Recall that $\boldsymbol{\theta}$ denotes a vector of parameters used in the calculation of the weights. We also denote $\boldsymbol{\theta}_0$ and $\hat{\boldsymbol{\theta}}$ as the corresponding vector of true parameters and estimators, respectively. Furthermore, we denote $\mathrm{b_{ij}}(\hat{\boldsymbol{\theta}})$ as a function of $\boldsymbol{X}_{i-1}$, $\boldsymbol{X}_{j-1}$, $\hat{\boldsymbol{\varepsilon}}_i$, and $\hat{\boldsymbol{\varepsilon}}_j$ from an initial fit, and other stochastic components. Similarly, we denote $\mathrm{b_{ij}}(\boldsymbol{\theta}_0)$ as a function of $\boldsymbol{X}_{i-1}$, $\boldsymbol{\varepsilon}_i$, $\boldsymbol{X}_{j-1}$, $\boldsymbol{\varepsilon}_j$, and no other stochastic components. It can be shown that, under model assumptions (13.1)–(13.3), (W1)–(W3), and (E1), AL can be established using $b_{ij} = \mathrm{b_{ij}}(\boldsymbol{\theta}_0)$. Analogously, it can be shown that, under model assumptions (13.1)–(13.3), (W1)–(W4), and (E3), the asymptotic distribution of $\boldsymbol{S}_\mathrm{n}(\boldsymbol{0})$ can be obtained using $b_{ij} = \mathrm{b_{ij}}(\boldsymbol{\theta}_0)$. The details are similar to those appearing in Terpstra and Rao (2001), and are therefore omitted for the sake of brevity.

13.4.1 Proof of Theorem 13.3.1

Heiler and Willers (1988) have shown that AL, AUL, and AUQ are equivalent in the context of linear regression. Their proof implies that linearity in parameters of the regression model and convexity of the dispersion function are sufficient conditions for this result to hold. Since the $\mathrm{VAR}_m(p)$ and $\mathrm{D_{WW}}(\cdot)$ satisfy these conditions, it suffices to establish AL.

To begin, let $\boldsymbol{\lambda} \in \mathbb{R}^{m^2p}$ be arbitrary but fixed and $T_n = \boldsymbol{\lambda}'(\boldsymbol{S}_\mathrm{n}(\boldsymbol{\Delta}) - \boldsymbol{S}_\mathrm{n}(\boldsymbol{0}))$. Thus, it suffices to show that $T_n + 2\boldsymbol{\lambda}'\boldsymbol{C}\boldsymbol{\Delta} = o_p(1)$. Then, it follows by definitions that

$$\begin{aligned} T_n &= \boldsymbol{\lambda}'(\boldsymbol{S}_\mathrm{n}(\boldsymbol{\Delta}) - \boldsymbol{S}_\mathrm{n}(\boldsymbol{0})) \\ &= n^{-\frac{3}{2}} \sum_{i<j}^{n} b_{ij}\boldsymbol{\lambda}'(\boldsymbol{x}_{j-1} - \boldsymbol{x}_{i-1})(\boldsymbol{u}(\boldsymbol{\varepsilon}_j - \boldsymbol{\varepsilon}_i - \boldsymbol{d}_{ijn}) - \boldsymbol{u}(\boldsymbol{\varepsilon}_j - \boldsymbol{\varepsilon}_i)) \\ &\stackrel{\text{def}}{=} n^{-\frac{3}{2}} \sum_{i<j}^{n} b_{ij}\boldsymbol{\lambda}'(\boldsymbol{x}_{j-1} - \boldsymbol{x}_{i-1})(\boldsymbol{u}_{\mathrm{ij}}(\boldsymbol{d}_{ijn}) - \boldsymbol{u}_{\mathrm{ij}}(\boldsymbol{0})), \end{aligned}$$

where $\boldsymbol{d}_{ijn} = (\boldsymbol{x}_{j-1} - \boldsymbol{x}_{i-1})'n^{-\frac{1}{2}}\boldsymbol{\Delta}$ and $\boldsymbol{u}_{\mathrm{ij}}(\boldsymbol{x}) = \boldsymbol{u}(\boldsymbol{\varepsilon}_j - \boldsymbol{\varepsilon}_i - \boldsymbol{x})$. Furthermore, we let

$$\begin{aligned} U_n^{\mathrm{AL}} &= n^{-\frac{3}{2}} \sum_{i<j}^{n} b_{ij}\boldsymbol{\lambda}'(\boldsymbol{x}_{j-1} - \boldsymbol{x}_{i-1})\boldsymbol{J}[\boldsymbol{u}_{\mathrm{ij}}(\boldsymbol{0})]\boldsymbol{d}_{ijn} \\ &= \left(\frac{n-1}{n}\right)\left(\frac{2}{n(n-1)}\right)\sum_{i<j}^{n} 2^{-1} b_{ij}\boldsymbol{\lambda}'(\boldsymbol{x}_{j-1} - \boldsymbol{x}_{i-1})\boldsymbol{J}[\boldsymbol{u}_{\mathrm{ij}}(\boldsymbol{0})](\boldsymbol{x}_{j-1} - \boldsymbol{x}_{i-1})'\boldsymbol{\Delta} \\ &\stackrel{\text{def}}{=} \left(\frac{n-1}{n}\right)\left(\frac{2}{n(n-1)}\right)\sum_{i<j}^{n} \mathrm{h_{AL}}(\boldsymbol{Z}_i, \boldsymbol{Z}_j), \end{aligned}$$

where $\boldsymbol{J}[\cdot]$ is the Jacobian operator and $\boldsymbol{Z}_t = (\boldsymbol{\varepsilon}_t', \boldsymbol{X}_{t-1}')'$.

Next, note that

$$
\begin{aligned}
|T_n - U_n^{\mathrm{AL}}| \\
&\leq n^{-\frac{3}{2}} \sum_{i<j}^{n} |b_{ij}\boldsymbol{\lambda}'(\boldsymbol{x}_{j-1} - \boldsymbol{x}_{i-1})(\boldsymbol{u}_{\mathrm{ij}}(\boldsymbol{d}_{ijn}) - \boldsymbol{u}_{\mathrm{ij}}(\mathbf{0}) - \boldsymbol{J}[\boldsymbol{u}_{\mathrm{ij}}(\mathbf{0})]\boldsymbol{d}_{ijn})| \\
&\leq n^{-\frac{3}{2}} \|\boldsymbol{\lambda}\| \sum_{i<j}^{n} b_{ij} \|\boldsymbol{x}_{j-1} - \boldsymbol{x}_{i-1}\| \, \| \boldsymbol{u}_{\mathrm{ij}}(\boldsymbol{d}_{ijn}) - \boldsymbol{u}_{\mathrm{ij}}(\mathbf{0}) - \boldsymbol{J}[\boldsymbol{u}_{\mathrm{ij}}(\mathbf{0})]\boldsymbol{d}_{ijn}\| \\
&= n^{-\frac{3}{2}} \|\boldsymbol{\lambda}\| \sum_{i<j}^{n} b_{ij} \|\boldsymbol{x}_{j-1} - \boldsymbol{x}_{i-1}\| \, \|\boldsymbol{d}_{ijn}\| \, \|\boldsymbol{d}_{ijn}\|^{-1} \| \boldsymbol{u}_{\mathrm{ij}}(\boldsymbol{d}_{ijn}) - \boldsymbol{u}_{\mathrm{ij}}(\mathbf{0}) - \boldsymbol{J}[\boldsymbol{u}_{\mathrm{ij}}(\mathbf{0})]\boldsymbol{d}_{ijn}\| \\
&\stackrel{\text{def}}{=} n^{-\frac{3}{2}} \|\boldsymbol{\lambda}\| \sum_{i<j}^{n} b_{ij} \|\boldsymbol{x}_{j-1} - \boldsymbol{x}_{i-1}\| \, \|\boldsymbol{d}_{ijn}\| \eta(\boldsymbol{d}_{ijn}) \\
&\leq n^{-2} m \|\boldsymbol{\lambda}\| \, \|\boldsymbol{\Delta}\| \sum_{i<j}^{n} b_{ij} \|\boldsymbol{X}_{j-1} - \boldsymbol{X}_{i-1}\|^2 \eta(\boldsymbol{d}_{ijn}).
\end{aligned}
$$

By taking expectations, and using assumption (E2) along with Lemma 2 (ii) of Arcones (1998), we have that

$$
\begin{aligned}
\mathrm{E}[|T_n - U_n^{\mathrm{AL}}|] &\leq n^{-2} m \|\boldsymbol{\lambda}\| \, \|\boldsymbol{\Delta}\| \sum_{i<j}^{n} \mathrm{E}[b_{ij} \|\boldsymbol{X}_{j-1} - \boldsymbol{X}_{i-1}\|^2 \eta(\boldsymbol{d}_{ijn})] \\
&\leq n^{-2} m \|\boldsymbol{\lambda}\| \, \|\boldsymbol{\Delta}\| \sum_{i<j}^{n} \mathbb{E}_{\mathrm{i,j}}[b_{ij} \|\boldsymbol{X}_{j-1} - \boldsymbol{X}_{i-1}\|^2 \eta(\boldsymbol{d}_{ijn})] + o(1) \\
&= m \|\boldsymbol{\lambda}\| \, \|\boldsymbol{\Delta}\| \left(\frac{n(n-1)}{2n^2} \right) \mathbb{E}_{\mathrm{i,j}}[b_{ij} \|\boldsymbol{X}_{j-1} - \boldsymbol{X}_{i-1}\|^2 \eta(\boldsymbol{d}_{ijn})] + o(1).
\end{aligned}
$$

By model assumptions (13.1)–(13.3) and the definition of the multivariate derivative, we have that $\boldsymbol{d}_{ijn} = \boldsymbol{o}_p(1)$ and $\eta(\boldsymbol{d}_{ijn}) = o_p(1)$. Furthermore, it can be shown that $\eta(\boldsymbol{d}_{ijn}) \leq (m^{\frac{1}{2}} + 1)\|\boldsymbol{\varepsilon}_j - \boldsymbol{\varepsilon}_i\|^{-1}$. Thus, by assumption (E2) and the Lebesgue Dominated Convergence Theorem, the expectation on the right hand side (RHS) is $o(1)$. Note that the remaining factor on the RHS is $O(1)$, which implies that $\mathrm{E}[|T_n - U_n^{\mathrm{AL}}|] = o(1)$. Using Markov's Inequality, it now follows that $T_n - U_n^{\mathrm{AL}} = o_p(1)$.

Therefore, it suffices to show that $U_n^{AL} + 2\boldsymbol{\lambda}'\boldsymbol{C}\boldsymbol{\Delta} = o_p(1)$. However, under assumption (W4), $(\frac{n}{n-1})U_n^{\mathrm{AL}}$ is a U-statistic with symmetric kernel $\mathrm{h}_{\mathrm{AL}}(\boldsymbol{Z}_i, \boldsymbol{Z}_j)$. Hence, it follows from assumption (E2) and Theorem 1 (ii) of Arcones (1998) that $U_n^{\mathrm{AL}} - \mathbb{E}_{\mathrm{i,j}}[\mathrm{h}_{\mathrm{AL}}(\boldsymbol{Z}_i, \boldsymbol{Z}_j)] = o_p(1)$.

Finally, we have that

$$\begin{aligned}\mathbb{E}_{i,j}[h_{AL}(\mathbf{Z}_i,\mathbf{Z}_j)] &= \mathbb{E}_{i,j}[2^{-1}b_{ij}\boldsymbol{\lambda}'(\boldsymbol{x}_{j-1}-\boldsymbol{x}_{i-1})\boldsymbol{J}[\boldsymbol{u}_{ij}(\mathbf{0})](\boldsymbol{x}_{j-1}-\boldsymbol{x}_{i-1})'\boldsymbol{\Delta}] \\ &= -2\boldsymbol{\lambda}'\left(\frac{1}{4}\mathbb{E}_{i,j}[b_{ij}(\boldsymbol{x}_{j-1}-\boldsymbol{x}_{i-1})\boldsymbol{J}[-\boldsymbol{u}_{ij}(\mathbf{0})](\boldsymbol{x}_{j-1}-\boldsymbol{x}_{i-1})']\right)\boldsymbol{\Delta} \\ &= -2\boldsymbol{\lambda}'\boldsymbol{C}\boldsymbol{\Delta},\end{aligned}$$

which completes our proof. □

13.4.2 *Proof of Theorem 13.3.2*

To obtain the asymptotic distribution of $\boldsymbol{S}_n(\mathbf{0})$, we follow the approach for U-statistics under absolute regularity proposed by Denker and Keller (1983). To begin, let $\boldsymbol{\lambda} \in \mathbb{R}^{m^2p}$ be arbitrary but fixed. Then, it follows from definitions that

$$\begin{aligned}\boldsymbol{\lambda}'\boldsymbol{S}_n(\mathbf{0}) &= n^{-\frac{3}{2}}\sum_{i<j}^{n} b_{ij}\boldsymbol{\lambda}'(\boldsymbol{x}_{j-1}-\boldsymbol{x}_{i-1})\,\boldsymbol{u}(\boldsymbol{\varepsilon}_j-\boldsymbol{\varepsilon}_i) \\ &= \left(\frac{n-1}{n}\right)n^{\frac{1}{2}}\left(\frac{2}{n(n-1)}\right)\sum_{i<j}^{n}2^{-1}b_{ij}\boldsymbol{\lambda}'(\boldsymbol{x}_{j-1}-\boldsymbol{x}_{i-1})\,\boldsymbol{u}(\boldsymbol{\varepsilon}_j-\boldsymbol{\varepsilon}_i) \\ &\stackrel{\text{def}}{=} \left(\frac{n-1}{n}\right)n^{\frac{1}{2}}U_n,\end{aligned}$$

where, under assumption (W4), U_n is a U-statistic with symmetric kernel $h(\mathbf{Z}_i,\mathbf{Z}_j) = 2^{-1}b_{ij}\boldsymbol{\lambda}'(\boldsymbol{x}_{j-1}-\boldsymbol{x}_{i-1})\,\boldsymbol{u}(\boldsymbol{\varepsilon}_j-\boldsymbol{\varepsilon}_i)$ and $\mathbf{Z}_t = (\boldsymbol{\varepsilon}_t', \boldsymbol{X}_{t-1}')'$. Next, we obtain

$$\begin{aligned}&\mathbb{E}_{i,j}[h(\mathbf{Z}_i,\mathbf{Z}_j)] = \\ &\iint 2^{-1}\boldsymbol{\lambda}'(\boldsymbol{x}_{j-1}-\boldsymbol{x}_{i-1})\left(\iint b_{ij}\,\boldsymbol{u}(\boldsymbol{\varepsilon}_j-\boldsymbol{\varepsilon}_i)\,dF(\boldsymbol{\varepsilon}_i)\,dF(\boldsymbol{\varepsilon}_j)\right)dG(\boldsymbol{X}_{i-1})\,dG(\boldsymbol{X}_{j-1}),\end{aligned}$$

where $G(\cdot)$ denotes the distribution function of $\boldsymbol{X}_t$. Note that, by definition of $\boldsymbol{u}(\cdot)$, and assumption (W4), it follows that

$$\iint b_{ij}\,\boldsymbol{u}(\boldsymbol{\varepsilon}_j-\boldsymbol{\varepsilon}_i)\,dF(\boldsymbol{\varepsilon}_i)\,dF(\boldsymbol{\varepsilon}_j) = -\iint b_{ij}\,\boldsymbol{u}(\boldsymbol{\varepsilon}_j-\boldsymbol{\varepsilon}_i)\,dF(\boldsymbol{\varepsilon}_i)\,dF(\boldsymbol{\varepsilon}_j),$$

$$\iint b_{ij}\,\boldsymbol{u}(\boldsymbol{\varepsilon}_j-\boldsymbol{\varepsilon}_i)\,dF(\boldsymbol{\varepsilon}_i)\,dF(\boldsymbol{\varepsilon}_j) = 0, \text{ and } \mathbb{E}_{i,j}[h(\mathbf{Z}_i,\mathbf{Z}_j)] = 0. \qquad (13.11)$$

Next, following the approach, we let

$$\mathrm{h}_1(\mathbf{Z}_i) = \iint \mathrm{h}(\mathbf{Z}_i, \mathbf{Z}_j)\, \mathrm{dF}(\boldsymbol{\varepsilon}_j)\, \mathrm{dG}(\boldsymbol{X}_{j-1}) \text{ and}$$

$$\sigma_n^2 = \mathrm{E}\left[\sum_{i=1}^{n} \mathrm{h}_1(\mathbf{Z}_i)\right]^2.$$

By model assumptions (13.1) and (13.2), we have

$$\begin{aligned}\sigma_n^2 &= \mathrm{E}\left[\sum_{i=1}^{n} \mathrm{h}_1(\mathbf{Z}_i)\right]^2 \\ &= n\,\mathrm{E}[\mathrm{h}_1^2(\mathbf{Z}_i)] + 2\sum_{k=2}^{n}(n-(k-1))\,\mathrm{E}[\mathrm{h}_1(\mathbf{Z}_1)\,\mathrm{h}_1(\mathbf{Z}_k)]. \end{aligned} \tag{13.12}$$

Focusing attention on the expectation inside the summation we have that

$$\begin{aligned}&\mathrm{E}[\mathrm{h}_1(\mathbf{Z}_1)\,\mathrm{h}_1(\mathbf{Z}_k)] \\ &= \int \mathrm{h}_1(\mathbf{Z}_1)\,\mathrm{h}_1(\mathbf{Z}_k)\,\mathrm{dH_k} \\ &= \int \mathrm{h}_1(\mathbf{Z}_1)\left(\iint 2^{-1} b_{kj}\boldsymbol{\lambda}'(\boldsymbol{x}_{j-1}-\boldsymbol{x}_{k-1})\,\boldsymbol{u}(\boldsymbol{\varepsilon}_j-\boldsymbol{\varepsilon}_k)\,\mathrm{dF}(\boldsymbol{\varepsilon}_j)\,\mathrm{dG}(\boldsymbol{X}_{j-1})\right)\mathrm{dH_k} \\ &= \iint \mathrm{h}_1(\mathbf{Z}_1)\left(\iint 2^{-1} b_{kj}\boldsymbol{\lambda}'(\boldsymbol{x}_{j-1}-\boldsymbol{x}_{k-1})\,\boldsymbol{u}(\boldsymbol{\varepsilon}_j-\boldsymbol{\varepsilon}_k)\,\mathrm{dF}(\boldsymbol{\varepsilon}_j)\,\mathrm{dG}(\boldsymbol{X}_{j-1})\right)\mathrm{d\tilde{H}_k}\,\mathrm{dF} \\ &= \int \mathrm{h}_1(\mathbf{Z}_1)\int 2^{-1}\boldsymbol{\lambda}'(\boldsymbol{x}_{j-1}-\boldsymbol{x}_{k-1})\left(\iint b_{kj}\,\boldsymbol{u}(\boldsymbol{\varepsilon}_j-\boldsymbol{\varepsilon}_k)\,\mathrm{dF}(\boldsymbol{\varepsilon}_k)\,\mathrm{dF}(\boldsymbol{\varepsilon}_j)\right)\mathrm{dG}\,\mathrm{d\tilde{H}_k},\end{aligned}$$

where $\mathrm{H_k}(\cdot)$ denotes the joint distribution function of $\mathbf{Z}_1$ and $\mathbf{Z}_k$, and $\mathrm{\tilde{H}_k}(\cdot)$ denotes the joint distribution function of $\mathbf{Z}_1$ and $\boldsymbol{X}_{k-1}$. It now follows from (13.11) that $\mathrm{E}[\mathrm{h}_1(\mathbf{Z}_1)\,\mathrm{h}_1(\mathbf{Z}_k)] = 0 \;\forall\; k \in \{2, 3, \ldots, n\}$. Next, we focus attention back to Eq. (13.12), and have that

$$\begin{aligned} n^{-1}\sigma_n^2 &= \mathrm{E}[\mathrm{h}_1^2(\mathbf{Z}_i)] \\ &= 4^{-1}\boldsymbol{\lambda}'\,\mathbb{E}_{\mathrm{i.j.k}}[b_{ij}b_{ik}(\boldsymbol{x}_{j-1}-\boldsymbol{x}_{i-1})\,\boldsymbol{u}(\boldsymbol{\varepsilon}_j-\boldsymbol{\varepsilon}_i)\,\boldsymbol{u}(\boldsymbol{\varepsilon}_k-\boldsymbol{\varepsilon}_i)'(\boldsymbol{x}_{k-1}-\boldsymbol{x}_{i-1})']\boldsymbol{\lambda} \\ &= 4^{-1}\boldsymbol{\lambda}'\boldsymbol{\Sigma}\boldsymbol{\lambda}. \end{aligned}$$

Thus, by g.a.r. of $\{\mathbf{Z}_t\}$ and assumption (E4), it follows from Theorem 1 part (c) of Denker and Keller (1983, p. 507) that $n^{\frac{1}{2}}U_n \xrightarrow{D} \mathrm{N}(0, \boldsymbol{\lambda}'\boldsymbol{\Sigma}\boldsymbol{\lambda})$. Finally, since $\boldsymbol{\lambda}$ is arbitrary, it follows that $\boldsymbol{S}_{\mathrm{n}}(\mathbf{0}) \xrightarrow{D} \mathrm{N}_{\mathrm{m^2p}}(\mathbf{0}, \boldsymbol{\Sigma})$, which completes our proof. □

13.4.3 Proof of Theorem 13.3.3

To obtain the asymptotic distribution of $\hat{\boldsymbol{\beta}}_n$, we follow the approach for regression coefficients by minimization of dispersion functions proposed by Jaeckel (1972). To begin, let $\boldsymbol{\Delta} = n^{\frac{1}{2}}(\boldsymbol{\beta} - \boldsymbol{\beta}_0)$ and $\tilde{\boldsymbol{\Delta}}_n = \min_{\boldsymbol{\Delta}} \mathrm{Q_n}(\boldsymbol{\Delta})$. Then, by minimization of $\mathrm{Q_n}(\boldsymbol{\Delta})$ and Theorem 13.3.2, we have that

$$\begin{aligned}\tilde{\boldsymbol{\Delta}}_n &= n^{\frac{1}{2}}(\tilde{\boldsymbol{\beta}}_n - \boldsymbol{\beta}_0)\\ &= \frac{1}{2}\boldsymbol{C}^{-1}\,\boldsymbol{S}_{\mathrm{n}}(\mathbf{0}) \xrightarrow{D} \mathrm{N_{m^2p}}\left(\mathbf{0}, \frac{1}{4}\boldsymbol{C}^{-1}\boldsymbol{\Sigma}\boldsymbol{C}^{-1}\right).\end{aligned}$$

Next, we let $\hat{\boldsymbol{\Delta}}_n = n^{\frac{1}{2}}(\hat{\boldsymbol{\beta}}_n - \boldsymbol{\beta}_0)$. Thus, by Theorem 13.3.1 and the Jaeckel (1972) convexity argument, we have that $\hat{\boldsymbol{\Delta}}_n - \tilde{\boldsymbol{\Delta}}_n = o_p(1)$. Finally, it follows by definitions that

$$n^{\frac{1}{2}}(\hat{\boldsymbol{\beta}}_n - \boldsymbol{\beta}_0) \xrightarrow{D} \mathrm{N_{m^2p}}\left(\mathbf{0}, \frac{1}{4}\boldsymbol{C}^{-1}\boldsymbol{\Sigma}\boldsymbol{C}^{-1}\right),$$

which completes our proof. □

13.5 Monte Carlo Study

13.5.1 The Process

In this section, we study the behavior of several estimates, in particular the WW-estimates, via Monte Carlo simulation. For the sake of simplicity and computation time only the VAR$_2$(1) is considered. Thus, we define the core process as

$$\begin{aligned}\boldsymbol{Y}_t &= \boldsymbol{\Phi}_1\boldsymbol{Y}_{t-1} + \boldsymbol{\varepsilon}_t; \quad t = 0, 1, \ldots, T,\\ \boldsymbol{\varepsilon}_t &\overset{\text{iid}}{\sim} (1-\rho)\,\mathrm{N}_2(\mathbf{0}, \boldsymbol{\Sigma}_{\boldsymbol{\varepsilon}}) + \rho\,\mathrm{N}_2(\mathbf{0}, \boldsymbol{\Sigma}_{\rho}),\end{aligned}$$

where $\rho \in [0, 1)$, and $\boldsymbol{\Sigma}_{\boldsymbol{\varepsilon}}$, $\boldsymbol{\Sigma}_{\rho}$ are both positive definite. In addition, we define the observed process as

$$\begin{aligned}\boldsymbol{Y}_t^* &= \boldsymbol{Y}_t + \boldsymbol{Z}_t,\\ \boldsymbol{Z}_t &\overset{\text{iid}}{\sim} (1-\gamma)\boldsymbol{\delta}_0 + \gamma\,\mathrm{N}_2(\boldsymbol{\mu}_{\gamma}, \boldsymbol{\Sigma}_{\gamma}),\end{aligned}$$

where $\gamma \in [0, 1)$, $\boldsymbol{\delta}_0$ represents a bivariate point mass distribution at $\mathbf{0}$, $\boldsymbol{\mu}_{\gamma} \in \mathbb{R}^2$, and $\boldsymbol{\Sigma}_{\gamma}$ is positive definite.

Note that when $\gamma = 0$ and $\rho > 0$, the observed process reduces to the core process, representing Fox's (1972) Type II or Innovation Outlier (IO) model. As discussed by Rousseeuw and Leroy (1987, p. 275), this model produces *good* leverage points in the sense that they have relatively little impact on estimates. When $\rho = 0$ and $\gamma > 0$, the observed process yields Fox's Type I or Additive Outlier (AO) model. This model produces *bad* leverage points which can have a significant impact on many estimates. When $\gamma > 0$ and $\rho > 0$, the observed process corresponds to a combination of the IO and AO models, denoted I&AO for convenience. In the remaining case, when $\gamma = 0$ and $\rho = 0$, the observed process is bivariate normal.

13.5.2 The Estimators

For interpretation convenience, we group estimators according to their weighting schemes as: no weights, Mallows, and Schweppe labeled as 1-3, respectively. Furthermore, in what follows, we refer to weighted L_2-estimates, weighted L_1-estimates, and WW-estimates as those estimates obtained by minimizing the dispersion functions given in (13.4), (13.5), and (13.6), respectively. This Monte Carlo study will simulate, compute, and compare the performance of the following estimators:

OLS(1): OLS estimate based on the dispersion in (13.4) with $b_t \equiv 1$,
L1(1): L_1-estimate based on the dispersion in (13.5) with $b_t \equiv 1$,
WIL(1): Wilcoxon-estimate based on the dispersion in (13.6) with $b_{ij} \equiv 1$,
BL2(2): Weighted L_2-estimate based on the dispersion in (13.4) with b_t defined in (13.7),
BL1(2): Weighted L_1-estimate based on the dispersion in (13.5) with b_t defined in (13.7),
THL(2): WW-estimate based on the dispersion in (13.6) with b_{ij} defined in (13.7),
ML2(2): Weighted L_2-estimate based on the dispersion in (13.4) with b_t defined in (13.8),
ML1(2): Weighted L_1-estimate based on the dispersion in (13.5) with b_t defined in (13.8),
GR(2): WW-estimate based on the dispersion in (13.6) with b_{ij} defined in (13.8),
HBL2(3): Weighted L_2-estimate based on the dispersion in (13.4) with b_t defined in (13.9),
HBL1(3): Weighted L_1-estimate based on the dispersion in (13.5) with b_t defined in (13.9),
HBR(3): WW-estimate based on the dispersion in (13.6) with b_{ij} defined in (13.9),
TMNL2(3): Weighted L_2-estimate based on the dispersion in (13.4) with b_t defined in (13.10),

TMNL1(3): Weighted L_1-estimate based on the dispersion in (13.5) with b_t defined in (13.10),

TMNR(3): WW-estimate based on the dispersion in (13.6) with b_{ij} defined in (13.10).

For the estimators based on Schweppe weights, we obtained initial residuals using ML1-estimates. We used the Minimum Covariance Determinant to obtain the robust measures of center and covariance required by d(·).

13.5.3 The Simulation Settings

We selected the simulation settings to account for the impact on estimates due to the degree of stationarity of the time series, the level of contamination due to innovation outliers, the level of contamination due to additive outliers, and the magnitude of the additive outlier. With regard to stationarity, the following three coefficient matrices are considered:

$$\boldsymbol{\Phi}_1 \in \left\{ \begin{bmatrix} 0.10 & 0.03 \\ 0.01 & 0.05 \end{bmatrix}, \begin{bmatrix} 0.30 & -0.20 \\ -0.10 & 0.40 \end{bmatrix}, \begin{bmatrix} 1.20 & -0.50 \\ 0.60 & 0.30 \end{bmatrix} \right\},$$

representing three degrees of stationarity. The processes associated to these $\boldsymbol{\Phi}_1$ matrices have roots with their largest modulus at 0.1, 0.5 and 0.8, respectively. Therefore, we denoted them as *very stationary*, *moderate stationary*, and *close to non-stationary process*, respectively. Note that all of the coefficient matrices satisfy the stationarity assumption in (13.2). Furthermore, we evaluate the sensitivity of the estimates to innovative outliers by considering $\rho \in \{0, 0.05, 0.1\}$. Next, we evaluate the sensitivity of the estimates to additive outliers by considering $\gamma \in \{0, 0.05, 0.1\}$. Finally, we evaluate the impact of the magnitude of the additive outlier by considering $\boldsymbol{\mu}_\gamma \in \{(10, 13)', (100, 130)'\}$; denoted as *close* and *far* from the core process, respectively. The remaining parameters of the simulation are fixed. We generated 1000 replicates of size $T = 100$ and set

$$\boldsymbol{\Sigma}_{\boldsymbol{\varepsilon}} = \mathbb{I}_2, \quad \boldsymbol{\Sigma}_\rho = 16\,\mathbb{I}_2, \quad \text{and} \quad \boldsymbol{\Sigma}_\gamma = \mathbb{I}_2.$$

All estimates were computed using R (R Core Team 2013).

13.5.4 Simulation Results

We estimate the efficiencies of estimators based on the trace of the empirical mean square error (MSE) matrix. For comparison purpose, results have the actual trace of the empirical MSE matrix in the OLS entry, while the other estimators are presented relative to OLS so that entries larger than 1 indicate a more efficient estimator.

Table 13.1 shows results for the *very stationary* setting. Under multivariate normality, the OLS estimates prove to be the most efficient followed by the Wilcoxon and the HBR estimates at 0.96 efficiency. Under the AO model, the OLS estimates deteriorate and are out-performed by all estimates based on robust dispersions. In particular, the Wilcoxon, Theil, and HBR estimates show the best all-around performance. Under the IO model, the OLS estimate is resistant, yet the robust dispersions perform better. Unweighted dispersions perform best, followed by Schweppe weighted estimates, and last are the Mallows weighted estimates. Under the I&AO model, the AO effect is mitigated, yet the OLS estimates performance drops. Similar to the AO model, the estimates based on robust dispersions have an overall better performance when compared to the OLS estimates. In particular, the Wilcoxon, HBR, Theil, and L_1 estimates show the best overall performance.

Tables 13.2 and 13.3 show results for the *moderate stationary* and *close to non-stationary* settings, respectively. Results remain similar to those under the *very stationary* setting. In particular, HBR estimates show the best all-around performance under the *moderate stationary*. On the other hand, under the *close to non-stationary* setting, the HBR estimates have a blind spot when additive outliers that are *close* to the core process are present. Under the *close to non-stationary* setting with additive outliers that are *close* to the core process, the HBL1 has the best performance. That said, these simulation results suggest that Schweppe weighted estimators can achieve comparable efficiencies across different degrees of stationarity and the presence of outliers. Additive outliers seem to have the largest effect on estimators and are best handled by Schweppe weighted robust dispersions.

13.6 Conclusion

In this paper a Schweppe-based weighted Wilcoxon estimate for a multivariate autoregressive time series parameter was considered. In Sects. 13.3 and 13.4 asymptotic linearity properties and the asymptotic distribution of the estimate were derived. Tests of hypotheses as well as standard errors for confidence interval procedures can be formulated from such results. The Monte Carlo study in Sect. 13.5 compared several estimates that essentially fall into one of nine categories. These categories are a product of the dispersion function [i.e. (13.4), (13.5), or (13.6)] and the class of weights [i.e. none, Mallows, or Schweppe]. Generally speaking, estimates based on (13.6) outperformed analogous estimates based on (13.5) and (13.4) for time series containing outliers. Moreover, the study highlights the flexibility of Schweppe weights relative to Mallows weights. For example, under the IO models considered, the Wilcoxon [WIL(1)] estimate was the most efficient estimate, followed closely by the Schweppe-based WW-estimates [i.e. HBR(3) and TMNR(3)] and finally the Mallows-based WW-estimates [i.e. THL and GR]. On the other hand, the Schweppe and Mallows-based WW-estimates are much more efficient than the Wilcoxon estimate for AO and I&AO models, where outliers can

Table 13.1 Efficiency of estimators—very stationary $\boldsymbol{\Phi}_1$ matrix

ρ	0					0.05					0.1				
γ	0	0.05		0.1		0	0.05		0.1		0	0.05		0.1	
$\boldsymbol{\mu}_\gamma$	–	Close	Far	Close	Far	–	Close	Far	Close	Far	–	Close	Far	Close	Far
OLS(1)	0.04	0.16	13.75	0.26	24.35	0.04	0.12	8.46	0.17	15.88	0.04	0.10	5.58	0.15	10.60
L1(1)	0.77	3.93	315.41	6.59	393.40	1.10	4.05	273.39	5.37	459.20	1.50	3.89	198.38	5.41	371.37
WIL(1)	0.96	4.44	389.17	7.21	639.61	1.34	4.32	314.56	5.50	487.99	1.66	4.05	214.39	5.29	383.11
BL2(2)	0.77	0.67	0.67	0.61	0.62	0.59	0.54	0.59	0.53	0.51	0.53	0.54	0.50	0.51	0.48
BL1(2)	0.59	2.58	222.27	4.51	330.09	0.66	2.30	1.35	3.11	26.41	0.79	1.66	124.82	3.15	137.93
THL(2)	0.92	4.15	357.66	6.79	557.66	1.13	3.82	279.41	4.89	404.27	1.32	3.59	198.78	4.70	325.98
ML2(2)	0.95	0.59	0.54	0.62	0.55	0.66	0.48	0.42	0.45	0.44	0.55	0.43	0.34	0.43	0.35
ML1(2)	0.74	2.70	207.49	4.37	314.19	0.79	2.19	143.16	2.84	217.71	0.88	1.96	100.53	2.79	161.62
GR(2)	0.92	3.19	240.40	5.14	338.41	0.96	2.54	160.49	3.16	233.54	1.00	2.17	106.81	2.88	160.18
HBL2(3)	0.98	3.04	211.38	3.77	199.10	1.03	2.46	143.90	2.46	147.17	1.07	2.21	98.11	2.26	105.27
HBL1(3)	0.75	2.92	218.33	5.19	363.65	0.86	2.62	163.57	3.72	280.79	1.02	2.50	126.22	3.93	231.41
HBR(3)	0.96	4.20	281.81	7.21	466.14	1.27	4.35	236.49	5.77	399.43	1.61	4.38	197.82	5.98	347.91
TMNL2(3)	0.98	0.86	0.62	0.91	0.61	0.92	0.99	0.69	0.93	0.69	0.99	1.09	0.77	1.04	0.74
TMNL1(3)	0.75	2.96	210.25	5.07	321.78	0.92	2.87	167.97	3.87	259.80	1.19	2.90	138.77	4.18	227.67
TMNR(3)	0.94	3.47	248.75	5.88	351.36	1.13	3.33	196.40	4.32	293.78	1.39	3.30	156.90	4.50	243.40

Table 13.2 Efficiency of estimators—moderate stationary $\boldsymbol{\Phi}_1$ matrix

ρ	0					0.05					0.1				
γ	0	0.05		0.1		0	0.05		0.1		0	0.05		0.1	
$\boldsymbol{\mu}_\gamma$	–	Close	Far	Close	Far	–	Close	Far	Close	Far	–	Close	Far	Close	Far
OLS(1)	0.04	0.19	10.27	0.28	18.61	0.04	0.14	6.57	0.21	11.64	0.04	0.13	4.31	0.18	8.33
L1(1)	0.80	2.18	107.42	3.03	191.64	1.17	1.98	75.85	2.58	125.56	1.60	1.94	51.58	2.35	97.44
WIL(1)	0.98	2.30	114.21	3.08	199.83	1.39	2.04	77.98	2.59	129.29	1.76	1.96	52.45	2.32	97.88
BL2(2)	0.78	0.82	0.64	0.72	0.64	0.61	0.75	0.61	0.64	0.52	0.60	0.78	0.53	0.72	0.54
BL1(2)	0.58	1.72	5.08	2.82	200.17	0.72	2.53	101.62	2.68	126.91	0.87	2.52	44.04	2.66	2.67
THL(2)	0.94	3.29	145.81	3.82	209.65	1.21	3.02	104.61	3.18	136.90	1.48	2.97	77.31	2.98	106.26
ML2(2)	0.97	0.80	0.51	0.74	0.55	0.72	0.71	0.42	0.60	0.43	0.66	0.68	0.35	0.66	0.41
ML1(2)	0.77	3.27	173.52	4.44	277.50	0.88	2.94	127.26	3.65	183.05	1.03	2.87	97.97	3.56	157.78
GR(2)	0.94	3.82	202.23	4.87	296.78	1.05	3.30	141.11	3.75	190.60	1.18	3.06	102.94	3.52	158.51
HBL2(3)	0.99	3.49	176.42	3.44	183.99	1.11	3.02	126.08	2.68	121.07	1.24	2.80	92.33	2.60	102.28
HBL1(3)	0.78	3.38	187.06	4.62	322.73	0.94	3.20	146.24	4.19	234.44	1.19	3.40	122.82	4.09	219.09
HBR(3)	0.97	3.68	235.94	4.28	395.39	1.34	3.59	198.77	3.74	319.73	1.74	3.61	178.06	3.46	285.92
TMNL2(3)	0.99	0.96	0.57	0.91	0.61	0.98	1.11	0.66	0.94	0.63	1.10	1.26	0.68	1.12	0.73
TMNL1(3)	0.78	2.96	175.74	3.89	285.44	1.01	2.81	148.00	3.32	216.15	1.35	2.91	130.58	3.22	212.13
TMNR(3)	0.96	3.41	206.48	4.27	308.45	1.23	3.23	168.63	3.58	236.78	1.57	3.29	144.71	3.47	223.95

Table 13.3 Efficiency of estimators—close to non-stationary $\boldsymbol{\Phi}_1$ matrix

ρ	0					0.05					0.1				
γ	0	0.05		0.1		0	0.05		0.1		0	0.05		0.1	
$\boldsymbol{\mu}_\gamma$	–	Close	Far	Close	Far	–	Close	Far	Close	Far	–	Close	Far	Close	Far
OLS(1)	0.01	0.32	6.32	0.52	11.48	0.01	0.22	4.80	0.38	7.80	0.01	0.16	3.34	0.28	5.54
L1(1)	0.77	3.67	8.02	2.03	13.70	1.16	4.91	6.11	2.42	9.53	1.58	5.80	4.40	3.12	6.83
WIL(1)	0.96	2.12	8.02	1.48	14.71	1.36	2.48	6.03	1.50	9.63	1.73	2.92	4.31	1.67	6.88
BL2(2)	0.77	1.92	0.75	1.52	0.67	0.68	1.80	0.69	1.48	0.64	0.67	1.53	0.67	1.42	0.60
BL1(2)	0.57	10.37	196.09	11.16	243.38	0.78	7.75	171.46	9.61	3.09	1.00	8.28	172.94	8.54	174.57
THL(2)	0.91	10.93	199.32	6.74	137.39	1.27	9.69	193.47	6.11	120.55	1.62	8.44	171.89	5.79	92.10
ML2(2)	0.95	2.49	0.84	1.80	0.75	0.84	1.97	0.81	1.56	0.71	0.83	1.62	0.76	1.41	0.69
ML1(2)	0.73	14.45	293.41	15.12	424.57	0.98	11.95	298.08	12.35	379.81	1.27	10.53	255.57	10.40	352.86
GR(2)	0.91	14.64	335.56	10.80	464.84	1.17	10.96	324.43	7.16	380.53	1.46	8.70	265.76	5.44	342.76
HBL2(3)	0.98	11.77	298.47	6.88	288.07	1.24	8.83	272.92	5.03	238.50	1.47	6.58	219.36	3.81	198.46
HBL1(3)	0.75	16.56	314.24	20.21	507.43	1.07	14.98	338.87	18.56	482.83	1.44	13.55	295.97	14.14	454.68
HBR(3)	0.96	12.11	386.91	5.13	622.49	1.38	10.45	424.09	3.89	594.39	1.80	8.80	369.76	3.49	535.52
TMNL2(3)	0.98	2.76	0.98	1.94	0.83	1.05	2.61	1.17	1.94	0.95	1.15	2.25	1.15	1.81	1.02
TMNL1(3)	0.75	14.90	300.29	15.87	437.88	1.09	13.76	325.26	14.61	425.18	1.50	12.44	290.34	12.75	409.07
TMNR(3)	0.94	15.40	346.02	11.52	480.29	1.31	13.37	369.74	9.02	439.75	1.73	11.19	316.56	7.36	418.20

become points of high leverage. Hence, the use of Schweppe weights yields a robust estimate that is both efficient and robust simultaneously. This is an attractive feature as the types of outliers are rarely known in practice.

Acknowledgements We would like to thank two anonymous referees for providing helpful comments regarding the initial version of this paper.

References

Arcones M. A. (1998). The law of large numbers for U-statistics under absolute regularity. *Electronic Communications in Probability, 3*, 13–19.

Ben, M. G., Martinez, E. J., & Yohai, V. J. (1999). Robust estimation in vector autoregressive moving-average models. *Journal of Time Series Analysis, 20*(4), 381–399.

Boldin, M. V. (1994). On median estimates and tests in autoregressive models. *Mathematical Methods of Statistics, 3*(2), 114–129.

Brockwell, P. J. & Davis, R. A. (2002). *Introduction to time series and forecasting* (2nd ed.). New York: Springer.

Bustos, O. H., & Yohai, V. J. (1986). Robust estimates for ARMA models. *Journal of the American Statistical Association, 81*(393), 155–168.

Chang, W. H., McKean, J. W., Naranjo, J. D., & Sheather, S. J. (1999). High-breakdown rank regression. *Journal of the American Statistical Association, 94*(445), 205–219.

Croux, C., & Joossens, K. (2008). Robust estimation of the vector autoregressive model by a least trimmed squares procedure. In P. Brito (Ed.), *COMPSTAT 2008* (pp. 489–501). Heidelberg: Physica-Verlag HD.

Denker, M., & Keller, G. (1983). On U-statistics and v. Mises' statistics for weakly dependent processes. *Zeitschrift für Wahrscheinlichkeitstheorie und Verwandte Gebiete, 64*(4), 505–522.

Fox, A. J. (1972). Outliers in time series. *Journal of the Royal Statistical Society B, 34*(3), 350–363.

Fuller, W. A. (1996). *Introduction to statistical time series* (2nd ed.). New York: Wiley.

Handschin, E., Schweppe, F. C., Kohlas, J., & Fiechter, A. (1975). Bad data analysis for power system state estimation. *IEEE Transactions on Power Apparatus and Systems, 94*(2), 329–337.

Heathcote, C. R., & Welsh, A. H. (1988). Multivariate functional least squares. *Journal of Multivariate Analysis, 25*(1),45–64.

Heiler, S., & Willers, R. (1988). Asymptotic normality of R-estimates in the linear model. *Statistics, 19*(2), 173–184.

Hettmansperger, T. P., & McKean, J. W. (2011). *Robust nonparametric statistical methods, Monographs on statistics and applied probability* (2nd ed., Vol. 119). Boca Raton, FL: CRC Press.

Jaeckel, L. A. (1972). Estimating regression coefficients by minimizing the dispersion of the residuals. *Annals of Mathematical Statistics, 43*,1449–1458.

Krengel, U. (1985). *Ergodic theorems. de Gruyter studies in mathematics* (Vol. 6). Berlin: Walter de Gruyter & Co.

Li, W. K., & Hui, Y. V. (1989). Robust multiple time series modelling. *Biometrika, 76*(2), 309–315.

Lütkepohl, H. (1993). *Introduction to multiple time series analysis* (2nd ed.). Berlin: Springer.

Mallows, C. L. (1975). *On some topics in robustness*. Unpublished Memorandum, Bell Telephone Laboratories, Murray Hill, NJ.

McKean, J. W., & Sheather, S. J. (2009). Diagnostic procedures. *Wiley Interdisciplinary Reviews: Computational Statistics, 1*(2), 221–223.

Muler, N., & Yohai, V́.J. (2013). Robust estimation for vector autoregressive models. *Computational Statistics & Data Analysis, 65*, 68–79.

R Core Team. (2013). *R: A language and environment for statistical computing*. Vienna: R Foundation for Statistical Computing.

Reber, J. C., Terpstra, J. T., & Chen, X. (2008). Weighted L_1-estimates for a VAR(p) time series model. *Journal of Nonparametric Statistics, 20*(5), 395–411.

Rousseeuw, P. J., & Leroy, A. M. (1987). *Robust regression and outlier detection*. New York: Wiley.

Terpstra, J. T., McKean, J. W., & Naranjo, J. D. (2001). Weighted Wilcoxon estimates for autoregression. *Australian & New Zealand Journal of Statistics, 43*(4), 399–419.

Terpstra, J. T., & Rao, M. B. (2001). Generalized rank estimates for an autoregressive time series: A U-statistic approach. *Statistical Inference for Stochastic Processes, 4*(2), 155–179.

Terpstra, J. T., & Rao M. B. (2002). On the asymptotic distribution of a multivariate GR-estimate for a VAR(p) time series. *Statistics & Probability Letters, 60*(2), 219–230.

Theil, H. (1950). A rank-invariant method of linear and polynomial regression analysis (parts 1–3) *Koninklijke Nederlandse Akademie van Wetenschappen. Proceedings. Series A 53*, 386–392, 521–525, 1397–1412.

Chapter 14
Median Stable Distributions

Gib Bassett

Abstract When i.i.d. data follows a stable distribution the sample mean has the same distribution as the rescaled data. The normal with a scaling factor of $\sqrt{n}$, and the Cauchy with scaling factor of 1 are well known examples of mean stable distributions. This idea is extended to *median* stable distributions by requiring that the sampling distribution of the median be identical to the distribution of the rescaled data. The median's sampling distribution is a functional of the data's cdf so that the analysis of median stability involves solutions to functional equations (as opposed to sums of random variables). A few properties of median stable distributions are presented including their relation to the limiting distribution of the remedian.

Keywords Remedian • Iteration • Functional composition

14.1 Introduction: The Sample Median with Small (n = 3) Data

Consider the distribution of the sample median in the simplest case of $n = 3$ i.i.d observations. That is, X_1, X_2, X_3, are i.i.d with the common distribution function F; F will be sometimes referred to as the distribution of the data. The random variable with the cumulative distribution function (cdf) of the data is denoted by $X = X(F)$. Let $\hat{X} = \hat{X}(F)$ denote the sample median. The (sampling) cdf of $\hat{X}$ is denoted by $M(x) = M(x : F) = \Pr[\hat{X} < x]$.

The distribution of the sample median can be written as a functional depending on the F of the data,

$$M(x) = G(F(x)) \tag{14.1}$$

G. Bassett (✉)
Department of Finance, University of Illinois at Chicago, 2431 University Hall, 601 S. Morgan Street, Chicago, IL 60607, USA
e-mail: gib@uic.edu

© Springer International Publishing Switzerland 2016

R.Y. Liu, J.W. McKean (eds.), *Robust Rank-Based and Nonparametric Methods*, Springer Proceedings in Mathematics & Statistics 168,
DOI 10.1007/978-3-319-39065-9_14

where the G-function is,[1]

$$G(w) = 3w^2 - 2w^3, \quad w \in [0, 1]. \tag{14.2}$$

While the analysis of the sample mean is about sums of random variables, the analysis of the sample median has to do with the behavior of functionals like G.

The G function is depicted in Fig. 14.1 along with schematics showing how G turns F into M. G is seen to be continuous, increasing, with fixed points at $G(1/2) = 1/2$, $G(0) = 0$, and $G(1) = 1$, and $G(w) > w$ for $1/2 < w < 1$, $G(w) < w$ for $0 < w < 1/2$. (These features hold for the median G-functions with $n > 3$ that will be considered later.)

One consequence of the G properties is that the median of M and the median of F will always be the same; see Appendix. If the median of the data F is an interval $[\mu^-, \mu^+]$, then the median of M will be the same interval. If the median of F is a point, $\mu = \mu^+ = \mu^-$, then the median of M will also be μ. The sample median being median-unbiased is analogous to the sample mean being mean-unbiased.

With the median of F and M the same, from now on without loss of generality the data will be centered so that either its unique median is 0, or 0 is in the interval of medians, $[\mu^-, \mu^+]$.

Another feature following directly from the G-properties is that the sample median will always be more concentrated than the data around their common median.[2] That is, the probability that the sample median is in the interval, $[\mu^- - b, \mu^+ + a]$, $b > 0$, $a > 0$, is always greater than the probability that the data is in the interval; see Appendix. This is indicated in the figure by, $M(x) > F(x)$, $F(x) \in (1/2, 1)$, and $M(x) < F(x)$, $F(x) \in (0, 1/2)$. Among other things, this means the sample median will always be a better estimate of the population median than the estimate that throws away all but one of the observations.[3]

Finally, consider a rescaling of the sample median so that not only its location and scale, but also its shape is the same as the data. That is, for a rescaling factor $\lambda > 1$ there is an H for the data such that the rescaled distribution of the median is the same as H,

$$\lambda \hat{X}(H) \stackrel{d}{=} X(H). \tag{14.3}$$

[1] $\hat{X} < x$ if: (1) two-out-of-three, or (2) three-out-of-three of the X_i are less than x. Two-out-of-three has probability, $F(x)^2(1 - F(x))$, and can occur in three-choose-two equals three ways. Three-out-of-three can occur in one way, which has probability, $F(x)^3$. So, $M(x) = F(x)^3 + 3F(x)^2(1 - F(x)) = 3F(x)^2 - 2F(x)^3 = G(F(x))$. Note that $G(x)$ is the distribution of the median when the data is uniformly distributed on [0,1].

[2] Except in the trivial case in which there is no variation in the data around the median in which case $F = M$; see Appendix.

[3] Note that this is not true for the sample mean. Unless tail conditions on the data are ruled out, averaging the data can be worse than the estimate based on a single observation; for examples, see, Feller (1971), p. 172. If averaging is the alternative, the advice to never put all your eggs in one (observation) basket is a mistake.

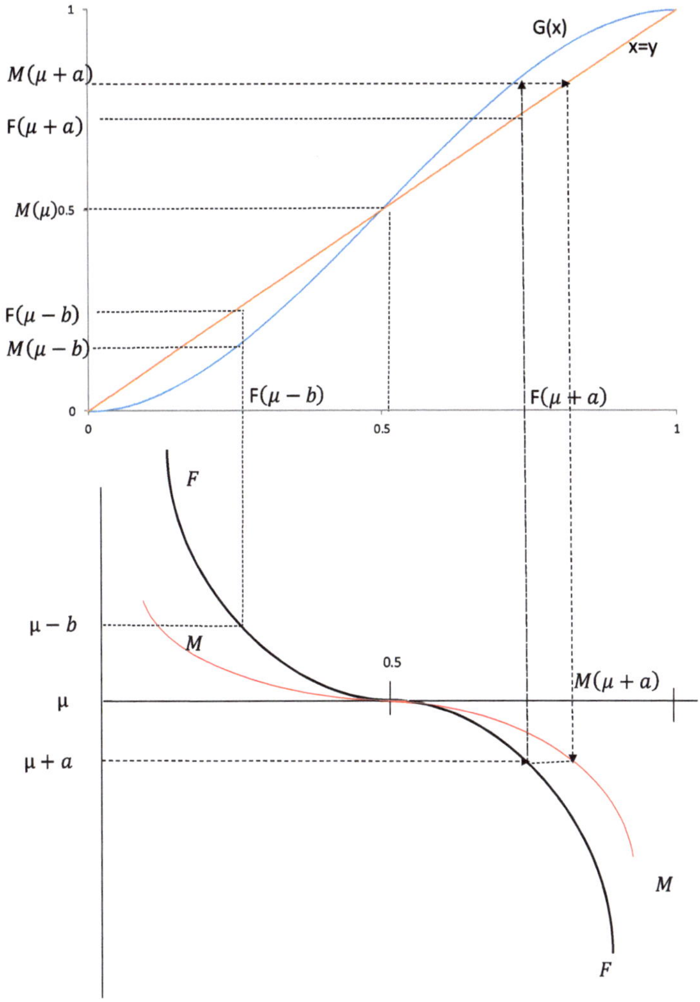

Fig. 14.1 Mapping from F to M, via G

Or, in terms of the cdfs, $M[x : H(\lambda^{-1}x)] = H(x)$; hence from (14.1), H solves the functional equation,

$$H(x) = G(H(\lambda^{-1}x)). \tag{14.4}$$

Such an H will be analogous to the normal distribution for the sample mean. With normal data (and $n = 3$) the sample mean (scaled by $\sqrt{3}$) has the same normal distribution. In a similar fashion, with H-data the sample median has the same H distribution. A picture illustrating the relationship between H and G is shown in Fig. 14.2.

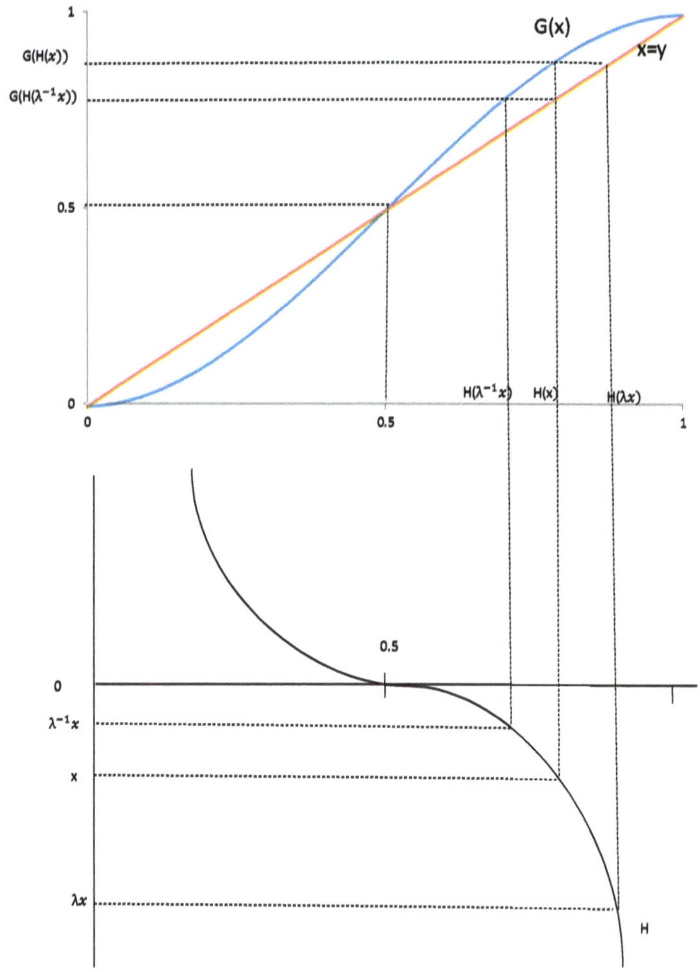

Fig. 14.2 The relationship between H and G

In the case of the sample mean the set of distributions that reproduce themselves in this fashion define the symmetric stable distributions; see e.g., Feller (1971), p. 169. The normal is best known. Other scaling factors lead to different distributions; for example, a Cauchy distribution for the data results in a Cauchy distribution for the sample mean, where the scaling factor is, $\lambda = 1$.

Similarly, a distribution H that satisfies (14.4) will be referred to as median stable. In this paper the idea of median stable distributions is introduced and an initial investigation is begun into what they look like. The simplest, $n = 3$ case is continued in Sect. 14.2, and serves as a template for $n > 3$ considered later.

Discussion Motivated by the sample mean, the classical presentation of symmetric stable distributions is in terms of sums of i.i.d random variables such that the rescaled sum has the same distribution as the common distribution of the summands. Rather than a definition in terms of "sums", a definition in terms of the sample mean lends itself to extensions of the idea of stable distribution. That is, when $\hat{X}(H)$ is the sample mean (given i.i.d data H), the definition of a (mean) stable distribution is (14.3). Substituting the sample median for the sample mean in this definition leads to the median stable distributions considered here.

One difference between the median and mean stable distributions concerns the way they depend on n, the number of observations. Mean stable distributions do not depend on—are the same—for all n: for $n = 3$ the normal is stable, and it is also stable for $n > 3$. In contrast, as discussed in Sect. 14.3, there are different median stable distributions for each n.[4]

Another contrast between mean and median stable distributions is that there are mean stable distributions with the λ-scaling factor less than one; the sample mean with fat-tailed data has to be scaled by a $\lambda < 1$ because it is more dispersed than the data. As mentioned above, since the sample median is more concentrated than the data, its scaling factor will always be greater than one.

14.2 More Small Data

When noting the dependence of the cdf H defined by (14.4) on λ, write, $H(x : \lambda)$. The next result summarizes general properties of $H(x : \lambda)$ that follow directly from the shape of G.

Theorem 14.1. *For each $\lambda > 1$, $H(x : \lambda)$ is symmetric, continuous, increasing, with $0 < H(x : \lambda) < 1$, and $H(1/2 : \lambda) = 0$. An $H(x : \lambda)$ determines a scale family of cdfs: if $H(x : \lambda)$ solves (14.4) then so does $H^*(x : \lambda) = H(\sigma x : \lambda)$, $\sigma > 0$. An $H(x : \lambda_0)$ determines all ($\lambda = \lambda_0^{1/\alpha}, \alpha > 0$), solutions to (14.4) via, $H(x : \lambda_0^{1/\alpha}) = H(x^\alpha : \lambda_0)$.*

For the last part, let $L(x) = H(x^\alpha : \lambda_0)$ so $L(x) = G(H(\lambda_0^{-1} x^\alpha : \lambda_0)) = G(H((\lambda_0^{-1/\alpha} x)^\alpha : \lambda_0)) = G(L(\lambda_0^{-1/\alpha} x))$, but this says $L(x)$ solves the functional equation with $\lambda = \lambda_0^{1/\alpha}$.

Note that since H is increasing the stable H's all have a unique median.

The derivative of H is; $h(x : \lambda) = g(H(\lambda^{-1}x))h(\lambda^{-1}x)\lambda^{-1}$, which at $x = 0$ says, $h(0 : \lambda) = g(0)h(0)\lambda^{-1} = sh(0)\lambda^{-1}$, so if the density is positive at the median the scaling parameter for the stable distribution will be $\lambda = s$. Since the solutions of (14.4) for any λ can be obtained from, $H(x : \lambda_0)$, we focus mostly on the case $\lambda_0 = s$ where the density of H is positive at $x = 0$.

[4] A median stable distribution with n observations, call it H_n, however, is a limiting distribution but, rather than the median, for the median-like (base n) remedian estimator; see Sect. 3 below and Rousseeuw and Bassett (1990).

It would be nice to have a way of going from the G function of (14.2) to an explicit formula for the cdfs implicitly defined by the functional equation (14.4). An explicit solution of the defining functional equation (14.4), even for this simplest $n = 3$ case, however is not evident. So as a starting point to understanding what H looks like, an approximation for $h(x : s)$ near zero is presented and compared to the normal distribution. This is followed by expressions for $H(x : s)$ in terms of functional composition, and for $H(x : s)$ as the limit of a sequence of cdfs that are related to the distribution of the remedian.

14.2.1 H(x : s), x Near Zero

Write the derivative of $G(w)$ as $g(w) = s[1-4(w-1/2)^2]$. The log of the derivative of the functional equation is then given by, $\log h(sx)) - \log(h(x)) = \log(1-4(H(x)-1/2)^2)$. Dividing by x^2 and taking limits as x goes to zero means the RHS is $-4h(0)^2$, while for the LHS,

$$\lim_{x\to 0} \frac{s^2 \log(h(x))}{x^2} - \frac{\log(h(sx))}{s^2x^2} = (s^2-1) \lim_{x\to 0} \frac{\log(h(x))}{x^2}.$$

So,

$$\lim_{x\to 0} \frac{\log(h(x))}{x^2} = \frac{-4h(0)^2}{s^2-1} = -3.2h(0)^2$$

Hence,

Theorem 14.2. $h(x : s) = \exp(-3.2x^2h(0)^2 + o(x))$.

This density whose value at $x = 0$ is 1 can be compared to the normal density whose value at 0 is also 1, or $n(x : 0, \sigma = 1/\sqrt{2\pi}) = \exp(-\pi x^2)$. Hence the stable median and normal densities near zero are both proportional to x^2 with the normal exponent being $\pi \approx 3.14$ whereas the median stable exponent is 3.20.

14.2.2 H as the Composition of Functions

Equation (14.4) "at" $s^{-1}x$ on the RHS says, $H(s^{-1}x) = G(H(s^{-2}x))$ which on substituting into the LHS gives $H(x) = G(G(H(s^{-2}x))) = G^{(2)}(H(s^{-2}x))$ where $G^{(2)}(x)$ is the composition of G two-times. Continuing in this fashion, $H(x) = G^{(k)}(H(s^{-k}x))$, $k = 1, 2, \ldots$. Further, since G is increasing its inverse is well-defined so that (14.4) says, $G^{-1}(H(x)) = H(s^{-1}x)$, and substituting as above gives, $H(x) = G^{(-k)}(H(s^kx))$, where $G^{(-k)}(w)$ is the composition of $G^{(-1)}(w)$,

k-times. Hence, for $k = \pm 1, \pm 2, \ldots$, a median stable distribution satisfies, $H(x) = G^{(k)}(H(s^{-k}x))$, or $H(s^k x) = G^{(k)}(H(x))$. This can be, in turn, extended to rational and then real values of k by defining fractional composition via, for example, $G^{(1/2)}$ as $G^{(1/2)}(G^{(1/2)}(w)) = G(w)$.

This maps out all solutions to the functional equation. Pick w_0 in $(1/2, 1)$, and $x_0 > 0$ so that

$$H(s^k x_0) = G^{(k)}(w_0).$$

This identifies the $H(x), x > 0$ in the scale family of cdfs satisfying (14.4)with $\lambda = s$ such that $H(x_0) = w_0$. (For $x < 0$, H is determined by symmetry, $H(x) = 1 - H(-x)$.)

The tail of $H(x)$, $x \to \infty$, is thus seen to depend on the rate at which $G^{(k)}(w_0) \to 0$, $k \to \infty$. This rate can be determined from a G-inequality involving monomials, which can be used to compute and bound $G^{(k)}(w_0)$; see Appendix. It gives,

Theorem 14.3.

$$\lim_{x \to \infty} x^{-\alpha} \log H(-x) = -c$$

where $0 < c < \infty$*, and* $s^\alpha = 2$.

Proof. See Appendix.

This limit means a median stable distribution has exponential, not fat, tails.

14.2.3 H as the Limit of $H_k(x)$

Consider $H(x) = G^{(k)}(H(s^{-k}x)$, since $H(s^{-k}x) = 1/2 + h(0)s^{-k}x + o(s^{-k}x)$, define the sequence of cdfs, $H_k(x) = G^{(k)}(1/2 + s^{-k}x)$, $|s^{-k}x| < 1/2$, $k = 1, 2, \ldots,$. The limit of the sequence of cdfs is the median stable distribution in which, $h(0) = 1$. [5]

Table 14.1 in Appendix shows the values of $H_k(x)$ for various x and k. For comparison, it also shows the values of the normal distribution whose density at zero is 1. The comparison shows $H_k(x)$ is very close to normal even for moderate k. [But the limit is not a normal distribution because the normal does not satisfy the functional equation (14.4).]

[5] $H_K(x)$ is the distribution of the remedian (base 3) with $n = 3^K$ data that are uniform on [0, 1]; see Rousseeuw and Bassett (1990).

14.3 Stable Median Distributions and the Remedian

A stable distribution for the mean does not depend on—is the same—given any number of observations. In addition, the limiting distribution of the sample mean with arbitrary, not necessarily stable, data will necessarily be a stable distribution. The property of the mean that makes its associated stable distributions the same for any number of observations is its recursive property in which the mean of means is the mean. For the regular median, as is well-known, the median of medians is not the median, and as a result median stable distributions are n-dependent. The recursive property for the mean however does hold for the median-like, remedian estimator.[6] As a result, the stable distribution for the (base b) remedian (which is the median stable distribution for b observations) is the same for all $n > b$.

The remedian is defined recursively as the median of medians. Its base-3 version is the ordinary median given n = 3 observations. For $n = 3^2$ write the data as a 3×3 array X_{ij}, $i = 1, \ldots, 3, j = 1, \ldots, 3$. The base-3 remedian for $n = 3^2$ is given by, $\hat{X}_{3^2} = \hat{X}_3(\hat{X}_{1.}, \ldots, \hat{X}_{3.})$, where $\hat{X}_{i.} = \hat{X}_3(X_{i1}, \ldots, X_{i3})$, $i = 1, \ldots, 3$. That is, the $n = 3^2$ version is the same as $\hat{X}_3$, but on a set of "data" that are themselves medians. In a similar fashion the estimate for $n = 3^k$ is recursively defined as the median of the "data", $(\hat{X}_{1\hat{X}_{(k-1)}}, \hat{X}_{2\hat{X}_{(k-1)}}, \hat{X}_{3\hat{X}_{(k-1)}})$.

The cdf of the remedian given i.i.d. data with cdf F is, $Pr[\hat{X}_{3^k}(F) < x] = G^{(k)}(F(x))$, see Rousseeuw and Bassett (1990).

Similar to the definition of mean and median stable distributions, define a remedian stable distribution so that the scaled-by-λ sampling distribution of the estimate is the same as the data. Let $H_n(x : \lambda_n)$ be notation for such a remedian stable distribution.

Since the remedian of remedians is the remedian, an H_n that is $\hat{X}_n$-stable with n observations will be also stable with n^2 observations. That is, $H_{n^2}(x : \lambda_{n^2}) = H_n(x : \lambda_n^2)$. The stable distribution is the same, and the scaling parameter for n^2 observations is just $\lambda_{n^2} = \lambda_n^2$, namely the squared value of the scaling parameter with n observations; see Appendix.

14.4 Median Stable for $n > 3$

Let $\hat{X}_r$ denote the median given $n = 2r + 1$, i.i.d F-distributed observations. The sampling distribution of $\hat{X}_r$, like the $r = 1$ case discussed above, is a functional of the data, $G_r(F)$. It is convenient to write this G_r function as a transformation of its $G = G_1$ version in Eq. (14.2).

[6]This contrasts with generalizing the median (the 0.5 quantile) via its M-estimator representation as in Bassett and Koenker (1978), Koenker and Bassett (1978).

Write the G function of (14.2) as, $G(w) = \int_0^w g(t)dt$, and consider the transformation to a new G function by raising the integrand to the power r, and scaling the result so that it integrates to 1:

$$G_r(w) = \frac{\int_0^w g(t)^r dt}{\int_0^1 g(t)^r dt} = C_r^{-1} \int_0^w g(t)^r dt$$

where $C_r = \int_0^1 g^r(t)dt = 4^r s^r \int_0^1 [t(1-t)]^r dt = 4^r s^r \frac{\Gamma^2(r+1)}{\Gamma(2r+2)} = \frac{r!r!}{(2r+1)!}$. The sampling cdf of the median is given by, $M_r(x) = G_r(F(x))$; for this and many additional results about the median see e.g., David (1981).

Let the median stable distributions for $n = 2r + 1$ be denoted by $H_r(x)$; as in section 14.1 they are defined by a functional equation given by $H_r(x : \lambda_r) = G_r(H_r(\lambda_r^{-1}x))$.

Consider the stable cdfs with positive density at the median. As in the $r = 1$ case this entails $\lambda_r = s_r$ so that the functional equation is: $H_r(x : s_r) = G_r(H(s_r^{-1}x))$ where $s_r = g_r(1/2)$, where $g_r(x)$ denotes the derivative of $G_r(x)$.

The previous results regarding H near zero and H in the tails extend readily to the $r > 1$ case:

Theorem 14.4.

$$h(x : s) = \exp(-\alpha(r)x^2 h(0)^2 + o(x))$$

where, $\alpha(r) = \frac{-4r}{s^2-1}$.

Proof. See Appendix.

Theorem 14.5.

$$\lim_{x\to\infty} x^{-\alpha} \log H(-x) = -c$$

where, $0 < c < \infty$, *and* α *satisfies,* $s^\alpha = r + 1$.

Proof. See Appendix.

Finally, given data with a positive density at the median we know the limiting distribution of the median;

$$\lim_{n\to\infty} M_n(n^{-1/2}x : F(x)) = G_n(F(n^{-1/2}x)) = N(x : 0, \sigma_M)$$

where $\sigma_M = (2f(0))^{-1}$. In terms of the G function this means

$$\lim_{r\to\infty} G_r(N(s_r^{-1}x : 0, \sigma)) = N(x : 0, \sigma)$$

(Verify noting

$$\lim_{n\to\infty} s_n^{-1}\sqrt{n} = \sqrt{\pi/2}$$

and $N((\pi/2)^{1/2}x : 0, \sigma_M) = N(x : 0, \sigma_M))$. This just means that the normal distribution solves the stable median functional equation as, $n = 2r + 1 \to \infty$.

Appendix

1. Median unbiased for any F.

   μ is a median of F if; (i) $F(\mu + 0) \geq 1/2$ and, (ii) $F(\mu - 0) \leq 1/2$. Since G is increasing with fixed point at $1/2$, $F(\mu + 0) \geq 1/2$, implies $G(F(\mu + 0)) \geq G(1/2) = 1/2$, or $M(\mu + 0) \geq 1/2$. On the other side, $F(\mu - 0) \leq 1/2$ implies $G(F(\mu - 0)) < G(1/2) = 1/2$, or $M(\mu - 0) \leq 1/2$. Hence μ will also be a median of M.
2. No variation in the data around the median.

   The sample mean and data have the same cdf in the trivial case in which there is no variability in the data; there is probability 1 point mass at 0; $\Pr(X = 0) = 1$. The analogous situation for the median occurs when the data is equal to "the" median with probability 1. As in the mean case, this trivially occurs when $\Pr(X = 0) = 1$. But it also occurs when $\Pr(X = \mu^- \text{or } X = \mu^+) = 1$ which means, $\Pr(X = \mu^-) = \Pr(X = \mu^+) = 1/2$. In this case $F = M$, the cdfs are the same.
3. The sample mean is more concentrated around the median than the data.

   Let $a > 0$ such that, $1/2 < F(\mu^+ + a + 0) < 1$. Since $G(x) > x$ for $1/2 < x < 1$, $1/2 < F(\mu^+ + a + 0) < G(F(\mu^+ + a + 0)) < 1$. But $G(F(\mu^+ + a + 0)) = M(\mu^+ + a + 0)$, so, $1/2 < F(\mu^+ + a + 0) < M(\mu^+ + a + 0) < 1$. Similarly, let $b > 0$ is such that $0 < F(\mu^- - b - 0) < 1/2$. Since $G(x) < x$ for $x < 1/2$, $0 < G(F(\mu^- + b - 0)) < F(\mu^- + b - 1) < 1/2$. Combining gives: for any $a > 0$ such that $1/2 < F(\mu^+ + a + 0) < 1$, $b > 0$ is such that $0 < F(\mu^- + b - 0) < 1/2$:

$$\Pr(\mu^- - b < \hat{X} < \mu^+ + a) > \Pr(\mu^- - b < X < \mu^+ + a).$$

4. Table 14.1
5. Stability for n^2.

   The LHS of the stability condition for n^2 is, $\lambda_{n^2}\hat{X}_{n^2}(H_{n^2})$. Make the substitution, $H_{n^2} = H_n$, $\lambda_{n^2} = \lambda_n^2$, or, $\lambda_n^2\hat{X}_{n^2}(H_n)$. Now use, $\hat{X}_{n^2} = \hat{X}_n(\hat{X}_{1.}, \ldots, \hat{X}_{n.})$ and linear homogeneity so that,

$$\lambda_n^2\hat{X}_{n^2}(H_n) = \lambda_n^2\hat{X}_n(\hat{X}_{1.}(H_n), \ldots, \hat{X}_{n.}(H_n)) = \lambda_n\hat{X}_n(\lambda_n\hat{X}_{1.}(H_n), \ldots, \lambda_n\hat{X}_{n.}(H_n)).$$

Table 14.1 $H_K(x)$ for alternative x and K

	$K = 1$	$K = 2$	$K = 5$	$K = 10$	$K = 15$	Normal
$x = 0.00$	0.5000	0.5000	0.5000	0.5000	0.5000	0.5000
$x = 0.10$	0.59941	0.59915	0.59896	0.59894	0.59894	0.59896
$x = 0.20$	0.69526	0.69330	0.69192	0.69179	0.69179	0.69193
$x = 0.30$	0.78400	0.77800	0.77395	0.77359	0.77358	0.77397
$x = 0.40$	0.86207	0.84981	0.84196	0.84126	0.84124	0.84199
$x = 0.50$	0.92593	0.90669	0.89492	0.89389	0.89387	0.89495
$x = 0.60$	0.97200	0.94818	0.93368	0.93241	0.93239	0.93371
$x = 0.70$	0.99674	0.97538	0.96033	0.95899	0.95896	0.96034
$x = 0.75$	1.00000	0.98435	0.96995	0.96862	0.96860	0.96994
$x = 0.80$		0.99076	0.97755	0.97628	0.97626	0.97753
$x = 0.90$		0.99769	0.98799	0.98692	0.98691	0.98796
$x = 1.00$		0.99976	0.99394	0.99312	0.99311	0.99391
$x = 1.10$		1.00000	0.99712	0.99655	0.99654	0.99709
$x = 1.50$			0.99992	0.99986	0.99986	0.99992

But $\lambda_n \hat{X}_{i.}(H_n) \stackrel{d}{=} X(H_n)$, so,

$$\lambda_n \hat{X}_n(\lambda_n \hat{X}_{1.}(H_n), \ldots, \lambda_n \hat{X}_{n.}(H_n)) \stackrel{d}{=} \lambda_n \hat{X}_n(X_1(H_n), \ldots, X_n(H_n)) \stackrel{d}{=} X(H_n).$$

The last step following from the fact that H_n is stable.

6. H, x near zero.

Suppressing the r-subscript, let G denote the G-function with $n = 2r + 1$ (as in Sect. 14.4), and $H(x)$ a median stable cdf with $\lambda = s$, s denoting the derivative of G at $1/2$. The derivative of the functional equation is $h(sx) = s^{-1}g(H(x))h(x)$, where $g(w) = s[1 - 4(w - 1/2)^2]^r$. Take the log of both sides so, $\log(h(sx)) - \log(h(x)) = r\log(1 - 4(H(x) - 1/2)^2)$. Divide both sides by x^2 and take limits as x goes to zero. The RHS is just $-4rh(0)^2$ and the LHS is

$$\lim_{x\to 0} \frac{s^2 \log(h(x))}{x^2} - \frac{\log(h(sx))}{s^2x^2} = (s^2 - 1) \lim_{x\to 0} \frac{\log(h(x))}{x^2}.$$

So,

$$\lim_{x\to 0} \frac{\log(h(x))}{x^2} = -\alpha(r)h(0)^2,$$

where $\alpha(r) = \frac{-4r}{s^2-1}$. When $r = 1$, $\alpha(1) = 3.2$ as in Theorem 14.2. As r increases, $\alpha(r)$ decreases and converges to π, the coefficient for the normal density.

7. Tails: $H(x), x \to -\infty$.

Suppressing the r-subscript, let G denote the G-function with $n = 2r + 1$ (as in Sect. 14.4), and $H(x)$ a median stable cdf with $\lambda = s$, s denoting the derivative of G at $1/2$.

Consider the following result for the tail of G. For $0 < t < 1/2$, write $t(1-t) = \omega t$ where, $1/2 < \omega < 1$. So,

$$G(w) = C^{-1} \int_0^w [6t(1-t)]^r dt = C^{-1}(6\omega)^r \int_0^w t^r dt =$$

$$C^{-1}(6\omega)^r \frac{w^{(r+1)}}{(r+1)}.$$

So, $G^{(k)}(w) = (A(r)w^{r+1})^{(k)}$ where $A(r) = \frac{C^{-1}(6\omega)^r}{r+1}$. Now verify,

$$(bx^\alpha)^{(k)} = b^{\frac{1-\alpha^k}{1-\alpha}} x^{\alpha^k}$$

so that,

$$\lim_{k\to\infty} \log \frac{(bx^{(\alpha+1)})^{(k)})}{\alpha+1} = \log x + \alpha^{-1} \log b$$

So,

$$\lim_{k\to\infty} \frac{\log G^{(k)}(w)}{(r+1)^k} = \lim_{k\to\infty} \frac{\log(A(r)w^{r+1})^{(k)}}{(r+1)^k} = \log x + r^{-1} \log A(r)$$

To prove the result, write $x = s^k x_0$, so

$$\lim_{x\to-\infty} x^{-\alpha} \log H(x) = \lim_{k\to\infty} (s^k w_0)^{-\alpha} \log H(-s^k x_0) = \lim_{k\to\infty} s^{-\alpha k} x_0^{-\alpha} \log G^{(k)}(w_0)$$

but, $s^\alpha = r + 1$ so that

$$= \lim_{k\to\infty} \frac{x_0^{-\alpha} \log(A(r)w^{r+1})^{(k)}}{(r+1)^k} = x_0^{-\alpha}(\log x + r^{-1} \log A(r)) = c.$$

References

Bassett, G. W., & Koenker, R. (1978). The asymptotic theory of least absolute error regression. *Journal of the American Statistical Association, 73*, 618–622.

David, H. A. (1981). *Order statistics*. New York: Wiley.

Feller, W. (1971). *An introduction to probability theory and its applications*. New York: Wiley.

Koenker, R., & Bassett, G. W. (1978). Regression quantiles. *Econometrica, 46*, 33–50.

Rousseeuw, P. J., & Bassett, G. W. (1990). The remedian: A robust averaging method for large data sets. *Journal of the American Statistical Association, 85*, 97–104.

Chapter 15
Confidence Intervals for Mean Difference Between Two Delta-Distributions

Karen V. Rosales and Joshua D. Naranjo

Abstract Traditional two-sample estimation procedures like pooled-t, Welch's t, and the Wilcoxon-Hodges-Lehmann are often used for skewed data and data inflated with zero values. We investigate how well these work compared to dedicated procedures that consider the specialized nature of the data.

Keywords Two-sample estimation • Confidence intervals • Skewed distribution • Zero-inflated data • Delta distribution

15.1 Introduction

Some data are inherently nonnegative and contain a large number of zeros. Aitchison (1955) first described a distribution that contains both zero and positive values in an application to household expenditures. Some households spend nothing on, say, children's clothing while others allocate high amounts that make the distribution skewed and approximately follow the lognormal curve. On marine surveys, data are frequently inflated with zeros. Pennington (1983) examined a series of ichthyoplankton surveys aimed at estimating the total egg production of Atlantic mackerel in the study region.

When zeros are mixed with lognormal positive values, this type of distribution is referred to as delta distribution (Aitchison 1955). One-sample confidence intervals for the mean of a delta distribution were investigated by Owen and DeRouen (1980), Pennington (1983), Zhou and Tu (2000a), Fletcher (2008), and Rosales (2009). Zhou and Tu (2000a) explored different methods of constructing confidence intervals for the mean of a delta distribution, including a bootstrap and two likelihood-based intervals. Fletcher (2008) investigated a profile-likelihood

K.V. Rosales
MMS Holdings, Inc., Canton, MI, USA
e-mail: krosales@mmsholdings.com

J.D. Naranjo (✉)
Department of Statistics, Western Michigan University, Kalamazoo, MI 49008, USA
e-mail: joshua.naranjo@wmich.edu

© Springer International Publishing Switzerland 2016

R.Y. Liu, J.W. McKean (eds.), *Robust Rank-Based and Nonparametric Methods*, Springer Proceedings in Mathematics & Statistics 168,
DOI 10.1007/978-3-319-39065-9_15

approach. Zhou and Tu (2000b) proposed a maximum likelihood-based method and a bootstrap method for constructing confidence intervals for the ratio in means of medical costs data that contained both lognormal and zero observations.

It remains unclear how well various two-sample confidence intervals work. For example, can we simply ignore the delta distribution structure of data and use traditional LS methods for estimating difference between means? Will more robust versions work better? In this paper, we focus on commonly used two-sample confidence intervals, and compare them to confidence intervals specifically derived under delta-distribution theory. We investigate how relative performance depends on sample size, proportion of zeros, the population means, and the population variances. In Sect. 15.2, we set up notation and terminology. In Sect. 15.3, we describe the confidence intervals included in the simulation study. In Sect. 15.4, we discuss results of a simulation study.

15.2 Notation and Terminology

Consider a population in which a proportion δ of the observations are zeros, and the non-zero values follow a lognormal distribution with parameters μ and σ^2. The population is said to have a Delta distribution, denoted as $\Delta(\delta, \mu, \sigma^2)$. We will index the populations of interest by $j = 1, 2$. Thus the jth population is said to have distribution $\Delta(\delta_j, \mu_j, \sigma_j^2)$, with mean κ_j and variance σ_j^2. The population mean and variance of the jth population are

$$\kappa_j = E[Y_j] = (1 - \delta_j)e^{\mu_j + \sigma_j^2/2} \tag{15.1}$$

$$\nu_j = Var[Y_j] = (1 - \delta_j)e^{2\mu_j + \sigma_j^2}(e^{\sigma_j^2} - (1 - \delta_j)) \tag{15.2}$$

Let $y_{1j}, \ldots, y_{n_j j}$ be a random sample from the jth population. Assume, without loss of generality, that the n_{j1} nonzero observations are listed first and the $n_{j0} = n_j - n_{j1}$ zero observations are listed last. For the nonzero observations let $x_{ij} = \log y_{ij}$ and

$$\hat{\delta}_j = n_{j0}/n_j \tag{15.3}$$

$$\hat{\mu}_j = \frac{\sum_{i=1}^{n_{j1}} \log y_{ij}}{n_{j1}} = \frac{\sum_{i=1}^{n_{j1}} x_{ij}}{n_{j1}} = \bar{x}_j \tag{15.4}$$

$$s_j^2 = \frac{\sum_{i=1}^{n_{j1}} (\log y_{ij} - \hat{\mu}_j)^2}{n_{j1} - 1} = \frac{\sum_{i=1}^{n_{j1}} (x_{ij} - \bar{x}_j)^2}{n_{j1} - 1} \tag{15.5}$$

Note that $\hat{\mu}_j$ and s_j^2 are simply the sample mean and variance of the *log-transformed nonzero* observations from the jth sample. The proportion of nonzero observations in the jth sample is $1 - \hat{\delta}_j$. Finney (1941) derived minimum-variance unbiased estimators for the lognormal mean and variance. Extending his results, Aitchison (1955) showed that the following is a minimum variance unbiased estimator of the mean of the Δ-distribution.

$$\hat{\kappa}_j = \begin{cases} \frac{n_{j1}}{n_j} e^{\hat{\mu}_j} G_{n_{j1}}\left(\frac{s_j^2}{2}\right) & \text{if } n_{j1} > 1 \\ \frac{x_{j1}}{n_j} & \text{if } n_{j1} = 1 \\ 0 & \text{if } n_{j1} = 0 \end{cases} \tag{15.6}$$

where $G_{n_{j1}}(t)$ is a Bessel function defined as,

$$G_{n_{j1}}(t) = 1 + \frac{n_{j1} - 1}{n_{j1}} t + \sum_{i=2}^{\infty} \frac{(n_{j1} - 1)^{2i-1} t^i}{n_{j1}^i (n_{j1} + 1)(n_{j1} + 3) \cdots (n_{j1} + 2i - 3) i!}$$

An estimate of asymptotic variance is given by Aitchison and Brown (1969)

$$\hat{v}_\infty(\hat{\kappa}_j) = \frac{e^{2\hat{\mu}_j} + S_j^2}{n_j} \left[\hat{\delta}_j (1 - \hat{\delta}_j) + \frac{(1 - \hat{\delta}_j)(2S_j^2 + S_j^4)}{2} \right] \tag{15.7}$$

Owen and DeRouen (1980) suggested confidence interval estimates based on these estimates of mean and variance. Pennington (1983) proposed an interval estimate using an alternative estimate of the variance, as follows:

$$\hat{v}_{pen}(\hat{\kappa}_j) = \begin{cases} \frac{n_{j1}}{n_j} e^{2\hat{\mu}_j} \left\{ \frac{n_{j1}}{n_j} G_{n_{j1}}\left(\frac{s_j^2}{2}\right) - \frac{n_{j1}-1}{n_j-1} G_{n_{j1}}\left(\frac{n_{j1}-2}{n_{j1}-1} s_j^2\right) \right\} & \text{if } n_{j1} > 1 \\ (\frac{x_{j1}}{n_j})^2 & \text{if } n_{j1} = 1 \\ 0 & \text{if } n_{j1} = 0 \end{cases} \tag{15.8}$$

15.3 Two-Sample Confidence Intervals

We are interested in confidence interval estimates for the difference between means $\kappa_1 - \kappa_2$ of two delta distributions. We first consider traditional least-squares confidence intervals based on Student's t-distribution, using either the pooled-SD version or the unpooled-SD Welch–Satterthwaite version. The pooled-t 100(1-α)% confidence interval is given by

$$\left[(\bar{y}_1 - \bar{y}_2) - t_{\alpha/2,df} S_p \sqrt{\frac{1}{n_1} + \frac{1}{n_2}},\ (\bar{y}_1 - \bar{y}_2) + t_{\alpha/2,df} S_p \sqrt{\frac{1}{n_1} + \frac{1}{n_2}} \right] \tag{15.9}$$

where $\bar{y}_j = \frac{1}{n_j}\sum_{i=1}^{n_j} y_{ij}$ is the sample mean for the jth sample, $t_{\alpha/2,df}$ is the upper percentile of the t-distribution, n_j is the sample size, $df = n_1 + n_2 - 2$, and S_p is the pooled standard deviation. We refer to this method as **Pooled-t** in the simulation study.

A 100(1-α)% confidence interval based on Welch's statistic is

$$\left[(\bar{y}_1 - \bar{y}_2) - t_{\alpha/2,\nu}\sqrt{\frac{s_1^2}{n_1} + \frac{s_2^2}{n_2}},\ (\bar{y}_1 - \bar{y}_2) + t_{\alpha/2,\nu}\sqrt{\frac{s_1^2}{n_1} + \frac{s_2^2}{n_2}}\right] \tag{15.10}$$

The degrees of freedom ν associated with this variance estimate is approximated using the Welch-Satterthwaite equation

$$\nu = \frac{(\frac{s_1^2}{n_1} + \frac{s_2^2}{n_2})^2}{\frac{s_1^4}{n_1^2(n_1-1)} + \frac{s_2^4}{n_2^2(n_2-1)}}$$

This method will be denoted as **Welch-t** in the simulation study.

Since the lognormal is right skewed, more robust alternatives might work better than the t-based methods. A rank-based alternative is the confidence interval based on the Wilcoxon rank sum test. See, for example, Hollander et al. (2014). The Wilcoxon interval may be computed as follows. Form all possible $(n_1)(n_2)$ pairwise differences $y_{h1} - y_{l2}$ between the first group and the second group. Let $O^{(1)}, O^{(2)}, \ldots, O^{(n_1 n_2)}$ denote these ordered differences. The Hodges-Lehmann point estimator of $\kappa_1 - \kappa_2$ is the median of these differences. A 100(1-α)% confidence interval is given by

$$\left(O^{(C_\alpha)},\ O^{(n_1 n_2 + 1 - C_\alpha)}\right) \tag{15.11}$$

where $C_\alpha = \frac{n_1(2n_2+n_1+1)}{2} + 1 - w_{\alpha/2}$, and $w_{\alpha/2}$ is an appropriate percentile of the rank sum distribution. For large samples, a normal approximation of C_α is given by

$$C_\alpha = \frac{n_1 n_2}{2} - Z_{\alpha/2}\left[\frac{n_1 n_2(n_1 + n_2 + 1)}{12}\right]^{1/2}$$

This method is denoted as **Wilcoxon** in the simulation study.

Both versions of the t-interval and the Wilcoxon interval ignore the zero-inflated nature of the data. One may construct a confidence interval based on Aitchison's minimum variance unbiased estimator $\hat{\kappa}$ and Pennington's estimator of the variance of $\hat{\kappa}$. A 100(1-α)% confidence interval for $(\kappa_1 - \kappa_2)$ is

$$(\hat{\kappa}_1 - \hat{\kappa}_2) \pm z_{\alpha/2}\sqrt{\hat{\upsilon}_{pen}(\hat{\kappa}_1) + \hat{\upsilon}_{pen}(\hat{\kappa}_2)} \tag{15.12}$$

where $\hat{\kappa}$ and $\hat{\upsilon}_{pen}$ are given in Eqs. (15.6) and (15.8), respectively. This method will be referred to as **MVUE1** in the simulation study.

An alternative confidence interval can be constructed based on the variance estimate from Aitchison and Brown (1969). This 100(1-α)% confidence interval for $(\kappa_1 - \kappa_2)$ is

$$(\hat{\kappa}_1 - \hat{\kappa}_2) \pm z_{\alpha/2}\sqrt{\hat{\upsilon}_\infty(\hat{\kappa}_1) + \hat{\upsilon}_\infty(\hat{\kappa}_2)} \tag{15.13}$$

where $\hat{\kappa}$ and $\hat{\upsilon}_\infty$ are given in Eqs. (15.6) and (15.7), respectively. We refer to this method as **MVUE2** for the rest of this dissertation.

In addition to the above confidence intervals, we propose two additional robust confidence intervals. Since the sample mean and the sample variance lack robustness, Al-Khouli (1999) proposed to directly replace $\hat{\mu}$ and s^2 in (15.4) and (15.5) with robust M-estimators to obtain robust estimators of κ and υ. In his simulation, using (T_H, S_b^2) in place of $(\hat{\mu}, s^2)$ seemed to work best, where T_H is the one-step Huber M-estimator of location and S_b^2 is a bi-weight A-estimator of scale.

Directly substituting T_H and S_b^2 in place of $\hat{\mu}$ and s^2 in (15.6) and (15.8), we get a robust version of the MVUE1 interval (15.12). The confidence interval is

$$(\hat{\kappa}_{M_1} - \hat{\kappa}_{M_2}) \pm z_{\alpha/2}\sqrt{\hat{\upsilon}_M(\hat{\kappa}_{M_1}) + \hat{\upsilon}_M(\hat{\kappa}_{M_2})} \tag{15.14}$$

where

$$\hat{\kappa}_{M_j} = \begin{cases} \frac{n_{j1}}{n_j} e^{T_{H_j}} G_{n_{j1}}\left(\frac{S_{b_j}}{2}\right) & \text{if } n_{j1} > 1 \\ \frac{x_1}{n_j} & \text{if } n_{j1} = 1 \\ 0 & \text{if } n_{j1} = 0 \end{cases}$$

and

$$\hat{\upsilon}_M(\hat{\kappa}_{M_j}) = \begin{cases} \frac{n_{j1}}{n_j} e^{2T_{H_j}} \left\{\frac{n_{j1}}{n_j} G_{n_{j1}}\left(\frac{S_{b_j}}{2}\right) - \frac{n_{j1}-1}{n_j-1} G_{n_{j1}}\left(\frac{n_{j1}-2}{n_{j1}-1} S_{b_j}\right)\right\} & \text{if } n_{j1} > 1 \\ (\frac{x_1}{n_j})^2 & \text{if } n_{j1} = 1 \\ 0 & \text{if } n_{j1} = 0 \end{cases}$$

This method is referred as **RMVUE1** in the simulation study.

Similarly, a robust version of the MVUE2 confidence interval (15.13) replaces $\hat{\mu}$ and s in Eqs. (15.6) and (15.7) with their robust versions. The confidence interval is

$$(\hat{\kappa}_{M_1} - \hat{\kappa}_{M_2}) \pm z_{\alpha/2}\sqrt{\hat{\upsilon}_\infty(\hat{\kappa}_{M_1}) + \hat{\upsilon}_\infty(\hat{\kappa}_{M_2})} \tag{15.15}$$

where

$$\hat{\kappa}_{M_j} = \frac{n_{j1}}{n_j} e^{T_{H_j}} G_{n_{j1}}\left(\frac{S_{b_j}}{2}\right)$$

and

$$\hat{v}_{\infty}(\hat{\kappa}_{M_j}) = \frac{e^{2T_{H_j}+S_{b_j}}}{n_j}\left[\hat{\delta}_j(1-\hat{\delta}_j) + \frac{(1-\hat{\delta}_j)(2S_{b_j}+S_{b_j}^2)}{2}\right]$$

We denote this method as **RMVUE2** in the simulation study.

15.4 Simulation

To assess the general performance and robustness of the interval estimators (15.9)–(15.15), we conducted a simulation study under various parameter combinations of the Δ-distribution. Performance of the different estimates will be assessed using the following criteria:

- Coverage Probability (CP): proportion of times that the 95 % confidence interval contains the true value of $\kappa_1 - \kappa_2$.
- Coverage Error (CE): absolute difference between the coverage probability and 95 %.
- Lower Error Rate (LER): proportion of times that the true value $\kappa_1 - \kappa_2$ falls below the interval
- Upper Error Rate (UER): proportion of times that the true value $\kappa_1 - \kappa_2$ falls above the interval
- Average Width (Width): average width of 95 % confidence interval

Note that all confidence intervals have confidence level set at 95 %. Ideally an estimation procedure will have CP=0.95, CE=0.0, LER=0.025, and UER=0.025. We also report the average width of each method. We evaluate performance at balanced sample sizes of 15 and 50. Ten thousand simulations are done for each combination of parameters and sample size.

Table 15.1 shows simulation results when the two delta distributions are the same. MVUE1 and RMVUE1 seem to do best, achieving narrower intervals without sacrificing coverage probability. Coverage probabilities all exceed 0.95, maybe due to overinflated standard error estimates because of skewness. The naive t-based intervals seem competitive, with reasonable width and coverage probability. The Wilcoxon interval has the shortest width.

Table 15.2 shows simulation results when $\delta_1 \neq \delta_2$. Again, MVUE1 and RMVUE1 seem to do best, with narrower intervals without sacrificing coverage probability. The naive t-based intervals remain competitive, with reasonable width and coverage probability. The Wilcoxon interval still has significantly shortest width but achieves this at the price of unacceptably low coverage probability, especially for larger differences in δ.

Table 15.3 shows simulation results when $\mu_1 \neq \mu_2$. MVUE1 and RMVUE1 still seem to do best, with RMVUE1 edging out MVUE1 in coverage probability

Table 15.1 95 % CI under equal distributions $\Delta_1(0.2, 0.5, 1)$ and $\Delta_2(0.2, 0.5, 1) : \kappa_1 - \kappa_2 = 0$

Method	Sample size	CP	CE	LER	UER	Width
Pooled-t	15	0.9609	0.0109	0.0196	0.0195	4.4816
Welch-t		0.9665	0.0165	0.0171	0.0164	4.5539
Wilcoxon		0.9681	0.0181	0.0168	0.0151	2.5595
MVUE1		0.9687	0.0187	0.0159	0.0154	4.3110
MVUE2		0.9888	0.0388	0.0056	0.0056	5.4533
RMVUE1		0.9684	0.0184	0.0163	0.0153	3.9451
RMVUE2		0.9904	0.0404	0.0049	0.0047	4.7320
Pooled-t	50	0.9561	0.0061	0.0224	0.0215	2.5258
Welch-t		0.9570	0.0070	0.0221	0.0209	2.5324
Wilcoxon		0.9779	0.0279	0.0114	0.0107	1.1137
MVUE1		0.9605	0.0105	0.0208	0.0187	2.4411
MVUE2		0.9739	0.0239	0.0139	0.0122	2.6163
RMVUE1		0.9700	0.0200	0.0161	0.0139	2.3764
RMVUE2		0.9805	0.0305	0.0114	0.0081	2.5332

and width. MVUE2 and RMVUE2 attain better coverage probabilities at the cost of significantly wider intervals. The naive procedures pooled-t and Welch-t are surprisingly competitive, with reasonable width and coverage probability. The Wilcoxon interval has unacceptably low coverage probability, especially for larger differences in μ.

Table 15.4 shows simulation results when $\sigma_1^2 \neq \sigma_2^2$. All intervals have problems maintaining close to 95 % coverage probability, especially for larger differences in σ^2.

The simulations show two notable features of Wilcoxon confidence intervals: they tend to be shorter and have low coverage probability. Wilcoxon intervals are a function of the ordered pairwise differences between the two samples [see e.g. Hollander et al. (2014)]. If (δ_1, δ_2) are both large, then enough pairwise differences are 0 regardless of the values of the positive observations. This seems to reduce length of the Wilcoxon interval more than the others. Low coverage probability may be a result of the Wilcoxon interval estimating the wrong parameter. The Wilcoxon point estimator is the median of pairwise differences, which is naturally a better estimate of the true median of differences (i.e. the *median of* $F_{Y_1-Y_2}$) rather than the difference in means $\kappa_1 - \kappa_2$. For example, given two distributions $\Delta(0.1, 0.5, 1)$ and $\Delta(0.5, 0.5, 1)$, the difference in means is $\kappa_1 - \kappa_2 = 1.0873$ while the median of the difference is $m = 0.7988$. In Table 15.5, we reassess the performance of Wilcoxon by looking at the percentage of time it contains the median of differences m instead of $\kappa_1 - \kappa_2$. The Wilcoxon 95 % interval coverage probability for $\kappa_1 - \kappa_2 = 1.0873$ are quite low at 0.8708 and 0.6734, respectively, but the coverage probability for $m = 0.7988$ are 0.9508 and 0.9479, respectively, as

Table 15.2 95 % CI under varying proportion of zeros δ

Method	Sample size	CP	CE	LER	UER	Width
$\Delta_1(0.2, 0.5, 1)$ *and* $\Delta_2(0.4, 0.5, 1) : \kappa_1 - \kappa_2 = 0.5437$						
Pooled-t	15	0.9619	0.0119	0.0152	0.0229	4.2539
Welch-t		0.9675	0.0175	0.0128	0.0197	4.3263
Wilcoxon		0.9369	0.0131	0.0150	0.0481	2.3203
MVUE1		0.9688	0.0188	0.0121	0.0191	4.0838
MVUE2		0.9872	0.0372	0.0037	0.0091	5.3487
RMVUE1		0.9649	0.0149	0.0149	0.0202	3.7071
RMVUE2		0.9887	0.0387	0.0035	0.0078	4.5085
Pooled-t	50	0.9561	0.0061	0.0174	0.0265	2.4063
Welch-t		0.9572	0.0072	0.0170	0.0258	2.4129
Wilcoxon		0.9059	0.0441	0.0064	0.0877	0.9932
MVUE1		0.9587	0.0087	0.0162	0.0251	2.3397
MVUE2		0.9750	0.0250	0.0098	0.0152	2.5297
RMVUE1		0.9665	0.0165	0.0145	0.0190	2.2688
RMVUE2		0.9779	0.0279	0.0100	0.0121	2.4350
$\Delta_1(0.1, 0.5, 1)$ *and* $\Delta_2(0.5, 0.5, 1) : \kappa_1 - \kappa_2 = 1.0873$						
Pooled-t	15	0.9596	0.0096	0.0107	0.0297	4.1678
Welch-t		0.9636	0.0136	0.0085	0.0279	4.2410
Wilcoxon		0.8708	0.0792	0.0051	0.1241	2.1206
MVUE1		0.9662	0.0162	0.0072	0.0266	4.0256
MVUE2		0.9857	0.0357	0.0024	0.0119	5.3849
RMVUE1		0.9636	0.0136	0.0109	0.0255	3.6579
RMVUE2		0.9878	0.0378	0.0029	0.0093	4.4512
Pooled-t	50	0.9575	0.0075	0.0133	0.0292	2.3900
Welch-t		0.9583	0.0083	0.0130	0.0287	2.3972
Wilcoxon		0.6734	0.2766	0.0007	0.3259	0.9820
MVUE1		0.9624	0.0124	0.0125	0.0251	2.3124
MVUE2		0.9776	0.0276	0.0073	0.0151	2.5068
RMVUE1		0.9707	0.0207	0.0119	0.0174	2.2398
RMVUE2		0.9815	0.0315	0.0074	0.0111	2.4070

found in the entry labeled W(for m). In fact, in all cases (see the rest of Table 15.5), as long as we measure the percentage of times that Wilcoxon interval contains the appropriate parameter m instead of $\kappa_1 - \kappa_2$, then the Wilcoxon has best coverage probability and narrowest width. Since the performance of MVUE2 and RMVUE2 trail MVUE1 and RMVUE1 in Tables 15.2, 15.3, and 15.4, they have been removed from Table 15.5 for space considerations.

Table 15.3 95 % CI under varying lognormal parameter μ

Method	Sample size	CP	CE	LER	UER	Width
$\Delta_1(0.2, 0, 1)$ *and* $\Delta_2(0.2, 0.5, 1)$: $\kappa_1 - \kappa_2 = -0.8556$						
Pooled-t	15	0.9366	0.0134	0.0580	0.0054	3.6641
Welch-t		0.9392	0.0108	0.0567	0.0041	3.7380
Wilcoxon		0.8280	0.1220	0.1701	0.0019	2.1170
MVUE1		0.9389	0.0111	0.0573	0.0038	3.5068
MVUE2		0.9678	0.0178	0.0314	0.0008	4.4297
RMVUE1		0.9444	0.0056	0.0510	0.0046	3.2131
RMVUE2		0.9722	0.0222	0.0268	0.0010	3.8499
Pooled-t	50	0.9466	0.0034	0.0447	0.0087	2.0790
Welch-t		0.9471	0.0029	0.0445	0.0084	2.0878
Wilcoxon		0.5473	0.4027	0.4526	0.0001	0.9866
MVUE1		0.9538	0.0038	0.0394	0.0068	2.0192
MVUE2		0.9657	0.0157	0.0310	0.0033	2.1633
RMVUE1		0.9669	0.0169	0.0277	0.0054	1.9697
RMVUE2		0.9759	0.0259	0.0208	0.0033	2.0995
$\Delta_1(0.2, 0, 1)$ *and* $\Delta_2(0.2, 0.9, 1)$: $\kappa_1 - \kappa_2 = -1.9252$						
Pooled-t	15	0.9018	0.0482	0.0949	0.0033	4.9543
Welch-t		0.9033	0.0467	0.0937	0.0030	5.0884
Wilcoxon		0.7171	0.2329	0.2821	0.0008	3.0142
MVUE1		0.9047	0.0453	0.0939	0.0014	4.7717
MVUE2		0.9391	0.0109	0.0602	0.0007	6.0067
RMVUE1		0.9147	0.0353	0.0824	0.0029	4.4050
RMVUE2		0.9464	0.0036	0.0529	0.0007	5.2732
Pooled-t	50	0.9246	0.0254	0.0709	0.0045	2.8552
Welch-t		0.9255	0.0245	0.0704	0.0041	2.8744
Wilcoxon		0.2817	0.6683	0.7183	0.0000	1.4666
MVUE1		0.9363	0.0137	0.0602	0.0035	2.7701
MVUE2		0.9477	0.0023	0.0505	0.0018	2.9663
RMVUE1		0.9542	0.0042	0.0423	0.0035	2.7066
RMVUE2		0.9630	0.0130	0.0349	0.0021	2.8846

15.5 Conclusion

Traditional two-sample estimation procedures like pooled-*t* and Welch *t* that require normal distribution are often used for skewed data and data inflated with zero values. Our simulations show that these naive nonrobust approaches do not do too badly compared to dedicated delta distribution procedures, in terms of coverage probabilities and interval width.

Among the dedicated approaches, we would recommend the MVUE1 and its robust version RMVUE1. The MVUE1 procedure is based on the mean estimator

Table 15.4 95 % CI under varying lognormal parameter σ^2

Method	Sample Size	CP	CE	LER	UER	Width
$\Delta_1(0.2, 0.5, 0.15)$ *and* $\Delta_2(0.2, 0.5, 1.0)$: $\kappa_1 - \kappa_2 = -0.7529$						
Pooled-t	15	0.8805	0.0695	0.1175	0.0020	3.1534
Welch-t		0.8826	0.0674	0.1157	0.0017	3.2449
Wilcoxon		0.7183	0.2317	0.2814	0.0003	2.2225
MVUE1		0.8894	0.0606	0.1095	0.0011	3.0699
MVUE2		0.9044	0.0456	0.0952	0.0004	3.7589
RMVUE1		0.8880	0.0620	0.1116	0.0004	3.0540
RMVUE2		0.9169	0.0331	0.0831	0.0000	3.5633
Pooled-t	50	0.9097	0.0403	0.0866	0.0037	1.8212
Welch-t		0.9103	0.0397	0.0862	0.0035	1.8366
Wilcoxon		0.2679	0.6821	0.7321	0.0000	1.0701
MVUE1		0.9246	0.0254	0.0721	0.0033	1.7831
MVUE2		0.9342	0.0158	0.0643	0.0015	1.8967
RMVUE1		0.9142	0.0358	0.0846	0.0012	1.8514
RMVUE2		0.9259	0.0241	0.0736	0.0005	1.9567
$\Delta_1(0.2, 0.5, 0.15)$ *and* $\Delta_2(0.2, 0.5, 2.0)$: $\kappa_1 - \kappa_2 = -2.1636$						
Pooled-t	15	0.7574	0.1926	0.2420	0.0006	6.9912
Welch-t		0.7651	0.1849	0.2347	0.0002	7.2802
Wilcoxon		0.3201	0.6299	0.6798	0.0001	2.8765
MVUE1		0.7892	0.1608	0.2106	0.0002	6.8989
MVUE2		0.8445	0.1055	0.1554	0.0001	11.7706
RMVUE1		0.6229	0.3271	0.3769	0.0002	4.2226
RMVUE2		0.7015	0.2485	0.2984	0.0001	5.4010
Pooled-t	50	0.8287	0.1213	0.1707	0.0006	4.4801
Welch-t		0.8308	0.1192	0.1686	0.0006	4.5326
Wilcoxon		0.0070	0.9430	0.9930	0.0000	1.2596
MVUE1		0.8768	0.0732	0.1232	0.0000	4.2748
MVUE2		0.8993	0.0507	0.1007	0.0000	5.0110
RMVUE1		0.5428	0.4072	0.4572	0.0000	2.6287
RMVUE2		0.5862	0.3638	0.4138	0.0000	2.8755

$\hat{\kappa}$ by Aitchison (1955) and the variance estimator by Pennington (1983). The RMVUE1 is similar to MVUE1 but uses M-estimates for the lognormal parameters μ and σ^2.

The Wilcoxon two-sample interval performed consistently badly, but only when it was asked to estimate the difference in means $\kappa_1 - \kappa_2$. When used to estimate the median of differences m, it performed very well in terms of coverage probability, and generally had the shortest interval width. Of course, usefulness of the Wilcoxon interval will depend more on whether the user wants to estimate the median of differences instead of the difference in means.

Table 15.5 95 % CI under varying parameters and sample size

Method	Sample Size	CP	CE	LER	UER	Width
Varying δ: $\Delta_1(0.1, 0.5, 1.0)$ *and* $\Delta_2(0.5, 0.5, 1.0)$						
$\kappa_1 - \kappa_2$ = *1.0873, m=0.7988*						
Pooled-t	15	0.9596	0.0096	0.0107	0.0297	4.1678
Welch-t		0.9636	0.0136	0.0085	0.0279	4.2410
Wilcoxon (for $\kappa_1 - \kappa_2$)		0.8708	0.0792	0.0051	0.1241	2.1206
Wilcoxon (for m)		0.9508	0.0008	0.0241	0.0251	2.1206
MVUE1		0.9662	0.0162	0.0072	0.0266	4.0256
MVUE2		0.9857	0.0357	0.0024	0.0119	5.3849
RMVUE1		0.9636	0.0136	0.0109	0.0255	3.6579
Pooled-t	50	0.9575	0.0075	0.0133	0.0292	2.3900
Welch-t		0.9583	0.0083	0.0130	0.0287	2.3972
Wilcoxon (for $\kappa_1 - \kappa_2$)		0.6734	0.2766	0.0007	0.3259	0.9820
Wilcoxon (for m)		0.9479	0.0021	0.0256	0.0265	0.9820
MVUE1		0.9624	0.0124	0.0125	0.0251	2.3124
RMVUE1		0.9707	0.0207	0.0119	0.0174	2.2398
Varying μ: $\Delta_1(0.2, 0, 1)$ *and* $\Delta_2(0.2, 0.9, 1)$						
$\kappa_1 - \kappa_2 = -1.9252$*, m=−0.8531*						
Pooled-t	15	0.9018	0.0482	0.0949	0.0033	4.9543
Welch-t		0.9033	0.0467	0.0937	0.0030	5.0884
Wilcoxon (for $\kappa_1 - \kappa_2$)		0.7171	0.2329	0.2821	0.0008	3.0142
Wilcoxon (for m)		0.9421	0.0079	0.0273	0.0306	3.0142
MVUE1		0.9047	0.0453	0.0939	0.0014	4.7717
RMVUE1		0.9147	0.0353	0.0824	0.0029	4.4050
Pooled-t	50	0.9246	0.0254	0.0709	0.0045	2.8552
Welch-t		0.9255	0.0245	0.0704	0.0041	2.8744
Wilcoxon (for $\kappa_1 - \kappa_2$)		0.2817	0.6683	0.7183	0.0000	1.4666
Wilcoxon (for m)		0.9335	0.0165	0.0343	0.0322	1.4666
MVUE1		0.9363	0.0137	0.0602	0.0035	2.7701
RMVUE1		0.9542	0.0042	0.0423	0.0035	2.7066
Varying σ^2: $\Delta_1(0.2, 0.5, 0.15)$ *and* $\Delta_2(0.2, 0.5, 2.0)$						
$\kappa_1 - \kappa_2 = -2.1636$*, m=0.0*						
Pooled-t	15	0.7574	0.1926	0.2420	0.0006	6.9912
Welch-t		0.7651	0.1849	0.2347	0.0002	7.2802
Wilcoxon (for $\kappa_1 - \kappa_2$)		0.3201	0.6299	0.6798	0.0001	2.8765
Wilcoxon (for m)		0.9565	0.0065	0.0236	0.0199	2.8765
MVUE1		0.7892	0.1608	0.2106	0.0002	6.8989
RMVUE1		0.6229	0.3271	0.3769	0.0002	4.2226
Pooled-t	50	0.8287	0.1213	0.1707	0.0006	4.4801
Welch-t		0.8308	0.1192	0.1686	0.0006	4.5326
Wilcoxon (for $\kappa_1 - \kappa_2$)		0.0070	0.9430	0.9930	0.0000	1.2596
Wilcoxon (for m)		0.9657	0.0157	0.0184	0.0159	1.2596
MVUE1		0.8768	0.0732	0.1232	0.0000	4.2748
RMVUE1		0.5428	0.4072	0.4572	0.0000	2.6287

The Wilcoxon interval is assessed for containing both $\kappa_1 - \kappa_2$ and the median of difference m

References

Aitchison, J. (1955). On the distribution of a positive random variable having a discrete probability mass at the origin. *Journal of the American Statistical Association, 50*(271), 901–908.

Aitchison, J., & Brown, J. (1969). *The lognormal distribution*. Cambridge: Cambridge University Press.

Al-Khouli, A. (1999). *Robust estimation and bootstrap testing for the delta distribution with applications in marine sciences*. Ph.D. dissertation, Texas A&M University.

Finney, D. J. (1941). On the distribution of a variate whose logarithm is normally distributed. *Journal of the Royal Statistical Society, Series B, 7*, 155–61.

Fletcher, D. (2008). Confidence intervals for the mean of the delta-lognormal distribution. *Environmental and Ecological Statistics, 15*(2), 175–189.

Hollander, M., Wolfe, D., & Chicken, E. (2014). *Nonparametric statistical methods*. Hoboken: Wiley.

Owen, W., & DeRouen, T. (1980). Estimation of the mean for lognormal data containing zeroes and left-censored values, with applications to the measurement of worker exposure to air contaminants. Biometrics, 36(4), 707–719.

Pennington, M. (1983). Efficient estimators of abundance, for fish and plankton surveys, *Biometrics, 39*(1), 281–286.

Rosales, M. (2009). *The robustness of confidence intervals for the mean of delta distribution*. Ph.D. dissertation, Western Michigan University.

Zhou, X. H., & Tu, W. (2000a). Confidence intervals for the mean of diagnostic test charge data containing zeros. *Biometrics, 56*(4), 1118–1125.

Zhou, X. H., & Tu, W. (2000b). Interval estimation for the ratio in means of log-normally distributed medical costs with zero values. *Computational Statistics and Data Analysis, 35*(2), 201–210.

Author Index

A
Abebe, A., 25, 61

B
Bassett, G., 249
Bathke, A.C., 121
Bilgic, Y., 61
Bindele, H., 25
Burgos, J., 227

D
Datta, S., 175

H
Harrar, S.W., 121
Hettmansperger, T.P., viii, 1

K
Kloke, J., 47, 61

L
Li, J., 209
Liu, R.Y., x, 209

M
Mathur, S., 175
McKean, J.W., x, 1, 61

N
Naranjo, J.D., 261

O
Oja, H., 189
Ozturk, O., 141

R
Rosales, K.V., 261

S
Sakate, D.M., 175
Sheather, S.J., viii, 101
Sun, Y., 141

T
Taskinen, S., 189
Terpstra, J.T., 81, 227

W
Wolfe, D.A., 163

© Springer International Publishing Switzerland 2016

R.Y. Liu, J.W. McKean (eds.), *Robust Rank-Based and Nonparametric Methods*,
Springer Proceedings in Mathematics & Statistics 168,
DOI 10.1007/978-3-319-39065-9

Author Index

A
Abbe, A., 18, 61

M

DOI 10.1007/978-3-319-39063-5

Subject Index

A
ACTG 320, 158–160
Additive outlier (AO) model, 240–242
Affine equivariant, 190–193, 195, 201
Affine invariance, 211
AIC, 31
AL. *See* Asymptotic linearity (AL)
AR(1) estimators, 48, 53, 55, 56, 58, 59, 92
AR model. *See* Autoregressive (AR) model
ART. *See* Asymptotic rank transform (ART)
Asymptotic linearity (AL), 10, 42, 86, 233, 242
Asymptotic quadraticity, 83, 86, 233
Asymptotic rank transform (ART), 126–128, 130, 131, 133
Asymptotic relative efficiency (ARE), 91
Asymptotic uniform linearity (AUL), 83, 86, 233
Asymptotic uniform quadraticity (AUQ), 83, 86, 233
AUL. *See* Asymptotic uniform linearity (AUL)
AUQ. *See* Asymptotic uniform quadraticity (AUQ)
Autoregressive (AR) model, 17, 82, 93, 227

B
Balanced rank set sampling, 169
Bartlett-Nanda-Pillai criterion, 126
BIC, 31, 32
Bifurcating autoregressive process, 81–90
Big data, 101–119
Bivariate data, 182
Bounded influence, 3, 13–17, 31, 93, 190, 192–195, 201
Breakdown, 3, 13–17, 26, 68, 70, 76, 104, 105, 190, 193, 203, 231
Breakdown point, 13–15, 190, 192, 193, 195

C
Calibration, 151–156
CFITS, 15, 16, 75
Cluster correlated data, 18–21, 47–59
Confidence intervals, 20, 47, 50, 68, 70, 72, 102, 105, 109, 111–114, 116–119, 164, 261–271
Convex hull, 212, 223
Cross validation, 30, 37, 40

D
Data depth, 209–225
Dell and Clutter model, 154, 156
Delta distribution, 261–271
Dempster's ANOVA type criterion, 126, 130
Depth-based ranking, 210, 216–217, 222, 224
Depth-*versus*-depth (DD) plot, 209–215, 221–225
Diagnostics, 7, 13, 15, 22, 49, 75, 123
Diagnostic testing, 175, 183
Direction vectors, 191
Dispersion function, 3, 4, 27, 49, 51, 54, 62, 64–66, 76, 144, 228–231, 235, 239, 240, 242
Distribution free, 2, 3, 148, 151, 153, 160, 172, 178

© Springer International Publishing Switzerland 2016

R.Y. Liu, J.W. McKean (eds.), *Robust Rank-Based and Nonparametric Methods*, Springer Proceedings in Mathematics & Statistics 168,
DOI 10.1007/978-3-319-39065-9

E

Efficiency, 2–4, 10–12, 15, 17, 18, 22, 26, 27, 32–36, 41, 50, 56, 57, 59, 70, 72, 76, 82, 91–93, 103–105, 150, 151, 183, 189–206, 210, 217, 231, 241–245
Elliptic distributions, 191, 197, 202, 225

F

Factor multivariate designs, 121–137
Functionals, 41, 69, 191–196, 200, 201, 203, 229, 234, 249–251, 253–259

G

GEEhbr estimators, 70–72
GEE models. *See* General estimating equations (GEE) models
GEERB estimators, 66–72, 75, 76
General estimating equations (GEE) models, 18, 61–78
Generalized joint rankings (GJR) estimators, 48, 52–57
Generalized Rank (GR) estimates, 229
GJR estimators. *See* Generalized joint rankings (GJR) estimators
GLS estimators, 59
G-properties, 250
Gradient, 5, 6, 13, 28, 30, 50, 64, 66, 76, 97, 233
GR estimates. *See* Generalized Rank (GR) estimates

H

HBR estimators, 15, 18, 242
Heterogeneity, 142
Hettmansperger and Randles (HR) estimator for location, 190, 192, 202
Hierarchical models, 22, 63, 71, 73–75
Hodges-Lehmann, 3, 264
Hotelling's T^2 test, 181

I

Inflated zero data, 261, 264, 269
Influence functions, 3, 13, 15, 31, 50, 93, 189–206
Innovation outlier (IO) model, 240, 242
Iterated reweighted rank-based estimators, 61–78

J

Joint rankings (JR) estimator, 48, 52
Judgment ordered statistic, 172

K

Knot values, 115, 116
k-step estimators, 193, 195, 198–200, 202, 203
k-step HR estimator, 193–196, 200, 202

L

LAD. *See* Least absolute deviations (LAD)
Lasso, 26, 27, 29, 32, 37, 39
Lawley-Hotelling criterion, 126
Least absolute deviations (LAD), 5, 14, 27, 31, 32, 34, 37, 81–99, 200
Least trimmed squares (LTS) estimators, 14, 102–105, 116
Linear hypotheses, 5, 17, 18, 21, 22
Longitudinal data, 62
LTS estimators. *See* Least trimmed squares (LTS) estimators

M

Mahalanobis depth, 217, 219, 225
Mallows weights, 91, 229–231, 242
Mann-Whitney, 62, 150, 156
Mardia test, 177, 181, 183
Maximum judgement order statistic, 172
Mean stable distribution, 153
Median, 2, 3, 5, 13, 37, 39, 40, 50, 65, 66, 71, 72, 103, 105, 109, 110, 142, 143, 145, 153, 158, 159, 167–173, 176, 182, 190, 192, 193, 195, 249–260, 264, 267, 270, 271
Median judgment order statistic, 172
Median stable distribution, 249–260
M estimators, 102–105, 109, 160, 190, 192, 193, 256, 265, 270
Minimum judgment order statistic, 172
MM estimator, 102, 104–105, 229
Moore-Penrose inverse, 126
Multivariate, 17, 18, 22, 42, 47, 98, 121–137, 176, 189–206, 209–225, 227–246
Multivariate LAD estimator, 200
Multivariate scale test, 209–225
Multivariate time series, 227–246
MVUE, 264–271

N

Negative binomial distribution, 166
Nested design, 18, 48, 63, 70, 71, 73, 74
Nonlinear models, 3, 16–22, 26, 47

O
Optimal scores, 12, 22, 49
Ordered restricted randomized design, 141–160

P
Partially sequential ranked set sampling, 163–173
Pearson-type distributions, 181
Penalized regression, 37
Perfect ranking, 143, 144, 148, 150, 151, 153–156
Permutation tests, 210, 215, 218–220, 225
Pooled-t, 263, 264, 267–271
Projection, 7, 54, 146

Q
Q test, 219–221, 224

R
Rank-based, 1–22, 26, 121–137, 264
Rank-based estimators, 5, 8, 12–16, 18, 20, 41, 47–59, 61–78, 109, 131, 135
Ranked set sampling, 144, 150, 163–173
Ranking error, 142, 143, 151–155
Ranks of angles, 177
Rank tests, 2, 3, 21, 210, 216, 217, 219, 220
Rank transform, 126–128, 130, 131, 133
Recursive property, 156
Regression spline models, 115, 116
Remedian, 253–256
Remedian stable distribution, 256
REML estimators, 56, 71, 72, 76
REML methods, 59
Repeated measures design, 41, 62, 71
Robust, 2, 4, 7, 8, 13–15, 19–21, 25–44, 48–50, 59, 62, 63, 65, 71, 72, 76, 82, 85, 137, 190, 192, 195, 199, 202, 217, 229–231, 241, 242, 246, 262, 264, 265, 269
Robustness, 3, 4, 13, 26, 47, 58, 59, 63, 76, 93, 193–195, 198, 202, 210, 265, 266
Robust regression, 101–119, 199
R package, `jrfit`, 19
R package, `rbgee`, 63
R package, `Rfit`, 8, 18
R package, `rlme`, 63

S
Sample depth, 216
SAS, 8, 102, 105, 109, 115
Scale curve, 223–225
Scale homogeneity, 210, 218–220, 224
Scaling factors, 250, 252, 253
Schweppe-type weights, 83, 231
Schweppe weights, 229–231, 241, 242, 246
Score function, 4, 8, 10, 11, 13, 17, 18, 27, 49, 50, 62, 65, 66, 68, 70, 76, 192, 198, 202
Sequential ranked set sampling, 163–173
S estimator, 195, 196, 198, 199
Signed-rank regression, 25–44
Simplicial depth, 211, 217, 219, 225
Spatial median, 190, 192, 193, 195, 206
Spatial signs, 190, 192, 199, 202
Spherical distributions, 191, 194, 201, 205
Stationarity, 241, 242
Studentized residuals, 7, 9, 20, 21, 37, 40, 59

T
TDBETAS, 15, 75
TDBETAS CFITS, 75
Time series, 17–18, 51, 82, 227–246
Two-sample test, 164
Tyler's M estimator, 192

U
Unweighted mean analysis, 126

V
Variable selection, 25–44
Variance-component estimators, 19, 20, 71, 72
Vector autoregressive (VAR) model, 227, 230
Visual detection, 210

W
Weighted L_1 estimators, 83, 229, 240, 241
Weighted Wilcoxon, 26, 227–246
Weighted Wilcoxon dispersion, 229, 231
Welch's t, 264, 267–271
Wilcoxon, 2, 3, 5, 8–17, 22, 26, 49, 50, 62, 65, 70–72, 76, 109, 114, 150, 156, 181, 183, 216, 227–246, 264, 266–271
Wilcoxon test, 2, 3, 156
Wilks's criterion, 126
WL1 estimators, 83, 92, 93

MIX
Papier aus verantwortungsvollen Quellen
Paper from responsible sources
FSC® C105338

If you have any concerns about our products,
you can contact us on
ProductSafety@springernature.com

In case Publisher is established outside the EU,
the EU authorized representative is:
Springer Nature Customer Service Center GmbH
Europaplatz 3, 69115 Heidelberg, Germany

Printed by Libri Plureos GmbH
in Hamburg, Germany